Natural Products

Microbial biofilm plays an important role in the life cycle of microorganisms surviving in diverse and harsh environments such as extremes of temperature, pH, salinity, nutrient scarcity, and the like. Biofilm formation is a survival strategy adopted by microorganisms that allows colonization of new niches through the dispersal of microbes from the microbial clusters embedded within an outer polymer layer produced by the microorganism itself. This layer comprises extracellular polymeric substances (EPS) that help the indwelling microbes to grow and divide in a protected environment, defending against invaders like antimicrobial agents, surfactants, biocides, and phagocytic cells of host organisms. Thus, the EPS matrix prevents antimicrobial resistance, which poses a threat to human health.

Natural Products: Alternative Therapeutics as Quorum Sensing Inhibitors provides a detailed and systematic review of the mechanism of quorum sensing in both Gram-positive and Gram-negative bacteria and alternative pharmacological developments as a potent solution to the rise of antimicrobial resistance during biofilm formation.

Features:

- A systematic overview of the mechanism of quorum sensing in the development of microbial biofilms
- Biofilm formation and its role in the emergence of antimicrobial resistance
- Recent pharmacological development of antibiofilm remedies involving biogenic compounds

A useful reference for researchers investigating the progression of events during microbial biofilm formation, starting from the genetic cascade regulating quorum sensing to the secretion of autoinducers and the design of safer methodologies for the successful eradication of microbial biofilm, this volume may also be used as a textbook for an undergraduate-level course in microbiology or microbial biotechnology.

Confluence of Research in Biotechnology for Human Welfare
Series Editor: Susmita Mukherjee
(*University of Engineering & Management, West Bengal, India*)

Natural Products: Alternate Therapeutics as Quorum Sensing Inhibitors
Edited by Moupriya Nag, Dibyajit Lahiri, Jaideep Banerjee and Taniya Roy Chowdhury

For more information, please visit series page: www.routledge.com

Natural Products

Alternate Therapeutics as Quorum Sensing Inhibitors

Edited by Moupriya Nag, Dibyajit Lahiri,
Jaideep Banerjee and Taniya Roy Chowdhury

CRC Press
Taylor & Francis Group
Boca Raton London New York

CRC Press is an imprint of the
Taylor & Francis Group, an **informa** business

First edition published 2024
by CRC Press
6000 Broken Sound Parkway NW, Suite 300, Boca Raton, FL 33487–2742

and by CRC Press
4 Park Square, Milton Park, Abingdon, Oxon, OX14 4RN

CRC Press is an imprint of Taylor & Francis Group, LLC

ISBN: 9781032292380 (hbk)
ISBN: 9781032292397 (pbk)
ISBN: 9781003300557 (ebk)

DOI: 10.1201/9781003300557

Typeset in Times
by Apex CoVantage, LLC

Contents

Preface.....xiii
Editor Biographies.....xv
List of Contributors.....xvii

Chapter 1 Introduction.....1

Dipro Mukherjee, Sayantani Garai, Sanchari Das, Moupriya Nag, Debasmita Bhattacharya, and Dibyajit Lahiri

1.1 Introduction.....1
1.2 Phytocompounds in Inhibiting the Quorum-Sensing Cascade.....1
1.3 AI-2 in Charge of Regulating QS.....2
1.4 QS Dominated by Peptide Secretion.....2
1.5 AHLS Are Responsible for Mediating QS.....3
1.6 Several Other Systems Mediating QS.....3
1.7 Microbiologically Synthesized Nanoparticles and Their Role in Silencing the Biofilm Signalling Cascade.....3
1.8 Microbial Exopolysaccharides for Synthesis of Nanoparticles.....4
1.9 Mechanism of Quorum-Sensing Inhibition Using Microbiogenic NPs.....5
1.10 NP-Associated Inhibition of Quorum-Sensing Cascade.....5
1.11 Plant-Mediated Synthesized Nanoparticles in Down-Regulating Quorum Sensing.....6
1.12 Fungal Components in Down-Regulation of Quorum Sensing.....7
1.13 Quorum Sensing and Antimicrobial Resistance.....8
1.14 Conclusion.....9
References.....9

Chapter 2 Quorum Sensing in Gram-Positive Bacteria.....13

Smaranika Pattnaik

2.1 Introduction.....13
2.2 Gram-Positive Bacterial Polysaccharides as the Basement of Biofilm Matrix.....14
2.3 MSCRAMMs as Adhesives in Biofilm Formation by Gram-Positives.....14

2.4 Cannibalism and Fratricide Mechanism of Gram-Positives to Form Biofilm 15
2.5 Quorum Sensing as Regulatory Mechanism for Biofilm Formation in Gram-Positives 15
2.6 QS System Operation by Gram-Positives 17
2.6.1 Quorum-Sensing System of *S. Aureus* 17
2.6.2 The ComX Quorum-Sensing System of *B. Subtilis* 18
2.6.3 Quorum-Sensing System of *E. Faecalis* 19
2.6.4 Quorum-Sensing System of *Clostridium difficile* 21
2.7 Conclusion 21
References 23

Chapter 3 Quorum Sensing in Gram-Negative Bacteria 25

Snehashis Koley, Ambar Bose, and Mandira Mukherjee
3.1 Introduction 25
3.2 Quorum Sensing in *Erwinia caratovora* 26
3.3 Quorum Sensing in *Agrobacterium tumefaciens* 27
3.4 Quorum Sensing in *Vibrio* Species 28
3.5 Quorum Sensing in *Aeromonas* Species 30
3.6 Quorum Sensing in *E. coli* 32
3.7 Quorum Sensing in *Helicobacter pylori* 35
3.8 Quorum Sensing in *Haemophilus influenzae* 36
3.9 Quorum Sensing in *Neisseria meningitidis* 36
3.10 Quorum Sensing in *Pseudomonas aeruginosa* 37
3.11 Quorum Sensing in *Salmonella* Species 38
3.12 Quorum Sensing in *Burkholderia cepacia* 40
3.13 Quorum Sensing in *Yersinia* Species 41
3.14 Quorum Sensing in *Brucella* Species 42
3.15 Quorum Sensing in *Campylobacter jejuni* 43
3.16 Conclusion 44
3.17 Acknowledgements 45
References 45

Chapter 4 Fundamentals of Bacterial Quorum Sensing and Antimicrobial Resistance 62

Tahseena Naaz, Vandana Singh, Subhasree Ray, Kalpana Sharma, Barun Kumar, and Soumya Pandit
4.1 Introduction 62
4.2 Antimicrobial Resistance Mechanism 64
4.3 Microbial Quorum-Sensing Mechanism and Its Regulation 67

4.4 Microbial Competition and Quorum Sensing 69
4.5 Microbial Regulation System by Quorum Sensing 71
4.5.1 Bacterial Efflux Pump Regulation by Quorum Sensing 71
4.5.2 Quorum Sensing in Regulation of Bacterial Biofilm Formation 72
4.5.3 Bacterial Secretion System Regulation by Quorum Sensing 73
4.6 New Strategy for Preventive Treatment of Microbial Resistance 74
4.7 Future Outlook 75
4.8 Challenges 76
4.9 Conclusion 76
References 76

Chapter 5 Phytocompound-Mediated Inhibition of Quorum-Sensing Cascade in Pathogenic Bacteria 79

M.P. Roy, M.D. Adhikari, and B.K. Tiwary

5.1 Introduction 79
5.2 Overview of QS Mechanism 81
5.2.1 QS Based on AHL Molecules 81
5.2.2 QS Based on Auto-Inducing Oligopeptides 82
5.2.3 QS Based on Production of AI-2 System 83
5.3 Phytocompounds with Inhibitory Effects against Quorum Sensing in Pathogenic Bacteria 83
5.3.1 Flavonoids 85
5.3.2 Alkaloids 87
5.3.3 Terpenoids and Limonoids 89
5.3.4 Phenol Derivatives 90
5.3.5 Essential Oils 90
5.4 Conclusions 92
References 95

Chapter 6 Microbiologically Synthesized Nanoparticles in Inhibition of Quorum Sensing 102

Drisha Roy, Moupriya Nag, Debasmita Bhattacharya, and Dibyajit Lahiri

6.1 Introduction 102
6.2 Quorum Sensing and Biofilm Formation 103
6.3 Quorum Quenching 104
6.4 Overview of Nanoparticles 105
6.4.1 Biosynthesis of Nanoparticles 106
6.4.2 Extracellular and Intracellular Synthesis of NPs..... 108

6.5 Bacteria-Mediated Synthesis of Nanoparticles 109
6.5.1 Metal NPs 109
6.5.2 Metal Oxides 110
6.5.3 Cyanobacteria 111
6.5.4 Actinobacteria 112
6.5.5 Algae and Microalgae 112
6.5.6 Viruses 113
6.5.7 Fungi and Yeast 114
6.6 Nanoparticles as Anti-QS Agents 115
6.6.1 Silver NPs 116
6.6.2 Gold NPs 117
6.6.3 Silicon Dioxide NPs 118
6.6.4 Titanium Dioxide NPs 118
6.6.5 Zinc Oxide NPs 119
6.6.6 Chitosan NPs 120
6.6.7 Other NPs 120
6.7 Demerits and Safety Concerns of Nanoparticles 124
6.8 Conclusion and Perspective for the Future 124
References 125

Chapter 7 Plant-Mediated Synthesized Nanoparticles in Down-Regulating Quorum Sensing 139

Kavita Singh, Avnish Kumar, and Vinita Gupta

7.1 Quorum Sensing 139
7.1.1 Introduction of Quorum Sensing 139
7.1.2 Types of Autoinducers 141
7.1.3 Mechanism of Quorum Sensing 141
7.1.4 Methods of Inhibition of Quorum Sensing 142
7.2 Methods of Inhibition of Quorum Sensing Using Nanoparticles 142
7.2.1 Inhibitors of Quorum Sensing 142
7.3 Study of Quorum-Sensing Inhibition 144
7.3.1 Mechanism of Inhibition by Silver Nanoparticle (Ag NPs) 144
7.3.2 Mechanism of Inhibition by Gold Nanoparticles 144
7.3.3 Mechanism of Inhibition by Zinc and Copper Nanoparticles 144
7.4 Plant Extracts Preparation Method 144
7.4.1 Maceration 144
7.4.2 Infusion 145
7.4.3 Digestion 145
7.4.4 Decoction 145
7.4.5 Percolation 145

7.4.6 Soxhlet Extraction ... 145
7.4.7 Microwave-Assisted Extraction ... 146
7.4.8 Ultrasound-Assisted Extraction ... 146
7.5 Preparation of Plant Extract-Based Nanoparticles and Their Characterization ... 146
7.6 Future of Plant-Based Nanoparticles ... 147
7.7 Conclusion ... 147
References ... 148

Chapter 8 Fungal Compounds in Down-Regulating Quorum Sensing ... 150

Barun Kumar, Dibyajit Lahiri, Soumya Pandit, Ankit Kumar, Kalpana Sharma, Prakriti Negi, and Moupriya Nag

8.1 Introduction ... 150
8.2 Quorum Sensing–Fungi ... 151
8.2.1 Regulation and Mechanism ... 152
8.2.2 *Vibrio fischeri* ... 152
8.2.3 Benefit of Quorum Sensing to *V. fischeri* ... 153
8.3 Origin and Development of Fungal Quorum Sensing ... 153
8.3.1 Bioengineering of Fungal QS ... 153
8.4 Down-Regulation of QS ... 155
8.4.1 Autoinducer ... 155
8.4.2 The Lux System ... 156
8.4.3 Biofilm ... 157
8.4.4 Molecular Response of Farnesol ... 158
8.4.5 Antimicrobial Resistance ... 160
8.5 Conclusion ... 161
References ... 161

Chapter 9 Antimicrobial Peptides: Sources, Types, and Antibiofilm Activity ... 165

Bhavini K, Abigail Fernandes, Vinothkannan Ravichandran, Dibyajit Lahiri, and Renitta Jobby

9.1 Introduction ... 165
9.2 Sources of Antimicrobial Proteins ... 166
9.2.1 AMPs from Mammals ... 167
9.2.2 AMPs from Plants ... 167
9.2.3 AMPs from Amphibians ... 167
9.2.4 AMPs from Insects ... 168
9.2.5 AMPs from Microorganisms ... 168
9.3 Structure of AMPs ... 168
9.3.1 α-Helical ... 168
9.3.2 β-Sheet ... 168
9.3.3 Loop ... 169

9.3.4 Extended Peptides ... 169
9.4 Types of AMPs ... 170
9.4.1 Antibacterial Peptides ... 170
9.4.2 Antiviral Peptides ... 170
9.4.3 Antifungal Peptides ... 170
9.4.4 Antiparasitic Peptides ... 171
9.5 Mechanism of Action of AMPs ... 171
9.6 AMPs and Biofilms ... 172
9.7 AMPs as Quorum Quenchers ... 173
9.8 Bioengineering Antimicrobial Peptides ... 174
9.9 Challenges in Manipulating QS ... 175
9.10 Conclusions and Future Perspectives ... 176
9.10.1 Acknowledgement ... 176
References ... 176

Chapter 10 Algal Compounds in Down-Regulating Quorum Sensing of Biofilm ... 183

Ashmita Samanta, Drisha Roy, Moupriya Nag, Dibyajit Lahiri, Debasmita Bhattacharya, and Rina Rani Ray

10.1 Introduction ... 183
10.2 The Tie-In between Biofilm and Quorum Sensing ... 184
10.3 Modes of Bacterial Quorum Sensing ... 184
10.3.1 Mechanism of Quorum Sensing in Gram-Positive and Gram-Negative Bacteria ... 185
10.4 Quorum Quenching ... 187
10.5 Biofouling ... 189
10.6 Mechanism of Biofouling ... 190
10.7 Prevention of Biofouling: Antifouling ... 191
10.8 Marine Antifouling from Algal Compounds ... 192
10.9 Quorum-Sensing Inhibition in the Marine Environment ... 192
10.10 Control of Biofouling by Quorum Quenchers ... 194
10.11 Conclusion ... 195
References ... 195

Chapter 11 Microbial Volatile Organic Compounds as Quorum Quenchers ... 200

Ashmita Samanta, Drisha Roy, Dibyajit Lahiri, Rina Rani Ray, Debasmita Bhattacharya, and Moupriya Nag

11.1 Introduction ... 200
11.2 Microbial Volatile Organic Compounds ... 201
11.2.1 Occurrence of Microbial Volatile Organic Compounds ... 201

11.2.2 Physiochemical Properties of Microbial Volatile Organic Compounds 202
11.2.3 Toxicokinetics of Microbial Volatile Organic Compounds 204
11.3 Antimicrobial Activity of Microbial Volatile Organic Compounds 205
11.4 Microbe Interactions in VOCs 206
11.4.1 Bacteria-Bacteria Interactions 206
11.4.2 Bacteria-Fungi Interactions 207
11.4.3 Fungi-Fungi Interactions 207
11.5 Quorum Sensing 208
11.5.1 Modes of Quorum Sensing 210
11.6 Quorum Quenching 210
11.6.1 Inhibiting AI Synthesis/Development of AI Adversary Molecules 211
11.6.2 Inhibition of the Autoinducer-Receptor Binding 212
11.6.3 Degradation of Autoinducers 212
11.6.4 Inhibition of Surface Adhesion Initiation 212
11.7 Quorum Quenching by VOCs 213
11.8 Conclusion 215
References 216

Preface

Biofilm is one of the predominant causes of various types of chronic infections that are associated with human beings and various animals. Biofilm is syntrophic associations of microbial cells that adhere to biotic and abiotic surfaces with the help of self-secreted produced extracellular polymeric substances (EPS). The formation of biofilm results from alteration of genetic expression within the planktonic cells that causes the adherence of the cells on the surface. Biofilm results in the development of various types of disorders, among them cystic fibrosis, chronic otitis media, valve endocarditis, and various types of wound-associated infections.

The bacterial species possess the ability to monitor and adapt to environmental changes with the help of cell-to-cell communication or cellular interaction known as quorum sensing (QS). QS is a density-dependent process and is based on contact associated with exchange of chemical substances, signalling associated with various types of electrical impulse, and chemical signalling. The interaction of the bacterial cells with the environment is a constant phenomenon by the process of cell-to-cell signalling for the identification of various available nutrients.

The mechanism of QS helps in the regulation of cellular virulence, competence, sporulation, and formation of the biofilm. Initially it was thought that the mechanism of QS is restricted to a few bacterial species, but it has been found that most bacterial and fungal species bring about cell-to-cell communication by this process. The molecules that are associated with cellular signalling are called autoinducers (AIs) and show variation among Gram-positive and Gram-negative bacterial cells. The Gram-negative bacterial cells use acyl-homoserine lactones (AHLs) for the purpose of cellular communication.

QS is responsible for the production of various types of virulence factors that include pyocyanin, lection, exotoxin A, and elastase in *Pseudomonas aeruginosa*, whereas lipases, haemolysins, enterotoxin, protein A, and fibronectins are produced by *Staphylococcus aureus*. The virulence factor developed in the bacterial cells dwelling within the biofilm helps in evading the immune system of the host, thus results in pathogenesis. The development of biofilm is a multistep process and involves the mechanism of gene transcriptions, which cause the conversion from planktonic to sessile counterparts. The sessile microcolonies result in the development of thick extracellular polymeric substances (EPS) that comprise exopolysaccharides, extracellular DNA (eDNA), proteins, lipids, and various other types of minerals that act as a physical barrier to the penetration of antibiotics and drugs, resulting in the development of multi-drug-resistant organisms. The mechanism of biofilm development occurs by the irreversible adhesion of sessile colonies on a surface followed by the production of QS molecules that help in the transportation of various substances within the biofilm, metabolism within the sessile microcolonies, EPS development, and metastasis of the sessile cells resulting in the spread of infections. Thus, the mechanism of QS followed by the formation of biofilm results in antibiotic resistance.

Traditionally, biofilm was addressed with combinatorial treatment of various types of antibiotics that exhibited the mechanism of killing. But the development of antibiotic resistance results in the failure of the conventional use of the drugs. This has resulted in the use of alternate therapeutics for the treatment of biofilm-associated infections. Consequently, the use of natural products as antibiofilm agents is currently increasing. This book therefore emphasises various types of natural compounds that have a significant role in inhibiting quorum sensing.

Editor Biographies

Dr. Moupriya Nag is presently Assistant Professor in the Department of Biotechnology, University of Engineering & Management, Kolkata. Her research interests include single-molecule biophysics of protein unfolding/refolding and protein aggregation kinetics and their mechanism of action, especially the amyloid proteins in neurodegenerative disorders. Recently, her research has focussed on functional amyloids in bacteria and their relation in forming biofilms. Her work deals with developing novel antimicrobial and antibiofilm compounds from natural sources including plants, microbes, and green synthesized nanoparticles. She has some six years of teaching experience in Biophysics, Bioinformatics, Bioprocess, and Biotechnology, Enzyme Technology, Computing, Molecular Modelling and Drug Design. She has also published many research articles in peer-reviewed international journals and authored or coauthored numerous book chapters. She is a member of many scientific societies, among them the Indian Biophysical Society and Indian Science Congress Association.

Dr. Dibyajit Lahiri is at present Assistant Professor in the Department of Biotechnology, University of Engineering & Management, Kolkata. His research interest is specifically biofilm isolated from human prosthesis and its inhibition by various novel phyto and nano compounds, and he is keen to explore the molecular mechanism behind the removal of biofilm by natural compounds. His research arena also encompasses computational drug development by using bioinformatics. He is a life member of the Indian Science Congress Association and the Institution of Engineers (India).

Dr. Jaideep Banerjee received his PhD in Molecular, Cell, and Developmental Biology from the Ohio State University in Columbus, Ohio. He completed his postdoctoral training at the US Army Institute of Surgical Research in San Antonio, Texas, and at George Washington University, in Washington, D.C. Over the past 10 years, Dr. Banerjee has been involved with various aspects of wound healing research. He is an editor for *Advances in Wound Care* and *Current Stem Cell Research & Therapy*. He has worked as a consultant with Walter Reed Army Institute of Research and with Mike O'Callaghan Military Medical Center. Currently, he leads clinical strategy for biologics and regenerative medicine, working with the team of medical science liaisons in the Department of Clinical Affairs (Research and Development) at Smith and Nephew, Inc., a medical technology company.

Dr. Taniya Roy Chowdhury works in the Agricultural Research Service at the US Department of Agriculture. Her research is focused on soil systems at two scales: the broadly ecological and the complexly microbial. She studies the vital locus of biodiversity at the critical interfaces of soil, microbes, water and trace gas emissions, microbial underpinnings of biogeochemical changes with a focus on nutrient cycles (carbon, nitrogen, and iron), and specifically on anaerobic microbial processes, including methanogenesis. Ecosystems of interest include wetlands, permafrost, and converted agricultural lands. She is interested in studying connected urban waterways and tackling issues pertaining to environmental health. By applying a range of skills from microbiology, soil science, metabolomics, and high-throughput sequencing technologies, she explores the complex interactions between microorganisms and biotic/abiotic factors. Her ultimate goal is to detect ecological patterns within a spatial-temporal framework to address management and sustainability challenges. Her extended research interests encompass linking microbiome stability/phenotype to environmental stability/function; observing microbiome shifts at high-resolution spatial and temporal scales; and deciphering abiotic-biotic interactions driving biogeochemical changes at ecosystem scale.

Contributors

M.D. Adhikari
Department of Biotechnology
University of North Bengal, Darjeeling
West Bengal, India

Ankita Banerjee
Department of Biotechnology
University of Engineering & Management, Kolkata
West Bengal, India

Debasmita Bhattacharya
Department of Basic Science and Humanities
Institute of Engineering and Management, Salt Lake
University of Engineering and Managment, Kolkata

Ambar Bose
Department of Biochemistry and Medical Biotechnology
Calcutta School of Tropical Medicine
West Bengal, India

Sanchari Das
Department of Biotechnology
University of Engineering & Management, Kolkata
West Bengal, India

Abigail Fernandes
Amity Institute of Biotechnology
Mumbai, Maharashtra, India
and
Centre of Excellence in Astrobiology
Amity University
Mumbai, Maharashtra, India

Sayantani Garai
Department of Biotechnology
University of Engineering & Management, Kolkata
West Bengal, India

Vinita Gupta
Department of Chemistry
Dr. Bhimrao Ambedkar University
Agra, India

Renitta Jobby
Amity Institute of Biotechnology
Amity University
Mumbai, Maharashtra, India
and
Centre of Excellence in Astrobiology
Amity University
Mumbai, Maharashtra, India

Bhavini K
Amity Institute of Biotechnology
Amity University
Mumbai, Maharashtra, India

Snehashis Koley
Department of Biochemistry and Medical Biotechnology
Calcutta School of Tropical Medicine
West Bengal, India

Ankit Kumar
Department of Life Science
Sharda University
Greater Noida, Uttar Pradesh, India

Avnish Kumar
Department of Biotechnology
Dr. Bhimrao Ambedkar University
Agra, India

Barun Kumar
Department of Life Science
Sharda University
Greater Noida, Uttar Pradesh, India

Dibyajit Lahiri
Department of Biotechnology
University of Engineering & Management, Kolkata
West Bengal, India

Dipro Mukherjee
Department of Biotechnology
University of Engineering & Management, Kolkata
West Bengal, India

Mandira Mukherjee
Department of Biochemistry and Medical Biotechnology
Calcutta School of Tropical Medicine
West Bengal, India

Tahseena Naaz
Department of Life Science, School of Sciences and Basic Research
Sharda University
Noida, Uttar Pradesh, India

Moupriya Nag
Department of Biotechnology
University of Engineering & Management, Kolkata
West Bengal, India

Prakriti Negi
Department of Life Science
Sharda University
Greater Noida, Uttar Pradesh, India

Soumya Pandit
Department of Life Science
Sharda University
Greater Noida, Uttar Pradesh, India

Smaranika Pattnaik
Department of Biotechnology and Bioinformatics
Sambalpur University
Burla, Odisha, India

Dantuluri Veerabhadra Raju
Department of Life Science
Sharda University
Greater Noida, Uttar Pradesh, India

Vinothkannan Ravichandran
Amity Institute of Biotechnology
Amity University
Mumbai, Maharashtra, India

Rina Rani Ray
Department of Biotechnology
Maulana Abul Kalam Azad University of Technology
Kalyani, West Bengal, India

Subhasree Ray
Department of Life Science
Sharda University
Greater Noida, Uttar Pradesh, India

Drisha Roy
Department of Biotechnology
University of Engineering & Management, Kolkata
West Bengal, India

M.P. Roy
Department of Microbiology
Ananda Chandra College
Jalpaiguri, West Bengal, India

Ashmita Samanta
Department of Biotechnology
University of Engineering & Management, Kolkata
West Bengal, India

Kalpana Sharma
Department of Life Science
Sharda University
Greater Noida, Uttar Pradesh, India

Kavita Singh
Department of Chemistry
Dr. Bhimrao Ambedkar University
Agra, India

Vandana Singh
Department of Allied Health Sciences
Sharda University
Greater Noida, Uttar Pradesh, India

B.K. Tiwary
Department of Microbiology
North Bengal St Xavier's College, Rajganj
Jalpaiguri, West Bengal, India

1 Introduction

Dipro Mukherjee, Sayantani Garai, Sanchari Das, Moupriya Nag, Debasmita Bhattacharya, and Dibyajit Lahiri

1.1 INTRODUCTION

Phytochemicals are mainly obtained from plants, usually as secondary metabolites that are formed by substituting oxygen or from phenols. Such secondary metabolites offer a variety of useful properties which include antibacterial properties. Phenolic acids, quinones, phenolics, flavonoids, saponins, coumarins, tannins, terpenoids, and alkaloids are among the major groups of antimicrobial chemicals chiefly extracted from plants [Mohamed et al., 2014; Livermore, 2011]. All such compounds may possess distinct antibiofilm behaviour owing to the variations in their chemical constitutions as well as structure. Polyphenols, terpenoids, phytosterols, and a variety of other chemicals present in plants have been revealed to possess significant quorum-quenching properties. Furthermore, antibiofilm actions have also been reported for certain substances which include oridin, edible fruit extract, naringenin, cinnamaldehyde, salicylic acid, ursolic acid, and methyl eugenol [Wu et al., 2015]. The QS mechanism is triggered by autoinducers (AIs) that are synthesized, secreted, and assimilated by the bacteria in their surrounding environment, their concentration being influenced by the number of bacterial cells. Extracellular signalling elements and AIs significantly raise in the neighbouring medium in accordance with cellular density, and therefore are subsequently utilized during intercellular communication [Newman et al., 2003]. These microbes have a crucial role in regulating gene expression among the bacterial cells in the population, affecting a wide range of bacterial responses. QS also has the ability to regulate a variety of physiological activities in bacterial cells, including motility, luminescence, pathogenicity, genetic competence development, biofilm development, sporulation, fluorescence, peptide antibiotic biosynthesis, and proteolytic enzyme synthesis and release. Researchers have recently discovered that an autoinducing mechanism is required for the synthesis of signal molecules, and that this process varies between Gram-positive and Gram-negative bacteria [Deep et al., 2011].

1.2 PHYTOCOMPOUNDS IN INHIBITING THE QUORUM-SENSING CASCADE

Certain chemical signalling molecules, commonly referred to as bacterial pheromones or autoinducers, are produced by bacterial cell to facilitate QS. Such signals

DOI: 10.1201/9781003300557-1

are released during the growth of bacterial population and continues till it achieves a threshold concentration that causes certain genes to be activated or repressed [Kabir et al., 2010]. A stimulatory quantity of such substances may accumulate only when a particular, sufficient number of cells, known as a quorum, is accessible [Krishnan et al., 2012]. The following fundamental parts can be viewed as the foundation of the QS system: 1) signal synthase; 2) signal response regulator; 3) signal receptor; 4) regulated genes (that constitute the "QS regulon"); 5) autoinducers; and 6) signal synthase. Compounds with antibiofilm properties are usually based on certain characteristics, for example, inhibiting cell attachment and its adhesion, suppressing the formation of polymers, reducing the generation of pathogenic factors, and disrupting the synthesis of ECM [Singh et al., 2009].

1.3 AI-2 IN CHARGE OF REGULATING QS

Both Gram-positive and Gram-negative bacteria have this quorum-sensing mediated biofilm-forming pathway. Chemically AI-2 is also recognized as furanosyl borate diester or tetrahydroxy furan. Based on the density as well as concentration around the threshold, this molecule has the ability to flip between intra- and interspecies communication [Rocha-Estrada et al., 2010]. The precursor peptide molecules of autoinducers are changed in Gram-positive bacteria and shipped out of the bacterial cell mostly through the transporter complex known as ATP-binding cassette (ABC). Whenever the threshold point is reached by the peptide AI concentration, the sensor kinase protein is activated, the protein regulating the response is phosphorylated, and this phosphorylated protein binds to the target promoter, resulting in gene regulation by QS. On the other hand, in the case of Gram-negative bacteria, AIs are released and freely disseminated from the cell. When the AI concentration reaches a certain threshold, a positive feedback loop is created, causing increased AI biosynthesis. The AI concentration continues to increase with the growth of the bacterial cell population, finally reaching a certain peak which further leads to the diffusion of these molecules into the bacterial cells to regulate particular gene transcription, triggering the synthesis and release of several virulence factors as well as helping in biofilm development [Guan et al., 2000].

1.4 QS DOMINATED BY PEPTIDE SECRETION

Gram-positive bacteria participate in QS using certain molecules produced by biosynthesizing oligopeptides. Autoinducer peptides (AIPs) of this kind are synthesized within the extracellular environment and have a high binding affinity for the histidine membrane receptor kinase [Ditu et al., 2011]. Molecular expression of AIP is dependent on the family of genes encoding AgrD. Upon expression of the AIPs, thiolactone alters the conformation of AgrB proteins bound to membranes, allowing them to be effluxed into the circumambient media as oligopeptides. Once the concentration of extracellular AIP reaches a certain threshold level, the AgrC receptor gets coupled with these compounds [Chan et al., 2011].

1.5 AHLS ARE RESPONSIBLE FOR MEDIATING QS

RhlR-I and LasR-I systems have a role in QS modulation in certain Gram-negative bacteria, like *Pseudomonas aeruginosa*. Lactone synthase, also known as LuxI, is responsible for the biosynthesis of the AHL molecule. The extracellular medium is where this enzyme is discharged. The AHL concentration found within the extracellular medium reaches a critical concentration at this stage when the cell density is boosted or quorum is established, and subsequently the AHLs are diffused further into the bacterial cells, thereby causing an interaction engaging with transcriptional regulators in the process [Hayek et al., 2013]. C4-HSL and 3OC12-HSL are two types of AHL molecules generated by *P. aeruginosa*. Each kind of AHL molecule interacts to a distinct transcriptional regulator, which includes RhlR for C4-HSL and LasR for 3OC12-HSL. Several transcriptional regulators, including toxA, LasB, and LasI, are activated by the complexes generated as a result. The AHL molecule concentration influences the activation of the regulators belonging to the genes of certain families [Castillo-Juárez et al., 2013]. AHLs, via stimulating the expression of QS genes, regulate physiological processes [Kang et al., 2016]. As a result, most AHLs discovered to date are composed of an acyl chain comprising an even count of carbon atoms extending 4–14 C atoms in length and coupled to the homoserine lactone moiety [Chong et al., 2011]. The protein family, particularly LuxI and LuxR, are involved in the AHL-regulated QS system. LuxR is known to inhibit or activate some gene transcription, such as that of the virulent genes. LuxI is essentially involved to develop AHLs; however, LuxR is known to suppress or stimulate specific gene transcription, including that of the virulent genes [Davies et al., 1998].

1.6 SEVERAL OTHER SYSTEMS MEDIATING QS

Occasionally, a natural alteration in the QS system might be seen. Certain bacteria strains re unable to synthesize self-inducing molecules or peptides. Instead, they react to certain auto-induced signals that are synthesized by several other bacterial cells. *Escherichia coli*, for instance, contains SdiA (a molecule homologous to LuxR), one of the signalling molecules which are activated by the corresponding signals from several other bacteria. Besides that, *Burkholderia cepacia* interacts with *P. aeruginosa* QS signals using cystic fibrosis signal molecules. [Vandeputte et al., 2011].

1.7 MICROBIOLOGICALLY SYNTHESIZED NANOPARTICLES AND THEIR ROLE IN SILENCING THE BIOFILM SIGNALLING CASCADE

Health and hygiene professionals are increasingly using nanoparticles (NPs) designed especially for penetration into living systems. As compared to physical and chemical methods for synthesizing NPs, microbial enzymes/proteins have been increasingly used as reducing agents over the past few decades. This is considered an efficient, ecologically friendly, and cost-effective alternative. The bacterial growth can be efficiently inhibited with the help of several metal nanoparticles which have the ability to penetrate through the exopolysaccharides that constitute the biofilm matrix, thus altering the quorum-sensing gene cascades, which subsequently halt the cellular communication mechanism, finally inhibiting biofilm synthesis. Nowadays, antibiotics are being replaced with numerous nanoparticles for treating several bacterial infections. Healthcare professionals often consider nanoparticles synthesized microbiologically to be an improved and safe approach for treating several antibacterial infections. It has also been documented that hyperthermia of tumours can be clinically treated using silver nanoparticles (Ag NPs). Several applications of silver nanoparticle have also been observed in the domain of chemical sensors medical imaging, wireless electronic logic memory chips, computer transistors, drug delivery, and catalysis. The nanoparticles can be efficiently stabilized using several biogenic sources which involve fungi, bacteria, and numerous plant parts. Exceptionally, fungi have been documented as playing an important role in utilizing enzymes present both intra- and extracellularly with a cell in the area of nanoparticle synthesis.

1.8 MICROBIAL EXOPOLYSACCHARIDES FOR SYNTHESIS OF NANOPARTICLES

Extracellular polymeric substances (EPS) nowadays are being employed as an effective substitute for the production of numerous metal nanoparticles microbiologically, as they are critically involved in surface adhesion and cellular communication. They are capable of synthesizing several nanoparticles via metal ions reduction, and also contribute to stabilizing nanoparticles by serving as a capping agent. EPS comprises several groups of organic compounds such as noncarbohydrate components, N-acetyl-D-glucosamine, and N-acetyl D-galactosamine, that subsequently affect their cation (primarily metal ions) interaction. In both Gram-positive and Gram-negative bacteria, during manufacture of nanoparticles, EPS comprises several functional groups which have the ability to synthesize and reduce via the process of chelation and capping. This further promotes size regulation, modifying the conformation of NPs, and finally particle dispersion. Numerous studies have also showed documentation mentioning that succinoglycan bacteria have been utilized during the production of silver nanoparticles. Another category of EPS is known as curdlan. It chiefly consists of the repeated units of (1, 3)-β-D-glucan coupled with β-(1, 3)-glycosidic bonds and is primarily synthesized by specific bacterial cells such as *Rhizobium* species, *Alcaligenes faecalis,* and *Agrobacterium* species. These are chiefly employed to synthesize and stabilize numerous nanoparticles. The silver ions were reduced more effectively due to the perseverance of groups comprising mainly hydroxyl and carboxyl having negative charge.

Dextran is chiefly synthesized by *Streptococcus mutans* and *Leuconostoc mesenteroides,* which are recognized as specific strains of lactic acid bacteria. It is a complex with branched glucan and notably consists of residues of glucose which remain coupled with α-(1,6) glycosidic bonds. Dextran is considered an essential element of EPS which facilitates the production of grapheme NPs. Bacterial biofilms comprising the strains of *Delftia acidovorans* and *Cupriavidus metallidurans* are predominantly employed in the synthesis of gold nanoparticles. Gold nanoparticles are capable of biofilm inhibition achieved by modifying the surface chemistry, the hydrophobicity, and finally the interaction between the lipids and proteins predominantly present on the cell membrane of a bacterium. The nanoparticles penetrate through the bacterial membrane and interact with its structural components, finally leading to biofilm disintegration.

1.9 MECHANISM OF QUORUM-SENSING INHIBITION USING MICROBIOGENIC NPs

As QS inhibitors, NPs act by influencing the process of cellular communication or by inhibiting the signals involved in QS. They hinder the synthesis of a wide range of signalling molecules and prevent the formation of a molecule-receptor complex. Consequently, the signal transduction cascade comes to a halt. Silver nanoparticles have been observed to consist of a durable antimicrobial activity, thus playing an important role in the process of quorum quenching.

A wide range of metal nanoparticles, including Au NPs, TiO_2, ZnO, and SiO_2, obtained from several strains of microbes, have the capability to potentially inhibit quorum-sensing cascade and subsequently inhibit biofilm development.

1.10 NP-ASSOCIATED INHIBITION OF QUORUM-SENSING CASCADE

Quorum sensing (QS), a noncytotoxic bacterial cell-to-cell communication mechanism, is the primary approach to monitor the proliferation of sessile populations of bacterial cells present in biofilm. The effectiveness of treating any chronic infection solely depends on developing an appropriate and efficient approach for conveying molecules to the respective target cells. The nanoparticles play an essential role in inhibiting quorum sensing, subsequently inhibiting biofilm development. NPs have been proven to be a widely recognized carrier employed mainly for delivering drugs, retaining the capability to suppress microbial cell proliferation and subsequently eliminating any harmful organisms. The capacity of ZnO NP to suppress the NorA efflux pump in *Staphylococcus aureus* has been discovered. ZnO NP, when combined with a specific antibiotic named ciprofloxacin, has been observed to improve the zone of inhibition of *E. coli* and *S. aureus* from 22% to 27%, respectively. Several studies have also demonstrated the fact that coating iron oxide NPs with polyacrylic acid improves the mechanism of efflux inhibition, subsequently restricting the growth of *Mycobacterium smegmatis*.

1.11 PLANT-MEDIATED SYNTHESIZED NANOPARTICLES IN DOWN-REGULATING QUORUM SENSING

A recent study demonstrated that gold nanoparticles coupled with a crude ethnobotanical extract (Au NPs + Ces) showed antibacterial activities against *P. aeruginosa* and thus they can be used to eliminate any microbial growth in biofilm. In total, 11 samples of crude plant extracts were mixed with gold nanoparticles and were employed against the expression of two QS genes, lasR and rhlR. The 10 plant species were *Hydra vulgaris, Dillenia philippinensis* leaves and bark, *Cassia alata, Persicaria odorata* leaves, *Eleusine indica* roots, *Mikania micrantha, Cymbopogon winterianus* leaves, *Urena lobata* leaves, *Phyllanthus urinaria,* and *Diplazium esculentum*, found in the region of Ilongot, Philippines. These were used for analysing the gene expression of RNA of *P. aeruginosa* PNMC 1335. It was observed that the expression of the gene rhlR showed significant down-regulation when treated with gold nanoparticles coupled with the crude extracts obtained from *H. vulgaris; D. esculentum;* from the leaves of *M. micrantha, Ceiba pentandra, E. indica, and U. lobata*; and from the bark of *P. odorata*, establishing the fact that they possess certain active QS inhibitors that can effectively lower the mechanism of biofilm formation. Down-regulation of both the genes *lasR* and *rhlR* was also observed in plants *C. alata, P. urinaria*, and *D. philippinensis*. However, the gene *lasR* showed sudden up-regulation in *D. esculentum*, in the leaves of *U. lobata* and *C. pentandra*, and in the roots of *E. indica*. Inhibiting the expression of two QS genes *lasR* and *rhlR*, respectively, has been known to display an antagonistic influence on the development of biofilm formation. By damaging tiny molecules in QS systems, QS circuits could be effectively disabled, subsequently reducing the bacterial pathogenicity [Schuster et al., 2003]. Transcription regulators LasR and RhlR must bind to their respective autoinducers, 3-oxo-dodecanoyl-homoserine lactone and butanoyl-homoserine lactone, to trigger the transcription of several characteristics associated with the virulence of the particular bacteria [Gopu et al., 2015].

A study also demonstrated that when the two strains of *Serratia marcescens*, the clinical isolate FJ584421 and ATCC 14756, as well as the two strains of *Proteus mirabilis*, MTCC 425 and ATCC 7002, when treated with the aqueous extract of *Piper betle* coupled with silver nanoparticles exhibit significant anti-QS effect and also inhibit the development of biofilm. The PbAg NPs significantly hinder the production of EPS in *S. marcescens* and *P. mirabilis*, both of which are well-known uropathogens causing urinary tract infections. Upon analyzing the gene expression in both pathogens, it was observed that several genes in both *S. marcescens* and *P. mirabilis* showed epic down-regulation [Ouyang et al., 2016; Corral-Lugo et al., 2016].

FlhD, which is known to be one of the two components of FlhDC operon, showed down-regulation. This gene is responsible for regulating the swarming motility in both *P. mirabilis* and *S. marcescens*. Flagellar biosynthetic protein B (FlhB) plays a crucial role in developing the rod structure in the flagellar

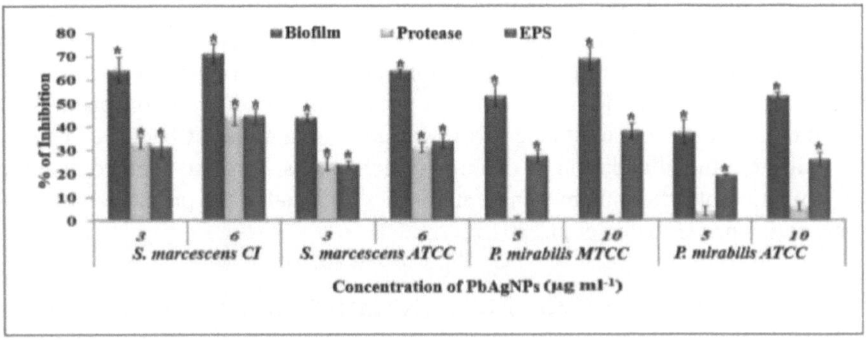

FIGURE 1.1 Demonstrating the impact of PbAg NPs on biofilm development, EPS, and protease in the two uropathogens.

basal body in the uropathogen *P. mirabilis*. Similarly, down-regulation was also observed in the RsbA, which focuses on regulating the biofilm development and the synthesis of EPS (Figure 1.1) [Vikram et al., 2010].

In *S. marcescens* down-regulation was also observed in the expression of three genes, fimA, fimC, and bsmB. FimA is a Type 1 fibrial protein and FimC is a Type 1 fimbriae periplasmic chaperone protein (Figure 1.1). In *S. marcescens* both of these proteins are responsible for surface adhesion and also assist the pathogen to develop colonies on biotic as well as abiotic surfaces. BsmB, on the other hand, is essential for synthesizing FimA [Vikram et al., 2012].

1.12 FUNGAL COMPONENTS IN DOWN-REGULATION OF QUORUM SENSING

P. aeruginosa is a pathogen mainly infecting individuals having a weakened immune system, like patients suffering from cystic fibrosis. A study showed that 33 extracts collected from 50 *Penicillium* species possesses QS inhibitory (QSI) components. Microorganisms which are pathogenic as well as resistant to different antibiotics lead to a serious public health issue all over the world. But the various genes that are related to antibiotic resistance are not confined to the clinic [Vikram et al., 2011]. They are actually widely available among the various bacterial populations inside the bacterial environment. For example, antibiotic synthesizing soil bacterium expresses the resistance genes that encode mechanisms of intrinsic resistance. Because of the development of antibiotic resistance in pathogens, most important reserves are of the resistance genes [Savka et al., 2015; Pejin et al., 2015; Burt et al., 2014]. It should be noted that the location of the factors that determine resistance in soil and in environmental bacteria implies a threat to human health. The location of these factors also affects their mobilization ability into new hosts and the different expression patterns expressed by them under various instances. For instance, transfer into plasmids and into integrins inside various pathogenic bacteria might be viewed as a huge danger sign [Majtan et al., 2014; Lu et al., 2014]. The selective pressure of various human activities also results in an increase of determinant factors in various populations of bacteria. This is why there is an urgent and increasing need to understand the distribution patterns of determinant factors of resistance in the various populations of bacteria. This in turn will help the elucidation of different mechanisms of resistance and also the determination of different environmental issues that lead to the promotion of antibiotic resistance [Fidaleo et al., 2015].

1.13 QUORUM SENSING AND ANTIMICROBIAL RESISTANCE

Every microorganism can develop resistance against drug molecules. Fungi develop antifungal resistance. Viruses develop antiviral resistance. Protozoa develop antiprotozoal resistance. Lastly, bacteria develop antibiotic resistance [Husain et al., 2013]. Antimicrobial resistance is the capacity of any microorganism to develop resistance against the effects of any kind of antibiotic treatment. Let us consider an example. *Lactobacillus fermentum, E. faecalis, Streptococcus* species including *S. mutans, Actinomyces naeslundii, Actinomyces viscosus, Prevotellabuccae, and Propionibacterium acidifaciens* and other bacterial species are seen as biofilm-producing bacteria and also they are sensitive to the antibiotics like penicillin G, amoxicillin, clindamycin, gentamicin, vancomycin, tetracycline, doxycycline, fosfomycin, rifampicin, ciprofloxacin, and moxifloxacin [Okusa et al., 2014]. This resistance arises from one of three mechanisms. The different genes responsible for the production of this enzyme that protects the bacterial cell from various antibiotics are inside the cell. The enzyme β-lactamase inhibits the activity of the primary group of antibiotics – for instance, penicillin by denaturation of the β-lactam ring which is located in penicillin [Adonizio et al., 2008]. The first β-lactamase enzyme underwent the process of identification inside *E. coli* before the secretion of the penicillin used in medicine. As well, families of antibiotics like aminoglycoside and aminocyclitol undergo inhibition by any element of antimicrobial resistance located inside the bacteria [Persson et al., 2005]. When the resistance increases because of the mutation effects inside the chromosome, this replaces replication, transcription, and translation processes which lead to different types of alteration in protein synthesis and forms mosaic protein by the process of transformation [Alvarez et al., 2012]. Because of horizontal as well as vertical gene transfer, the novel genetic component is injected inside the same or different species or genera. When the mutation takes place inside the nucleotide sequence, it gives rise to a novel generation of the antimicrobial resistance genes – for instance, a mutation occurring due to substitution of the nucleotide base of 23S rRNA, causes an alteration in ribosomal proteins L4 and L22, giving rise to macrolide resistance. Resistance can arise spontaneously or as a result of some random mutation [Zaki et al., 2013]. But increased use of the antimicrobial substances is encouraging in the process of selection of different mutations that can cause resistance [Song et al., 2012; Chaudhari et al., 2014].

1.14 CONCLUSION

Biofilm formation in bacteria is governed by a quorum sensing mechanism that is a complex circuit of signalling molecules whose outflow/inflow within the EPS matrix is a strictly regulated phenomenon. The QS mechanism is triggered by small molecules known as autoinducers (AIs) that are synthesized and secreted by the bacteria depending on the cell population density in the neighbourhood. These signalling molecules, as well as their receptors, are being categorized into three main parts: (i) autoinducer (AI)-2 that promotes interspecies communication in Gram-negative and Gram-positive bacteria, (ii) Autoinducing peptides or oligopeptides that include 5–34 amino acid residues and are mainly used in Gram-positive bacteria for promoting their intercellular communication [Olivero-Verbel et al., 2014; Alvarez et al., 2014]. Studies have also showed that the bulk of such peptides are discharged by particular systems and transformed post-translationally in various manners so that they can be detected by several other bacterial cells through the use of membrane-embedded receptors and finally (iii) N-acyl homoserine lactones known as AHLs released by Gram-negative bacteria possesses variable length and degrees of acyl side chain carrying the purpose of supervising their population density in QS-mediated gene expression and regulation [Abdul Wahab et al., 2014; Ceylan et al., 2014]. This book consists of chapters that will highlight on the detailed mechanism of biofilm formation via QS signalling molecules (AIs, AHLs, AIPs), varied survival strategies developed against antimicrobial agents, and alternative therapeutics developed for fighting antimicrobial resistance (AMR). This includes chapters focusing on phytocompounds, green synthesized nanoparticles, and fungal-derived bioactive compounds as alternative green approaches for fighting the persistent biofilms and sheds light on the plausible solution to AMR.

REFERENCES

Abdul Wahab NA, Zain MS, Kader J, Radzi SM, Noor HM. Study on anti-quorum sensing potential of selected local ulam in Malaysia. World J Pharm Pharm Sci. 2014;3:203–11.

Adonizio A, Kong KF, Mathee K. Inhibition of quorum sensing-controlled virulence factor production in Pseudomonas aeruginosa by South Florida plant extracts. Antimicrob Agents Chemother. 2008;52:198–203.

Alvarez MV, Moreira MR, Ponce A. Antiquorum sensing and antimicrobial activity of natural agents with potential use in food. J Food Saf. 2012;32:379387.

Alvarez MV, Ortega-Ramirez LA, Gutierrez-Pacheco MM, Bernal-Mercado AT, Rodriguez-Garcia I, Gonzalez-Aguilar GA, et al. Oregano essential oil-pectin edible films as anti-quorum sensing and food antimicrobial agents. Front Microbiol. 2014;5:699.

Burt SA, Ojo-Fakunle VT, Woertman J, Veldhuizen EJ. The natural antimicrobial carvacrol inhibits quorum sensing in Chromobacterium violaceum and reduces bacterial biofilm formation at sub-lethal concentrations. PLoS One. 2014;9:e93414.

Castillo-Juárez I, García-Contreras R, Velázquez-Guadarrama N, Soto-Hernández M, Martínez-Vázquez M. Amphypterygium adstringens anacardic acid mixture inhibits quorum sensing-controlled virulence factors of Chromobacterium violaceum and Pseudomonas aeruginosa. Arch Med Res. 2013;44:488–94.

Ceylan O, Ugur A, Sarac N. In vitro antimicrobial, antioxidant, antibiofilm and quorum sensing inhibitory activities of Bellis perennis L. J BioSci Biotechnol. 2014:35–42.

Chan KG, Atkinson S, Mathee K, Sam CK, Chhabra SR, Cámara M, et al. Characterization of N-acylhomoserine lactone-degrading bacteria associated with the Zingiber officinale (ginger) rhizosphere: Co-existence of quorum quenching and quorum sensing in Acinetobacter and Burkholderia. BMC Microbiol. 2011;11:51.

Chaudhari V, Gosai H, Raval S, Kothari V. Effect of certain natural products and organic solvents on quorum sensing in Chromobacterium violaceum. Asian Pac J Trop Med. 2014;7S1:S204–11.

Chong YM, Yin WF, Ho CY, Mustafa MR, Hadi AH, Awang K, et al. Malabaricone C from Myristica cinnamomea exhibits anti-quorum sensing activity. J Nat Prod. 2011;74:2261–4.

Corral-Lugo A, Daddaoua A, Ortega A, Espinosa-Urgel M, Krell T. Rosmarinic acid is a homoserine lactone mimic produced by plants that activates a bacterial quorum-sensing regulator. Sci Signal. 2016;9:ra1.

Davies DG, Parsek MR, Pearson JP, Iglewski BH, Costerton JW, Greenberg EP, et al. The involvement of cell-to-cell signals in the development of a bacterial biofilm. Science. 1998;280:295–8.

Deep A, Chaudhary U, Gupta V. Quorum sensing and bacterial pathogenicity: From molecules to disease. J Lab Physicians. 2011;3:4–11.

Ditu LM, Chifiriuc MC, Bezirtzoglou E, Voltsi C, Bleotu C, Pelinescu D, et al. Modulation of virulence and antibiotic susceptibility of enteropathogenic Escherichia coli strains by Enterococcus faecium probiotic strain culture fractions. Anaerobe. 2011;17:448–51.

Fidaleo M, Lavecchia R, Zuorro A. Antibacterial and anti-quorum sensing activities of selected Italian honeys against antibiotic-resistant pathogens. Online J Biol Sci. 2015;15:236–43.

Gopu V, Kothandapani S, Shetty PH. Quorum quenching activity of Syzygium cumini (L.) skeels and its anthocyanin malvidin against Klebsiella pneumoniae. Microb Pathog. 2015;79:61–9.

Guan LL, Onuki H, Kamino K. Bacterial growth stimulation with exogenous siderophore and synthetic N-acyl homoserine lactone autoinducers under iron-limited and low-nutrient conditions. Appl Environ Microbiol. 2000;66:2797–803.

Hayek SA, Gyawali R, Ibrahim SA. Antimicrobial natural products. In: Mendez-Vilas A, editor. Microbial Pathogens and Strategies for Combating Them: Science, Technology and Education. Vol. 2. Badajoz: Formatex Research Center; 2013. pp. 910–21.

Husain FM, Ahmad I, Asif M, Tahseen Q. Influence of clove oil on certain quorum-sensing-regulated functions and biofilm of Pseudomonas aeruginosa and Aeromonas hydrophila. J Biosci. 2013;38:835–44.

Kabir AH, Roy AG, Alam MF, Islam R. Detection of quorum sensing signals in gram-negative bacteria by using reporter strain CV026. Not Sci Biol. 2010;2:72–5.

Kang JE, Han JW, Jeon BJ, Kim BS. Efficacies of quorum sensing inhibitors, piericidin A and glucopiericidin A, produced by Streptomyces xanthocidicus KPP01532 for the control of potato soft rot caused by Erwinia carotovora subsp. atroseptica. Microbiol Res. 2016;184:32–41.

Krishnan T, Yin WF, Chan KG. Inhibition of quorum sensing-controlled virulence factor production in Pseudomonas aeruginosa PAO1 by Ayurveda spice clove (Syzygium aromaticum) bud extract. Sensors (Basel). 2012;12:4016–30.

Livermore DM. British Society for Antimicrobial Chemotherapy Working Party on the urgent need: Regenerating antibacterial drug discovery and development. Discovery research: The scientific challenge of finding new antibiotics. J Antimicrob Chemother. 2011;66:1941–4.

Lu J, Turnbull L, Burke CM, Liu M, Carter DA, Schlothauer RC, et al. Manuka-type honeys can eradicate biofilms produced by Staphylococcus aureus strains with different biofilm-forming abilities. Peer J. 2014;2:e326.

Majtan J, Bohova J, Horniackova M, Klaudiny J, Majtan V. Anti-biofilm effects of honey against wound pathogens Proteus mirabilis and Enterobacter cloacae. Phytother Res. 2014;28:69–75.

Mohamed GA, Ibrahim SR, Shaaban MI, Ross SA. Mangostanaxanthones I and II, new xanthones from the pericarp of Garcinia mangostana. Fitoterapia. 2014;98:215–21.

Newman DJ, Cragg GM, Snader KM. Natural products as sources of new drugs over the period 1981–2002. J Nat Prod. 2003;66:1022–37.

Okusa PN, Rasamiravaka T, Vandeputte O, Stévigny C, Jaziri ME, Duez P. Extracts of Cordia gilletii de wild (Boraginaceae) quench the quorum sensing of Pseudomonas aeruginosa PAO1. J Intercult Ethnopharmacol. 2014;3:138–43.

Olivero-Verbel J, Barreto-Maya A, Bertel-Sevilla A, Stashenko EE. Composition, anti-quorum sensing and antimicrobial activity of essential oils from Lippia alba. Braz J Microbiol. 2014;45:759–67.

Ouyang J, Sun F, Feng W, Sun Y, Qiu X, Xiong L, et al. Quercetin is an effective inhibitor of quorum sensing, biofilm formation and virulence factors in Pseudomonas aeruginosa. J Appl Microbiol. 2016;120:966–74.

Pejin B, Ciric A, Glamoclija J, Nikolic M, Sokovic M. In vitro anti-quorum sensing activity of phytol. Nat Prod Res. 2015;29:374–7.

Persson T, Hansen TH, Rasmussen TB, Skindersø ME, Givskov M, Nielsen J, et al. Rational design and synthesis of new quorum-sensing inhibitors derived from acylated homoserine lactones and natural products from garlic. Org Biomol Chem. 2005;3:253–62.

Rocha-Estrada J, Aceves-Diez AE, Guarneros G, de la Torre M. The RNPP family of quorum-sensing proteins in gram-positive bacteria. Appl Microbiol Biotechnol. 2010;87:913–23.

Savka MA, Dailey L, Popova M, Mihaylova R, Merritt B. Chemical composition and disruption of quorum sensing signaling in geographically diverse United States propolis. J Evid Based Complementary Altern Med. 2015;2015:472593–603.

Schuster M, Lostroh CP, Ogi T, Greenberg EP. Identification, timing, and signal specificity of Pseudomonas aeruginosa quorum-controlled genes: A transcriptome analysis. J Bacteriol. 2003;185:2066–79.

Singh BN, Singh BR, Singh RL, Prakash D, Sarma BK, Singh HB. Antioxidant and anti-quorum sensing activities of green pod of Acacia nilotica L. Food Chem Toxicol. 2009;47:778–86.

Song C, Ma H, Zhao Q, Song S, Jia Z. Inhibition of quorum sensing activity by ethanol extract of Scutellaria baicalensis Georgi. J Plant Pathol Microbiol. 2012;S7:1–4.

Vandeputte OM, Kiendrebeogo M, Rasamiravaka T, Stévigny C, Duez P, Rajaonson S, et al. The flavanone naringenin reduces the production of quorum sensing-controlled virulence factors in Pseudomonas aeruginosa PAO1. Microbiology. 2011;157:2120–32.

Vikram A, Jesudhasan PR, Jayaprakasha GK, Pillai BS, Patil BS. Grapefruit bioactive limonoids modulate E. coli O157:H7 TTSS and biofilm. Int J Food Microbiol. 2010;140:109–16.

Vikram A, Jesudhasan PR, Jayaprakasha GK, Pillai SD, Patil BS. Citrus limonoids interfere with Vibrio harveyi cell-cell signalling and biofilm formation by modulating the response regulator LuxO. Microbiology. 2011;157:99–110.

Vikram A, Jesudhasan PR, Pillai SD, Patil BS. Isolimonic acid interferes with Escherichia coli O157:H7 biofilm and TTSS in QseBC and QseA dependent fashion. BMC Microbiol. 2012;12:261.

Wu H, Moser C, Wang HZ, Høiby N, Song ZJ. Strategies for combating bacterial biofilm infections. Int J Oral Sci. 2015;7:1–7.

Zaki AA, Shaaban MI, Hashish NE, Amer MA, Lahloub MF. Assessment of anti-quorum sensing activity for some ornamental and medicinal plants native to Egypt. Sci Pharm. 2013;81:251–8.

2 Quorum Sensing in Gram-Positive Bacteria

Smaranika Pattnaik

2.1 INTRODUCTION

Gram-positive bacteria involved in chronic infections are notorious for their pan-drug resistance and biofilm production. The bacteria may be cocci, bacilli, and are spiral in shape (see Figure 2.1). Among the most significant disease-causing bacteria are strains of *Staphylococcus aureus*, *S. epidermidis*, *S. saprophyticus*, *S. haemolyticus*, *Streptococcus pyogenes*, *S. pneumoniae*, *S. mutans*, *Enterococcus faecalis*, *Bacillus subtilis*, *B. anthracis*, *Clostridium tetani*, *C. perfringes*, and *C. difficile*. The bacteria, being adaptive by nature, can make a niche in clinical as well as nosocomial sinks. The β-lactams, methicillins, and moreover the last-resort drug, vancomycin, are well resisted by the candidate bacteria, prolonging the infection. The Gram-positive response could be due to a thick layer of peptidoglycal intercalated with lipotiechoic acid. The overall architecture gives the bacterium rigidity, strength, and shape maintenance. Further, the hypervirulent bacteria are found to be associated with capsular polysaccharides, which enable the bacteria to evade the action of immunocytes as well as keep the bacteria away from drug molecule interaction. The polysaccharide can

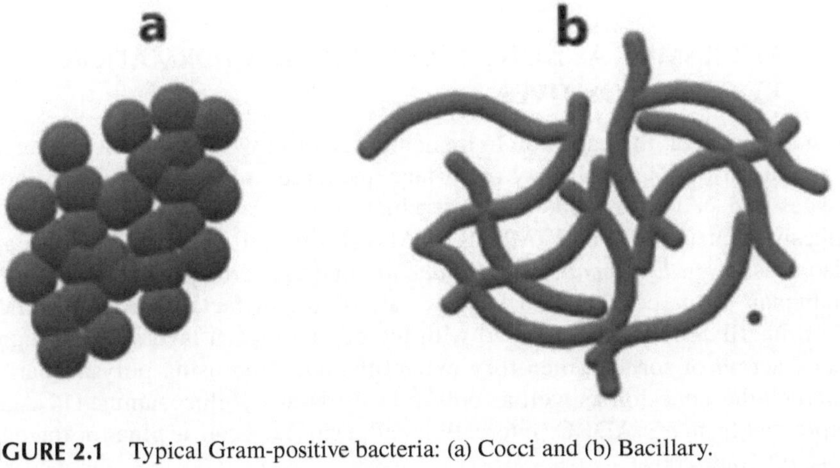

FIGURE 2.1 Typical Gram-positive bacteria: (a) Cocci and (b) Bacillary.

DOI: 10.1201/9781003300557-2

mimic the mammalian glycan benefit, so the host immune system may be biased (Cress et al. 2014).

Biofilm, a three-dimensional architectural product of a team of bacteria, acts as a barrier to the entry of antibiotics. The biofilm is a cumulative product of bacterial candidates; it aggregates as well as integrates biomolecules like extracellular DNA, protein and polysaccharides. The wall teichoic acid, covalently linked with peptidoglycan and the lipoteichoic acid, can significantly enhance the interbacterial adhesion process to initiate biofilm matrix aggregation (Balducci et al. 2023). Furthermore, the polysaccharides play the matrix role and can act like a glue so that the bacteria can adhere to one another by compromising shape and space (Limoli et al. 2015). Staphylococcal biofilm are constituted with the biomolecule poly-β (1–6)-N-acetylglucosamine (PNAG), partially deacetylated, positively charged (Arciola et al. 2015; Nguyen et al. 2020).

2.2 GRAM-POSITIVE BACTERIAL POLYSACCHARIDES AS THE BASEMENT OF BIOFILM MATRIX

The proportion of exopolysaccharide (EPS) (Lahiri et al. 2022) in biofilms can comprise approximately 50–90% of the total organic matter (Vu et al. 2009). Gram-positives like *Streptococcus pneumoniae* (Paton and Trappetti 2019) or *Staphylococcus* have capsular polysaccharide assembly, which has glycosidic linkage with the peptidoglycan layer via β-D-N-acetylglucosamine and can be knitted into biofilm (Larson and Yother 2017). The Gram-positive bacterial polysaccharides act as foundation of biofilm matrix, such as using polysaccharide intercellular adhesion by Staphylococcal candidates, or using flagellar adsorption to secreted exopolysaccharides by *Bacillus* strains (Ruhal and Kataria 2022), the bacterial cells go for the more complex biofilm formation.

2.3 MSCRAMMS AS ADHESIVES IN BIOFILM FORMATION BY GRAM-POSITIVES

In staphylococci, the adhesion to human tissue or indwelling medical devices is regulated by a large variety of surface anchored proteins that bind to host tissues and cells, referred to as microbial surface components recognizing adhesive matrix molecules (MSCRAMMs) (Foster 2019). These proteins are also present on *E. faecalis* and *E. faecium* and are required in the first step of human tissue colonization (Sava et al. 2010). In fact, the MSCRAMMs are primarily covalently attached with the peptidoglycan layers by the enzymatic action of sortage, then they extend the adhesion using polysachharide intercellular adhesion as well as poly-β (1–6)-N-acetylglucosamine (PNAG), expressed from *ica*-ADBC (intercellular adhesion) operon as glue for the purpose (O'Gara 2007).

2.4 CANNIBALISM AND FRATRICIDE MECHANISM OF GRAM-POSITIVES TO FORM BIOFILM

Cannibalism in *B. subtilis* and fratricide mechanism in *S. pneumoniae* to form biofilm has been reported by authors in the past (Claverys and Håvarstein 2007). The cannibalism and fratricide mechanism (also called stress response social behaviour) are deployed by Gram-positives. Under stress, the Gram-positive rods like *Bacillus* species undergo long-term sporulation, which is known as cannibalism (Hofler et al. 2016). The bacteria produce two types of toxins, namely sporulation-delaying protein and sporulation-killing factor. As the names suggest, the former protein allow the bacteria to defer sporulation until a favourable time, and the latter is used as a chemical weapon to kill the siblings. The possible reason behind this behaviour is to compensate for loss of food and nutrition during a period of scarcity. Likewise, the Gram-positive cocci, Staphylococcal, or Streptococcal cells undergo fratricide mechanism, an autolysis or suicidal act, as a result of which the sister cells commit self-immolation so the bacteria can overcome the unfavourable condition. In the clinical scenario, these phenomena are ubiquitous due to prescription of antibiotics. The Gram-positive bacteria opt for these mechanisms to maintain the viability of competent cells, with sacrifice of some sister cells. These mechanisms are also extremely useful for the aggregation of biofilms. The autolysed or deliberately killed bacterial cell extrudes biomolecules like protein and DNA. Lipids are engaged as extracellular constructive units for biofilm formation.

2.5 QUORUM SENSING AS REGULATORY MECHANISM FOR BIOFILM FORMATION IN GRAM-POSITIVES

Quorum sensing (QS) is directed towards biofilm-producing bacterial cells for the purpose of communication and coordination; they keep safe under the umbrella of biofilm. Hence, certain sensor molecules, known as quorum-sensing molecules (QSMs), are deployed to detect the degree and depth of cell density. There is a direct relationship between depth in density of bacteria and the functionality of quorum sensing (Darch et al. 2012). The molecules associated with QS are released from a set of bacterial cells to the extraneous environment, so that the QSMs can bind with cognate receptors of other cells to intensify the QS signal transduction pathway. The Gram-positives carry out post-translational modifications and express oligopeptides which are known as auto-inducer peptides (AIPs). AIPs can undergo frequent structural variations (Verbeke et al. 2017). Figure 2.2 shows the AIP-mediated quorum sensing in Gram-positive bacteria.

It is evident that the AIPs establish contact with bacterial surface membrane protein (BSMP), as shown in Figure 2.3, to carry forward the signalling pathway, to express both the cytoplasmic effector molecule and the bacterial internalized protein (BIP). BSMP and BIP are considered as QS regulators, of indirect and direct QS mechanisms. The BIPs are clustered and designated as an RRNPP protein family (R, Rap; R, Rgg; N, Npr; P, PlcR; P, PrgX).

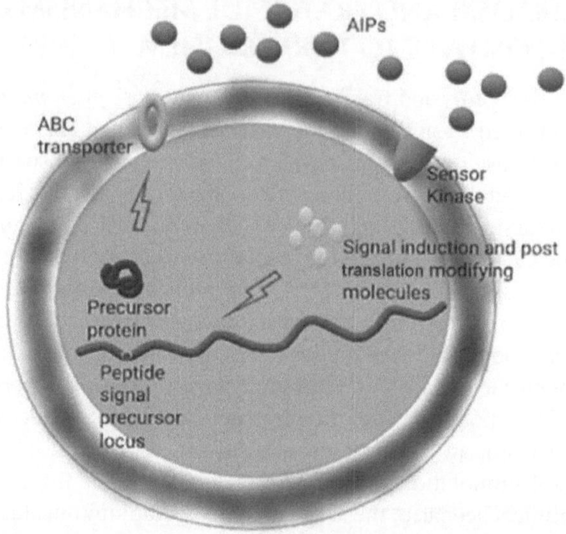

FIGURE 2.2 AIP-induced quorum sensing in Gram-positive bacteria. The signal peptide sequence is present in the peptide signal precursor locus. After being sensitized by signal, a post-translational modification event starts, and the genes express precursor protein, which is moved outside through the ABC transporter. The sensor kinase acts as a regulator.

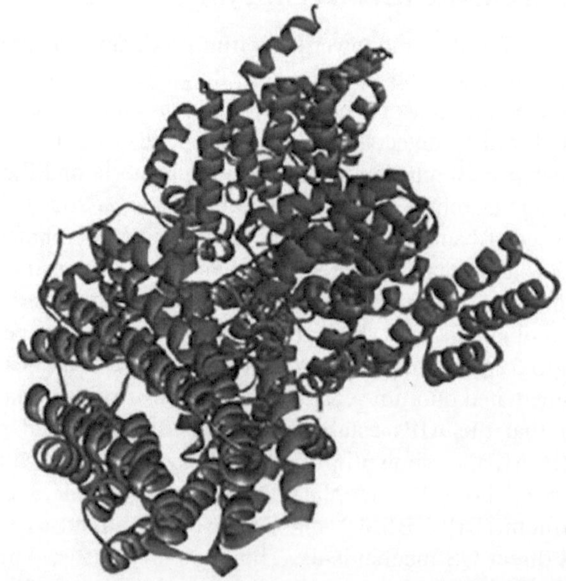

FIGURE 2.3 The interacting AIP (red) and bacterial surface membrane protein (green).

The specificity of these proteins lies with their common structural entity, the presence of tetratricopeptide repeats (TPRs). The TPRs are meant for protein-protein interactions, defining the BSMP-AIP protein interaction models. The Gram-positive AIP-binding QS receptors which are well studied are PrgX in. *faecalis* (Shi et al. 2005), and PlcR and NprR in *B. cereus* (Rocha et al. 2012; Perchat et al. 2011).

2.6 QS SYSTEM OPERATION BY GRAM-POSITIVES

The QS events are well operated by *S. aureus*, *S. epidermidis* and *Listeria monocytogenes*, reported as the agr system. The *fsr* QS system is active in *E. faecalis* and in the *B. subtilis* LuxS QS operating system. PlcRa, a quorum-sensing regulator from *B. cereus* and Rgg/small hydrophobic peptide quorum-sensing systems are evident in *S. pneumoniae*. Indole is also a QS molecule engaged in biofilm formation by *Micrococcus luteus*. *B. subtilis* cells are able to sense self-produced autoinducers, which gives rise to stronger quorum-sensing-mediated responses, in a process that can influence the generation of persisters during antibiotic treatment (Saha et al. 2018).

2.6.1 Quorum-Sensing System of *S. Aureus*

The agr QS system is composed of the response regulator AgrA and the sensor histidine kinase AgrC that in response to auto-inducing peptides expresses a regulatory RNA, RNAIII. At high cell density RNAIII mediates the transition from production of host matrix binding and immune evasion proteins to expression of a large number of extracellular toxins including the α-haemolysin encoded by *hla*. Within RNAIII itself the δ toxin is also encoded (Paulander et al. 2018).

The agr locus is composed of two divergent transcripts, RNAII (encoding agrB, agrD, agrC, and agrA) and RNAIII. RNAII is a polycystronic transcript of four genes, agrBDCA, which encodes a two-component signal transduction system (AgrCA), a 46-residue propeptide (AgrD), and a product (AgrB) involved in the processing and transport of the mature 7–9-residue-autoinducing peptide (AIP). RNAIII express the d-haemolysin gene hld. RNAIII is the key player of expression of a-haemolysin (*hla*) and protein A (*spa*) genes. There is a direct coordination between the suppression of cell wall–associated protein production and enhancement of secreted exoprotein production, linked with high cell density population. RNAII encodes four genes associated with AIP production and channelling the sensation of AIP. AgrD is the gene of autoinducing signal peptides, ArgB the locus for secretion of cyclic oligopeptides, and AgrA and AgrC, the two-component histidine kinase response regulator pair, are constitutively expressed in response to AIPs. Furthermore, AgrA and AgrC maintains the transcription of RNAII as well as RNAIII. Interestingly, the RNAIII transcript can act as a regulatory effector molecule in the agr QS system. In addition, RNA III encodes d-haemolysin. Figure 2.4 shows the agr QS system model of *S. aureus*.

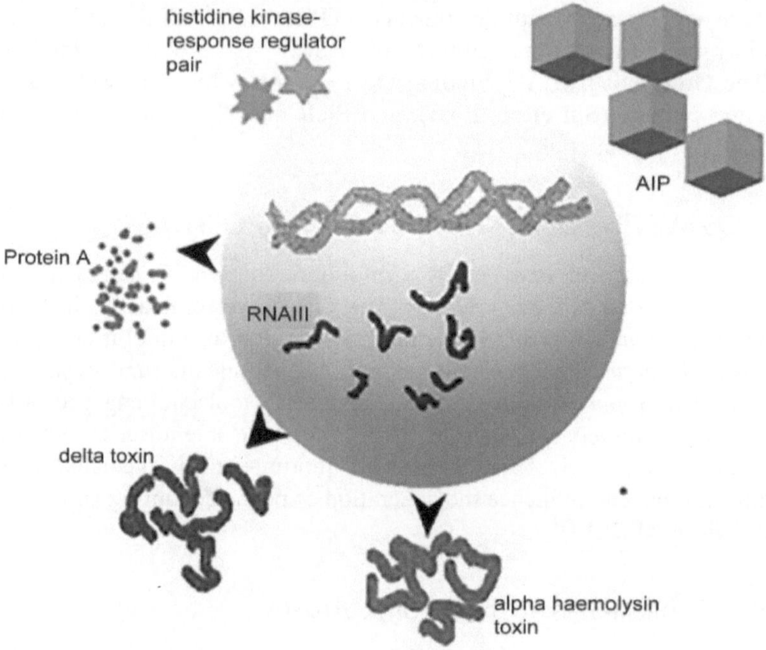

FIGURE 2.4 The agr QS system model of *S. aureus*. The AIP product concentration is directly linked to expression of RNA III and RNA II, which in turn express and expel alpha-haemolysin toxin and delta-toxin, respectively. In addition, protein A is also produced. Histidine kinase response regulator is present to coordinate and regulate the QS system.

2.6.2 The ComX Quorum-Sensing System of *B. Subtilis*

The ComX QS system of *B. subtilis*, an opportunistic bacterium that causes nosocomial infections, is notorious for promoting biofilm formation. ComX pheromone, the product of *comX* and *comQ* genes, is a 10-amino-acid peptide that acts as a modulator for gene expression in concurrence to cell population density (Comella 2004). It has been reported that to sense ComX pheromone, the cell-coupling is established through microchannels to share intracellular contents and electro-channels to decipher the metabolic status, respectively. Figure 2.5 shows the ComX QS model of *B. subtilis*. The coordination of different gene loci and their expressions in the form of processing and production of ComX, ComP, ComQ, proteases, pheromones, and polysaccharides for biofilm aggregation is shown. In addition, the expression of surfactin, which supplements the gliding movement of bacteria on surfaces, is also shown. It is pertinent to mention that this adaptation of bacteria towards gliding movement favours biofilm growth (Kalamara et al. 2018). Hence, the ComX QS system in *B. subtilis* is multimeric, both structurally and functionally.

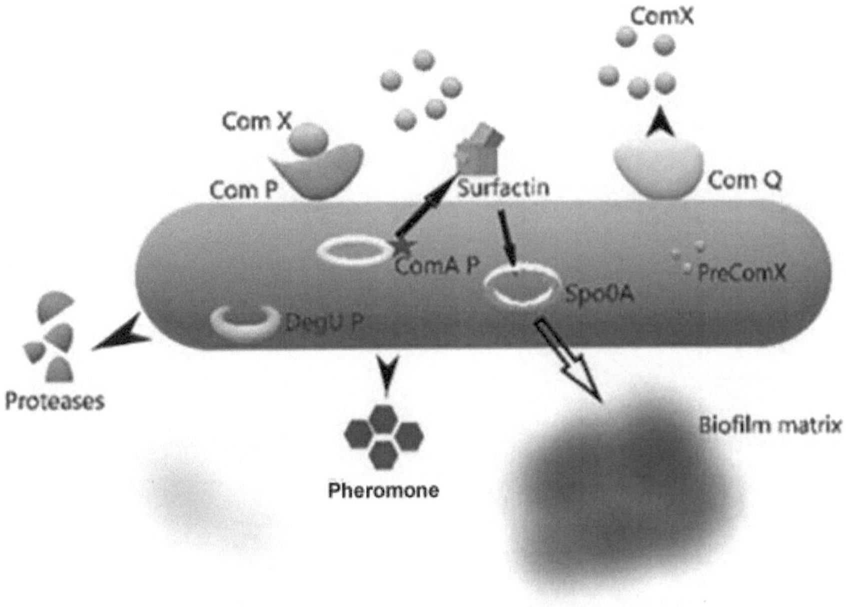

FIGURE 2.5 The ComX QS system model of *B. subtilis*. Prior to biofilm formation, the ComX is expressed and associated with ComP, giving a signal to DegUP (phosphorylated), which in turn expresses extracellular proteases. ComAP (phosphorylated) expresses the surfactin, a lipid-based molecule, for which bacteria attains the sliding movement. Surfactin in turn activates Spo0A, which encodes the biomolecules (polysaccharides) for biofilm matrix aggregation.

Structurally, also, there is attainment of polymorphotypes by the *Bacillus* candidate. The bacteria cell may be: (a) flagellated for swarming movement; (b) surfactin-coated for sliding movement; (c) competent cells; (d) mucoid cells; (e) cannibalistic cells; (f) miner cells; and (g) dead/ghost cells, as shown in Figure 2.6.

2.6.3 Quorum-Sensing System of *E. faecalis*

E. faecalis, previously called *S. faecalis*, is a major nosocomial pathogen. It is isolated from food, clinical specimens, and oral clefts (Anderson et al. 2016). *E. faecalis* possesses a number of different QS systems that enhance its pathogenesis within the host and also facilitate plasmid transfer from donor cells to recipient cells (see Figure 2.7).

In the horizontal transfer of plasmid and shutdown of pheromone, the QS molecule is correlated with transfer of antibiotic resistance genes and/or fitness genes to survive in stress-imposed conditions, thriving under the umbrella of biofilms (Xayarath and Freitag 2016). The Enterococci (*E. faecalis*) can also express *epa*

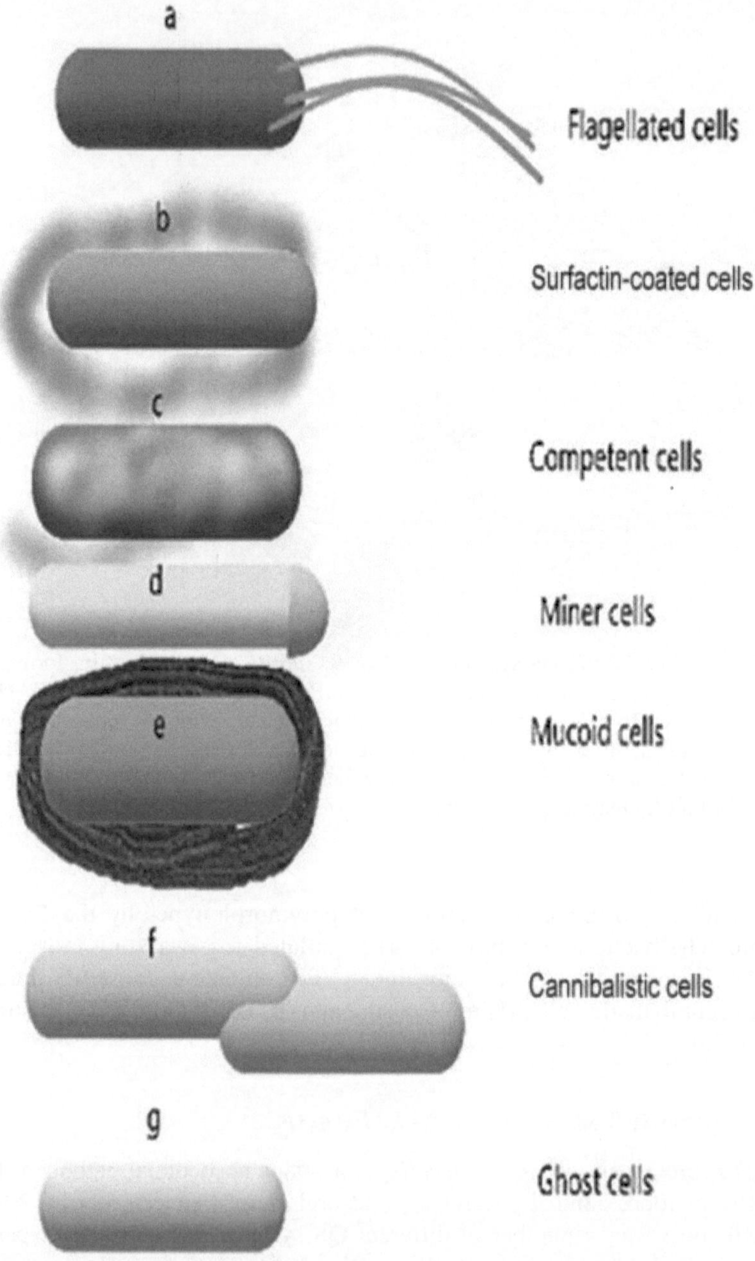

FIGURE 2.6 The morphotypes adapted by biofilm producing QS *B. subtilis* bacteria. (a) flagellated for swarming movement; (b) surfactin-coated for sliding movement; (c) competent cells; (d) miner cells; (e) mucoid cells; (f) cannibalistic cells; and (g) ghost (dead) cells.

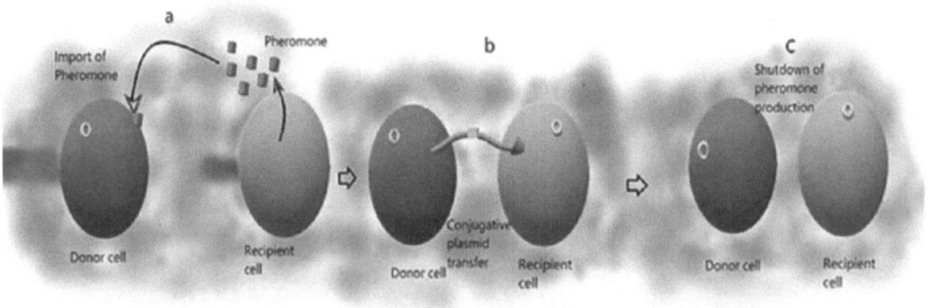

FIGURE 2.7 Pheromone production by recipient cell and import to donor cell (a) induces conjugative transfer of plasmid to recipient cell (b) followed by shutdown of pheromone production (c).

(enterococcal polysachharide antigen), which provokes phage intrusion as well as keeps bacteria fit. This is regarded as novel *Enterococcus*-phage interaction (Chatterjee et al. 2020).

2.6.4 Quorum-Sensing System of *Clostridium difficile*

C. difficile is an opportunistic Gram-positive, anaerobic gut bacteria that can create biofilm with other genera (*Bacteroides fragilis*). This is defined as an interbacterial quorum-sensing technique which is depicted in Figure 2.8. This bacteria produces the QS molecule AI-1, which can establish community QS by indulging the interbacterial interactions. LuxS-expressed AI-2 is also reported as a cross-species QS molecule. In addition, the bacteria is also involved in biofilm formation with utilization of eDNA (extracellular DNA), the output product of phage-mediated lysis (Slater et al. 2019). *C. difficile* also carries out QS and biofilm formation, in association with bacteriophage. The bacteriophages, causing lysis to the host bacteria, liberate the DNA, protein, and other the cellular inclusions, which form the biofilm in association with the phages (See Figure 2.9).

2.7 CONCLUSION

In a nutshell, Gram-positive bacteria meticulously operate a QS system with formation of biofilm by autoinducer molecules. The degree and mode of the QS operating system is typical with *Staphylococcus* species, *Bacillus* species, *Enterococcus* species, and *Clostridiodes* species, in context of utilization and acclimatization of biomolecules for the aggregation of biofilm matrix. It is important to mention that plasmids as well as phages also support the QS and form biofilm, as seen in *E. faecalis* and *C. difficile,* respectively. Hence, efforts should be made to block the QS system operated by candidate bacteria in the sequence of chronic and persistent infections. The QS molecule autoinducers may be targeted and the

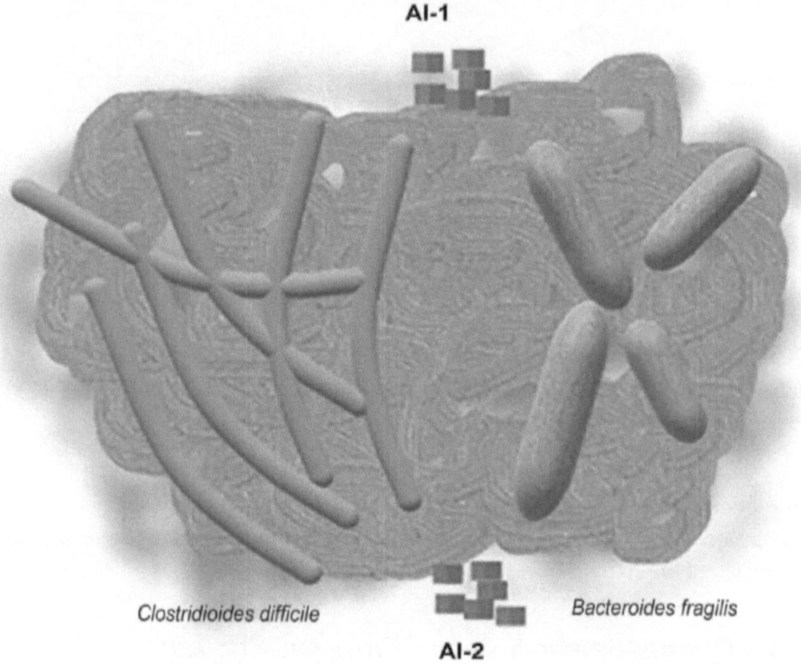

FIGURE 2.8 The interbacterial QS system operated by *C. difficile* and *B. fragilis*. Both AI-1 and AI-2 can indulge the QS and form biofilm under which they bind a common niche.

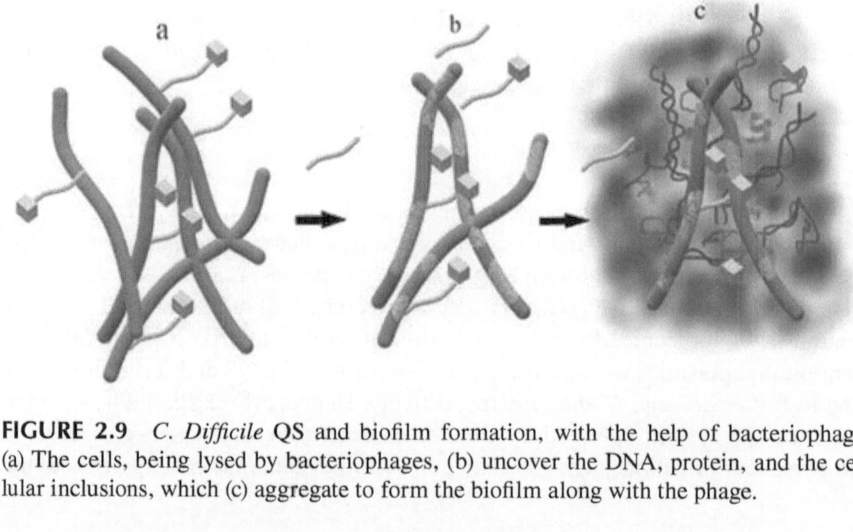

FIGURE 2.9 *C. Difficile* QS and biofilm formation, with the help of bacteriophage. (a) The cells, being lysed by bacteriophages, (b) uncover the DNA, protein, and the cellular inclusions, which (c) aggregate to form the biofilm along with the phage.

pertinent anti-autoinducers may be prescribed. In case of Staphylococcal QS, RNAII and RNAIII transcripts are more challenging, as the molecules encode the genes and act as player molecules also. The CRISPR-Cas sequence editing technologies are needed to block the QS signals.

REFERENCES

Anderson, Annette C., Daniel Jonas, Ingrid Huber, Lamprini Karygianni, Johan Wölber, Elmar Hellwig, Nicole Arweiler, Kirstin Vach, Annette Wittmer, and Ali Al-Ahmad. "*Enterococcus faecalis* from food, clinical specimens, and oral sites: prevalence of virulence factors in association with biofilm formation." *Frontiers in Microbiology* 6 (2016): 1534.

Arciola, Carla Renata, Davide Campoccia, Stefano Ravaioli, and Lucio Montanaro. "Polysaccharide intercellular adhesin in biofilm: structural and regulatory aspects." *Frontiers in Cellular and Infection Microbiology* 5 (2015): 7.

Balducci, Evita, Francesco Papi, Daniela Eloisa Capialbi, and Linda Del Bino. "Polysaccharides' structures and functions in biofilm architecture of antimicrobial-resistant (AMR) pathogens." *International Journal of Molecular Sciences* 24, no. 4 (2023): 4030.

Chatterjee, Anushila, Julia L.E. Willett, Uyen Thy Nguyen, Brendan Monogue, Kelli L. Palmer, Gary M. Dunny, and Breck A. Duerkop. "Parallel genomics uncover novel enterococcal-bacteriophage interactions." *MBio* 11, no. 2 (2020): e03120–e03119.

Claverys, Jean-Pierre, and Leiv S. Håvarstein. "Cannibalism and fratricide: mechanisms and raisons d'etre." *Nature Reviews Microbiology* 5, no. 3 (2007): 219–229.

Comella, Natalia. "Characterization of a transcriptional response to population density in Bacillus subtilis." PhD diss., Massachusetts Institute of Technology, 2004.

Cress, Brady F., Jacob A. Englaender, Wenqin He, Dennis Kasper, Robert J. Linhardt, and Mattheos A.G. Koffas. "Masquerading microbial pathogens: capsular polysaccharides mimic host-tissue molecules." *FEMS Microbiology Reviews* 38, no. 4 (2014): 660–697.

Darch, Sophie E., Stuart A. West, Klaus Winzer, and Stephen P. Diggle. "Density-dependent fitness benefits in quorum-sensing bacterial populations." *Proceedings of the National Academy of Sciences* 109, no. 21 (2012): 8259–8263.

Foster, Timothy J. "The MSCRAMM family of cell-wall-anchored surface proteins of gram-positive cocci." *Trends in Microbiology* 27, no. 11 (2019): 927–941.

Höfler, Carolin, Judith Heckmann, Anne Fritsch, Philipp Popp, Susanne Gebhard, Georg Fritz, and Thorsten Mascher. "Cannibalism stress response in *Bacillus subtilis*." *Microbiology* 162, no. 1 (2016): 164–176.

Kalamara, Margarita, Mihael Spacapan, Ines Mandic-Mulec, and Nicola R. Stanley-Wall. "Social behaviours by Bacillus subtilis: quorum sensing, kin discrimination and beyond." *Molecular Microbiology* 110, no. 6 (2018): 863–878.

Lahiri, Dibyajit, Moupriya Nag, Bandita Dutta, Ankita Dey, and Rina Rani Ray. "Bacterial extracellular polysaccharides in biofilm formation and function." In *Application of Biofilms in Applied Microbiology*, pp. 1–23. Academic Press, 2022.

Larson, Thomas R., and Janet Yother. "*Streptococcus pneumoniae* capsular polysaccharide is linked to peptidoglycan via a direct glycosidic bond to β-D-N-acetylglucosamine." *Proceedings of the National Academy of Sciences* 114, no. 22 (2017): 5695–5700.

Limoli, Dominique H., Christopher J. Jones, and Daniel J. Wozniak. "Bacterial extracellular polysaccharides in biofilm formation and function." *Microbial Biofilms* (2015): 223–247.

Nguyen, Hoai T.T., Thuan H. Nguyen, and Michael Otto. "The staphylococcal exopolysaccharide PIA–biosynthesis and role in biofilm formation, colonization, and infection." *Computational and Structural Biotechnology Journal* 18 (2020): 3324–3334.

O'Gara, James P. "ICA and beyond: biofilm mechanisms and regulation in *Staphylococcus epidermidis* and *Staphylococcus aureus*." *FEMS Microbiology Letters* 270, no. 2 (2007): 179–188.

Paton, James C., and Claudia Trappetti. "*Streptococcus pneumoniae* capsular polysaccharide." *Microbiology Spectrum* 7, no. 2 (2019): 7–2.

Paulander, Wilhelm, Anders Nissen Varming, Martin Saxtorph Bojer, Cathrine Friberg, Kristoffer Bæk, and Hanne Ingmer. "The agr quorum sensing system in *Staphylococcus aureus* cells mediates death of sub-population." *BMC Research Notes* 11 (2018): 1–5.

Perchat, Stéphane, Thomas Dubois, Samira Zouhir, Myriam Gominet, Sandrine Poncet, Christelle Lemy, Magali Aumont-Nicaise et al. "A cell–cell communication system regulates protease production during sporulation in bacteria of the Bacillus cereus group." *Molecular Microbiology* 82, no. 3 (2011): 619–633.

Rocha Jorge, Victor Flores, Rosina Cabrera, Adriana Soto-Guzmán, Giovana Granados, Eusebio Juaristi, Gabriel Guarneros, and Mayra De La Torre. "Evolution and some functions of the NprR–NprRB quorum-sensing system in the Bacillus cereus group." *Applied Microbiology and Biotechnology* 94 (2012): 1069–1078.

Ruhal, Rohit, and Rashmi Kataria. "Synthetic biology approaches for bioremediation of metals." In *Bioremediation of Toxic Metal (loid)s*, pp. 292–306. CRC Press, 2022.

Saha, Abhik, Kriti Arora, Andaleeb Sajid, and Gunjan Arora. "Cellular signaling in bacterial biofilms." *Implication of Quorum Sensing System in Biofilm Formation and Virulence* (2018): 81–109.

Sava, Irina G., Esther Heikens, and Johannes Huebner. "Pathogenesis and immunity in enterococcal infections." *Clinical Microbiology and Infection* 16, no. 6 (2010): 533–540.

Shi, Ke, C. Kent Brown, Zu-Yi Gu, Briana K. Kozlowicz, Gary M. Dunny, Douglas H. Ohlendorf, and Cathleen A. Earhart. "Structure of peptide sex pheromone receptor PrgX and PrgX/pheromone complexes and regulation of conjugation in Enterococcus faecalis." *Proceedings of the National Academy of Sciences* 102, no. 51 (2005): 18596–18601.

Slater, Ross T., Lucy R. Frost, Sian E. Jossi, Andrew D. Millard, and Meera Unnikrishnan. "*Clostridioides difficile* LuxS mediates inter-bacterial interactions within biofilms." *Scientific Reports* 9, no. 1 (2019): 9903.

Verbeke, Frederick, Severine De Craemer, Nathan Debunne, Yorick Janssens, Evelien Wynendaele, Christophe Van de Wiele, and Bart De Spiegeleer. "Peptides as quorum sensing molecules: measurement techniques and obtained levels in vitro and in vivo." *Frontiers in Neuroscience* 11 (2017): 183.

Vu, Barbara, Miao Chen, Russell J. Crawford, and Elena P. Ivanova. "Bacterial extracellular polysaccharides involved in biofilm formation." *Molecules* 14, no. 7 (2009): 2535–2554.

Xayarath, Bobbi, and Nancy E. Freitag. "When being alone is enough: noncanonical functions of canonical bacterial quorum-sensing systems." *Future Microbiology* 11, no. 11 (2016): 1447–1459.

3 Quorum Sensing in Gram-Negative Bacteria

Snehashis Koley, Ambar Bose,
and Mandira Mukherjee

3.1 INTRODUCTION

Quorum sensing (QS) was defined as the regulation of gene expression in response to variations in cell-population density. The different steps in QS included (i) synthesis and release of extracellular signal molecules known as autoinducers (AI); (ii) activation of a specific receptor present in the cytoplasm or in the membrane on reaching a certain threshold concentration of the AIs; and (iii) regulation of expression of genes in the QS regulon (1). Thus quorum sensing empowers cells in a population to function in harmony and conduct the different cellular activities collectively.

Gram-negative bacteria synthesize and release acyl-homoserine lactones (AHLs) or other chemical molecules whose production depends on S-adenosylmethionine (SAM) as a substrate which freely diffuses across the inner and outer membranes. These chemical compounds, the autoinducers (AIs), and are grouped as AI-1 type. In addition, another type of AI, called AI-2, which is a furanosyl borate diester, was detected in *Vibrio harveyi,* a free-living bioluminescent Gram-negative marine bacterium (2). A third, autoinducer-3 (AI-3), derived from threonine dehydrogenase (Tdh) products and "abortive" tRNA synthetase reactions, was first detected in enterohaemorrhagic *Escherichia coli* (EHEC), which regulated its virulence (3).

Incidence of QS was observed among intraspecies as well as interspecies Gram-negative bacteria. Co-culture systems of *V. harveyi* and *E. coli* demonstrated that AI-2 production by one species regulated gene expression in the other (4–6). Moreover, in another study, *P. aeruginosa* cell culture supernatant was found to be capable of mediating cell death of *E. coli* through a QS-facilitated process (7).

QS, in most of the Gram-negative bacteria, utilizes a LuxI (AI synthase)/LuxR (transcriptional regulator) type of system (1). There are varied homologues of the AI synthase (LuxI type) and the transcriptional regulator across the different species of the Gram-negative bacteria which actively used QS, suggesting a high degree of evolutionary conservation. Unlike quorum sensing by homoserine lactone autoinducers, AI-2 is a one-of-a-kind, 'universal' signal that might be employed by a wide range of bacteria for intra- and interspecies communication. In *E. coli*, *Salmonella typhimurium*, *V. harveyi*, and *V. cholerae*, evidence

DOI: 10.1201/9781003300557-3

suggested that the biosynthetic pathway and biochemical intermediates in AI-2 production were identical, which further supported the idea that QS mediated by AI-2 production facilitated both intra- and interspecies communication (6) In this chapter focus on the quorum sensing circuits in the different species of disease-causing Gram-negative bacteria, their mechanism of synthesis, and action at the molecular level.

3.2 QUORUM SENSING IN *ERWINIA CARATOVORA*

E. carotovora, E. amylovora, and *E. herbicola* are the three major groups of bacteria that were identified as the common etiologic agent in plant infections. They are economically important because they caused diseases in a variety of commercial crops, including potato, carrot, celery, rhubarb, onion, cucumber, and pineapple. Although the microbes were commonly found as part of epiphytic flora, insect vectors increased their transmission. Some *Erwinia* species were identified as insect pathogens and opportunistic animal diseases (8). *E. chrysanthemi, E. carotovora* subspecies *carotovora* (Ecc), *atroseptica* (Eca), and *betavascolorum* (Ecb) were among the soft-rot *Erwinia* species. Exoenzymes such as pectin methylesterases, pectate lyases (Pels), pectin lyases, polygalacturonases, cellulases (Cels), and proteases (Prts) were produced in abundance by these microbes, resulting in maceration of plant tissue (9, 10). Exoenzyme expression is critical for organisms' pathogenic success; disruption of genes encoding individual exoenzymes frequently resulted in a reduction of virulence in plants (11, 12). Two distinct secretory pathways transferred exoenzymes to the external environment: a single-step (type I) secretory pathway for Prts export and a two-step (type II) secretion system for Pels and Cels secretion. The products related to Ecc were encoded by at least 13 genes which were required for translocation of the periplasmic form of the enzymes to outside the cell via the latter pathway (13).

According to Vieira et al. (14), quorum sensing regulated *evf* transcription in Ecc using the acyl-homoserine lactone (AHSL) signal synthase ExpI and AHSL receptors ExpR1 and ExpR2. They discovered that Ecc load in the gut was influenced by *evf* that was controlled by quorum sensing. However, a study also discovered a second AHSL receptor (ExpR2) and demonstrated a unique quorum sensing mechanism in Ecc, in which ExpR2 functioned in tandem with ExpR1 to suppress virulence gene expression. They showed that addition of AHSLs relieved the repression, which appeared to be mediated in part by the negative regulator RsmA. They also discovered that ExpR2 also could detect AHSLs with varied acyl chain lengths. The *expI/expR1* double mutant revealed the ability to respond to a variety of AHSLs (15). Furthermore, other research showed that Ecc-infected larvae experienced developmental delay as a direct result of *evf* control using quorum sensing. Finally, they showed that during plant infection, *evf* was coexpressed with plant cell wall–degrading enzymes in a QS-dependent manner. Overall, these findings revealed that Ecc used quorum sensing to regulate both pectolytic enzyme and Evf synthesis.

3.3 QUORUM SENSING IN *AGROBACTERIUM TUMEFACIENS*

The pathogen *A. tumefaciens* caused crown gall tumours in plants by transferring the oncogenic DNA to the nucleus of its host. *A. tumefacines* used opines, secreted by the tumours, as a source of nutrition and a chemical signal that induced the conjugal transfer of Ti plasmids to non-Ti-plasmid-containing *Agrobacterium* cells within the tumour. Octopines and agrocinopines induced the transfer of octopine-type and nopaline-type Ti plasmids (16, 17). The two mechanisms were coordinately regulated in octopine-type Ti plasmids by an activator named OccR (18–20), which regulated the expression of LuxR homologue, TraR, and other essential genes involved in octopine catabolism (20). Key QS regulators included TraI and TraR encoded by Ti plasmid, as *traR* was located on a 14-gene, octopine-induced operon (*occ* operon) (21). AI pheromone 3-oxo-C_8-Homoserine Lactone ($3OC_8HSL$) was synthesized by TraI bound to TraR (22, 23). TraR then activated a number of operons which bind to the palindromic *tra* box (containing four 18 bp inverted repeat operator sequences, *tra* box I, II, III and IV). *tra* boxes were found in approximately 65 nucleotides upstream of class I-type promoter region and in approximately 45 nucleotides upstream of class II-type promoter region (24). *traR* gene also had been reported to be self-regulatory (21). A single class II-type promoter activated by tra box I, *traCDGyci* and *traAFBH* operons were transcribed in different ways, and coded for replication machineries, which were involved in the conjugative processing of Ti plasmid (25, 26). At the *oriT* region of the Ti plasmid, the proteins TraA, TraC and TraD were hypothesized to form a relaxosome that could suppress the expression of *traCDGyi* and *traAFBH* operons (26). The promoter of *traI-trabBCDEJKLFGHI* operon, though, was a QS regulated class-II promoter distinguished by the presence of a *tra* box II. The *trb* genes were similar to type IV secretion systems and encoded a mating pair formation pathway for Ti plasmid transfer (27). TrbJ and TrbK were two proteins expressed by these genes that worked together and established an entry-exclusion mechanism that prevented conjugation events between donor and recipient *A. tumifaciens* cells with similar Ti plasmids (28). TraR mediated up-regulation of the operons, namely *traCDGyi*, *traAFBH,* and *traI-trbCDEJKLFGHI*, that led to the induction of the conjugation of the Ti plasmid, which was consistent with the gene function. The modulation of *traI* expression by TraR, on the other hand, resulted in a positive feedback loop that amplified through an increase in 3OC8HSL production (20, 29).

Another LuxR homologue, TrlR or TraS, had also been detected on octopine-type Ti plasmids. It had immense similarity with TraR in terms of sequence, though a frameshift mutation in TrlR that resulted in a deletion of the carboxy-terminal domain. TrlR inhibited TraR activity in a dominant-negative manner, due to the presence of TraR-TrlR heterodimers (30, 31). Mannopine could stimulate *trlR* expression in cells only when grown in presence of succinate and glutamine, which suggested a nutrient-dependence of the conjugal transfer activity in the bacteria (30, 31).

TraM, encoded on both octopine and nopaline-type Ti plasmid, was identified as a TraR anti-activator (32, 33). This 11-kDa protein functioned as 3OC8HSL-independent and bound to the carboxy-terminal domain of TraR, and prevented the LuxR homologue from binding to the target promoter sequence (34, 35). TraM considerably contributed to the quorum-dependent dimension by delaying the ability of TraR to transactivate target genes (36). A *traM* defective strain was reported to be QS-active independent of cell population density (37). Moreover, TraM was also found to be essential for the existence of *A. tumifaciens*' QS 'off' state (38). Another aspect of TraR-mediated *traM* regulation was that TraR production must eventually outpace TraM synthesis, or else QS activation would be blocked indefinitely. This was supported by the fact that TraM was uniquely transcribed from a mildly active promoter associated with a *tra* box IV system (39).

The Ti-plasmids had a unique replication and partitioned strategy encoded in a single locus designated *repABC*. RepA and RepB were assumed to be important in the stable partitioning of plasmid into daughter cells, whereas RepC was required for DNA replication (40). In both octopine and nopaline type plasmids, TraR stimulated the expression of operon repABC, which was correlated with vegetative replication (41, 42). The expression of the operon was controlled by three different TraR-dependent (*repAP1*, *repAP2*, and *repAP3*) promoter and one TraR-independent (*repA4*) promoter (43). The Ti plasmid replication related to cell division was assumed to be mediated by promoter *repAP4*, although it was also auto-repressed by RepA and RepB. RepAP4 was also found downstream of *repAP1*, *repAP2*, and *repAP3*. As a result, auto-repression of *repAP4* might prevent activation of TraR-dependent promoters. Furthermore, the virulence proteins VirA and VirG could increase *repABC* expression, which suggested that the regulation of this operon was complicated and might be sensitive to diverse physiological conditions (44, 45).

3.4 QUORUM SENSING IN *VIBRIO* SPECIES

Vibrio species are rod-shaped Gram-negative bacteria found in freshwater, estuarine, and marine habitats. Despite their genetic diversity, they originated from aquatic and marine environments, and favoured warm, brackish water. Their abundance in nature was inclined to reflect ambient temperatures. *V. cholerae* causes cholera, a severe diarrhoeal illness that can be deadly if left untreated. Cholera is disseminated by contaminated water and person-to-person contact. Noncholera *Vibrio* species (for example, *V. parahaemolyticus*, *V. alginolyticus*, and *V. vulnificus*) caused vibriosis, which was acquired via consumption of undercooked infected seafood. With the exception of *V. vulnificus*, an opportunistic pathogen with a high mortality rate caused wound infections that developed septicaemia. Moreover, noncholera bacteria of the *Vibrio* species could induce a variety of clinical symptoms such as mild to self-limiting gastroenteritis (46).

Vibrio species use QS, and reports suggested that they express varied AI types (AI-1, AI-2, CAI-1). A common two-component LuxO/LuxU (VanO/VanU) phosphorelay cascade was identified in *Vibrio* species (47). The released AIs bind to their respective receptors intracellularly to regulate the expression of genes involved in different cellular pathways. The DNA binding domain of the key regulatory protein regulated the transcription of QS AI synthases and several other functional genes. There are mainly two types of key regulators: intracellular AHL receptors (LuxR, VanR, and VfqR) and AHL-independent receptors (VanT, LitR, HapR, AphA, SmcR), which directly reacts to QS activation without binding to the AHLs (47–49).

S-adenosyl-L-methionine (SAM) and a loaded acyl carrier protein were the two substrates required for AHL synthases, which belonged to the acyltranferase class of enzymes (ACP). They catalyzed the formation of an amide bond between the ACP-carrying fatty acid and the preliminary amine group of SAM. Lactonization of the SAM-acyl adduct resulted in the release of the AHL and 5-methylthioadenosine, which formed the homoserine lactone ring (MTA) (50–53). Different variations of the *V. fischeri* QS systems, such as the LuxI synthase homologues VanI or VfqI and AinS synthase homologues LuxM, VanM, or opaM, might be found in *Vibrio* species (54–58). The signalling pathways of *Vibrio* species are almost comparable, with two or three components: a membrane and/or intracellular receptor (LuxN/LuxR), a two-component phosphorelay protein (LuxU/O), and a key regulatory factor. LuxM synthases were substantially more common in *Vibrio* species than LuxI synthases. When *luxM* and *luxI* genes from various *Vibrio* species were compared, DNA sequence identities for *luxM* ranged from 19.9% to 98.9% and *luxI* ranged from 42.4% to 100%. *luxM* phylogeny was very similar to multilocus sequence analysis or *16s rRNA* gene phylogenies, which suggested that both systems had inextricably intertwined throughout evolution (59). Furthermore, it was also found that *luxI* and *luxM* were preserved at the clade level, with 96% to 100% similarity within a single species.

The LuxP/Q sensor kinase, the two-component phosphorelay proteins LuxU/O, and a specified key regulator, HapR for *V. cholerae* and Smc for *V. vulnificus* (60, 61) were reported. LuxP was the membrane receptor, while LuXQ was the accompanying kinase which used a phosphorylation cascade to transport information from LuxU to LuxO. The phosphorylation cascade and key regulator were shared by the various QS mechanisms in most *Vibrio* species (47). Unlike the *luxI* and *luxM* genes, *luxS/vanS* was a well-conserved gene in *Vibrio* due to its universal feature. The participation of *luxS* in the activated methyl cycle explained its conservation (62). LuxS was engaged not just in QS, but also in protein synthesis and sulphur metabolism, according to physiological research, which raised doubt about the role of the AI-2 molecule in this bacteria (62–64). Moreover, LuxP-mediated QS in *Vibrio* species regulated expression of virulence factor genes, biofilm production, and various other cellular functions (Figure 3.1).

The cholera autoinducer-1 (CAI-1) molecule was first found in *V. cholerae*. S-3-hydroxytridecan-4-one, or CAI-1, was a α-hydroxyketone (65). CAI-1 interacted with its receptor (Cholerae QS Sensor; CqsS), and the two QS pathways shared an intracellular signalling mechanism (47). It was later reported that in *V. harveyi* a similar chemical was generated by a CqsA homologue which was detected by CqsS (66). The *cqsA* gene was found in all *Vibrio* species and was not exclusive to *V. cholera* and *V. harveyi*.

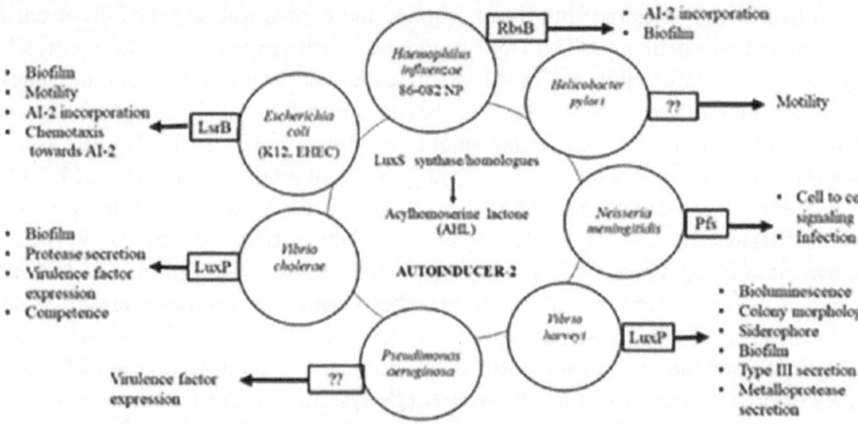

FIGURE 3.1 Autoinducer-2 (AI-2) mediated quorum sensing. AI-2 receptors and their regulation in *E. coli* (K12, EHEC), *Haemophilus influenza, Helicobacter pylori, Neisseria meningitides, V. cholera, Pseudomonas aeruginosa, V. harveyi.* Boxes represent the respective receptors of the AI-2 in each bacteria.

3.5 QUORUM SENSING IN *AEROMONAS* SPECIES

A. hydrophila, a Gram-negative, motile, nonspore-forming, facultative anaerobic bacteria is the primary etiologic agent in *Aeromonas*-induced gastroenteritis. *Aeromonas* species are found in varying amounts in drinking water, aquatic ecosystems, sewage, and foods such as shellfish, raw milk, chicken, vegetables, and meats including lamb, veal, pork, and ground beef. Two main families of *Aeromonas* were identified; i) motile, mesophilic species that might infect humans; and ii) nonmotile, psychrophilic organisms that often exclusively infected fish. *Aeromonas* infections were grouped into four categories: gastrointestinal tract syndromes; wound infections and connective tissue infections; bloodborne dyscrasias; and a broad category that encompassed a variety of less common diseases and infectious processes (67).

AHL molecules were synthesized by the LuxI type enzyme known as AhyI in *A. hydrophila* and AsaI in *A. salmonicida* (68). In this bacteria, AHLs interacted with AhyR, a cytoplasmic homologue of LuxR. AhyR was a transcriptional regulator of target genes including the AhyI gene (68, 69). During the exponential phase, AhyR and AhyI interacted in the activation loop of the system of AI-1, which led to the autoamplification phenomenon termed 'autoinduction'. AHLs readily diffused in and out of the cell membrane of *Aeromonas*, which accumulated in the cell surroundings of the bacteria. The autoinduction phenomenon was repressed in the stationary phase over an exogenous AHL concentration threshold, but intracellular communication occurred between two bacterial cells and it was the sole active phenomenon observed in *A. hydrophila* (70). A study showed that

AhyI was synthesized throughout the exponential phase but was degraded during the post-exponential phase. AhyR up-regulated AhyI expression and increased AHL production throughout the exponential phase. During the post-exponential phase, on the other hand, AhyR suppressed AhyI expression (71). Furthermore, the AI-2 synthase LuxS and the AI-3 response regulator QseB promoted *ahyRI* transcription (72).

The AI-1 QS system might vary by bacterial species and might include a variety of LuxRI homologues. Because LuxRI homologues were found in all 73 examined strains, spanning the known variety in the genus *Aeromonas*, only an AI-1 system was characterized in *Aeromonas*, and it was present in every *Aeromonas* strain (73). In *A. hydrophila*, the genes *ahyI* and *ahyR* from the locus *ahyRI* encoded this system, which were transcribed divergently together with a 62-bp intergenic region (68, 71). Other *Aeromonas* species had homologues for these genes as well (e.g., *A. salmonicida*, *A. molluscorum*, *A. veronii*, *A. media*, and *A. diversa*). In the intergenic region, 10-bp upstream of the *ahyI* promoter, a potential AhyR binding site was discovered (71). In *Aeromonas*, activation of the AI-1 system was linked to increased biofilm development (74, 75) and pathogenicity (Figure 3.2). Indeed, the AhyRI system increased the expression of a variety of virulence factors, which included metalloproteases, serine proteases, haemolysin, amylase, DNAse, S-layer formation, and pigment synthesis (69, 75–77). The transition between two metabolic pathways was likely regulated by type 1 QS. *Aeromonas* could convert from mixed acid fermentation to butanediol fermentation, which generated fewer acid molecules, to prevent deadly acidification of the medium in the late development phase (78).

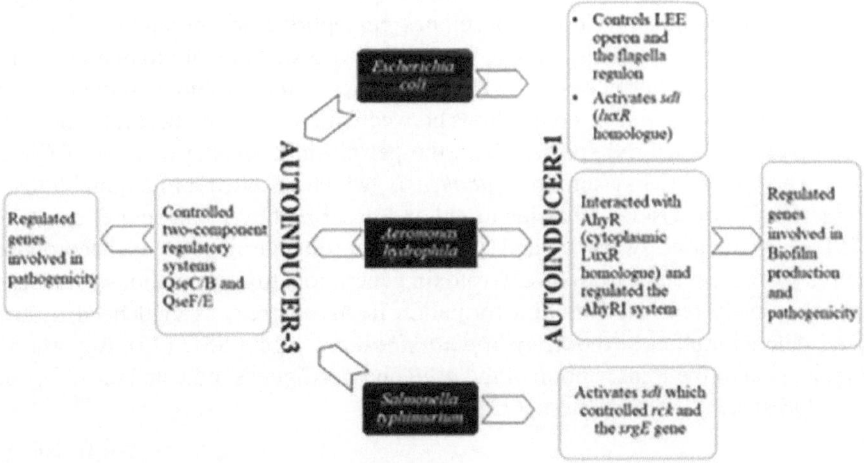

FIGURE 3.2 Autoinducer-1(AI-1) and autoinducer-3 (AI-3) mediated quorum sensing. Regulatory effect of AI-1 and AI-3 controlling pathogenicity in *E. coli*, *A. hydrophila*, *S. typhimurium*.

Bioassays and high-performance liquid chromatography (HPLC) analysis of *A. hydrophila* AH-1 and *A. salmonicida* type strain NCIMB 1102^T revealed that N-butanoyl homoserine lactone (C4-HSL) and N-hexanoyl homoserine lactone (C6-HSL) were the major autoinducers produced by aeromonads, with C4: C6 ratio of 70:1 (68). In another investigation, all *A. hydrophila* (n=20) and *A. sorbia* (n=2) clinical strains generated either C4-HSL, C6-HSL or both, with 10 strains producing an extra AHL, a suspected N-pentanoyl homoserine lactone (C5-HSL) (79). Reports indicated that *A. caviae* YL12 produced comparable results to the afore-mentioned species, however presence of other AHLs was not thoroughly described (80–83). The results of the HPLC analysis corroborated the bioassay results.

Exogenous AHLs with long chains, such as 10–14 carbons HSL, inhibit the *Aeromonas* AI-1 system *in vitro*, resulting in lower exoprotease synthesis (69). The 3-oxo-C10-HSL generated in *V. anguillarum* consistently suppressed *A. sulmonicida* and *A. hydrophila* protease activity (84). Similarly, the synthetic 2(5H)-furanone, which was generated from the competitive inhibitor of AHL produced by the marine algae *Delisea pulchra*, inhibited C4-HSL and C6-HSL, which were often produced by the aeromonads (85).

The genomes of *A. hydrophila* ATCC 7966^T and *A. hydrophila* SSU contained AI-2 synthase homologue LuxS and signal transformation enzymes (AI-2 sensor kinase/phosphatase LuxQ, phosphorelay protein LuxU, regulatory protein LuxO), as well as a LitR encoding gene, a homologue for the transcriptional regulator of *V. cholerae*, HapR and Lit (86). The expression of the locus *ahyRI* inhibited *luxS* transcription (86) *ΔluxS* deletion mutants of the *A. hydrophila* strain SSU revealed the role of AI-2 in the QS circuit.

The two-component QseBC system, in which QseC was the sensor kinase and QseB was the response regulator, transmitted a hormone-like signal known as the type 3 autoinducer (AI-3) system. Since QseC was also a bacterial adrenergic receptor for the eukaryotic host hormones epinephrine and norepinephrine and was engaged in interkingdom cross-signalling, AI-3 was thought to operate similarly to eukaryotic hormones (87). The gastrointestinal microbiota produced AI-3 in the context of symbiotic connections between microbiota and host (88). Several Gram-negative bacterial species share the periplasmic sensing domain of QseC (88). The QseBC QS system in *A. dhakensis*, which encoded a functional homologue for *E. coli* QseBC, was identified and described by Khajanchi et al. 2012 (75). AI-3 increased swarming and swimming motility, virulence (e.g., haemolytic activity, protease and cytotoxic enterotoxin generation, toxicity in mouse models), and negatively regulated biofilm formation in *Aeromonas*, as evidenced by the *qseB* deleted mutant of the *A. hydrophila dhakensis* strain SSU (75). Reports on suppression of the transcription of the *qseB* and *qseC* genes indicated the effect of AI-1 QS system in this bacteria (72).

3.6 QUORUM SENSING IN *E. COLI*

E. coli is one of the most common causes of bacterial illnesses in both humans and animals. *E. coli* is the primary causative agent of clinical infections such as enteritis, urinary tract infection, septicaemia, infant meningitis, etc. In pets and farm animals, *E. coli* was frequently linked to diarrhoea. Antimicrobial resistance posed complications to the therapeutic options caused by *E. coli*. Multidrug-resistant *E. coli* strains developed severe complications and were mismanaged, which caused a major global public health concern (89).

The regulation of gene expression by external signals was very well studied in *E. coli*. Growth of *E. coli* in conditioned media stimulated expression of the *rpoS* gene, which encoded the alternative sigma factor 38. The expression of *rpoS* was shown to be reduced in a *thrA, metL, lysC* triple mutant that lacks homoserine synthesis, but it was restored by addition of exogenous homoserine lactone. Extracellular factors regulated a number of other *E. coli* genes that included *sdiA* and *ftsQAZ*, which were involved in cell division (90, 91).

QS in *E. coli* was first demonstrated by Surette and Bassler (92). The *V. harveyi* quorum-sensing signals were particularly useful in tracking the autoinducers in *E. coli*. The density-sensing systems of *V. harveyi*, also known as signalling system 1 and signalling system 2, consisted of a sensor and a cognate autoinducer. It was shown that system 1 was made up of sensor 1 and responded to autoinducer 1 (AI-1), while system 2 was made up of sensor 2 and detected autoinducer 2 (AI-2) (93). The *V. harveyi* system 1 autoinducer (AI-1) was a species-specific system with great precision. It had been identified as hydroxybutanoyl-L-homoserine lactone, and its synthesis was controlled by the genes *luxL* and *luxM* (94). The system II, on the other hand, was not species specific. It has been postulated as a 'universal' signal for interspecies communication (95). According to another study, the LuxS protein was the AI-2 synthase, and AI-2 was synthesized in three enzymatic steps from S-adenosylmethionine (96). Unlike quorum sensing using homoserine lactone autoinducers, AI-2 was a one-of-a-kind, 'universal' signal that may be employed by a wide range of bacteria for intra- and interspecies communication. Furthermore, AI-2 (or AI-2-like) activity mediated by *luxS* gene was found in nearly all pathogenic and nonpathogenic *E. coli* strains (92) (see Figure 3.1).

The crystal structure of an AI-2 sensor protein, LuxP, that usually made a complex with the autoinducer, was described by Chen et al. (97). LuxP belonged to the periplasmic binding proteins (98), a broad protein family that binds a variety of ligands by clamping them between two domains (99). LuxP, a hybrid two-component sensor kinase consisting of a periplasmic sensor domain and cytoplasmic histidine kinase and response regulatory domains, was involved in AI-2 detection in conjunction with the inner membrane protein LuxQ (100), a hybrid two-component sensor kinase which consisted of a periplasmic sensor domain and a cytoplasmic histidine kinase and response regulatory LuxQ that operated as a kinase at low cell density and cross-phosphorylated histidine residues within the histidine kinase domain (101). The phosphorylation of LuxO was the result of a cascade of phosphotransfer events. Phospho-LuxO repressed the transcriptional activator LuxR indirectly (102). LuxQ was changed from a kinase to a phosphatase by interacting with AI-2-bound LuxP at high cell density. As a result of dephosphorylation of LuxQ, phosphoryl groups moved backwards through the AI-2 signal transduction pathway, which depressed LuxR. LuxR activated the transcription of *lux* operon that also encoded a luciferase enzyme (103). *E. coli* had an operon (formerly known as the *b1513* operon) that was homologous to the *lsr* operon of *Salmonella enterica* serovar *Typhimurium*. AI-2 activated the expression of the *lsr* operon in the *E. coli lsr*-lacZ fusion, which revealed that the *lsr* operon was necessary for AI-2 import and processing, very similar to *S. enterica* serovar *Typhimurium* (104). The AI-2 activity was maximum in the mid-exponential phase in the presence of particular favoured carbon sources, according to an analysis of AI production. The AI-2 signal in *E. coli*, however, was diminished when the cells entered stationary phase, unlike other known quorum-sensing systems (105). The levels of AI-2 production and degradation had been demonstrated to be influenced by a number of environmental conditions. Rapid logarithmic growth, preferred carbon sources, low pH, and/or high osmolarity boosted AI-2 production, whereas stationary phase, a lack of preferred carbon source, neutral pH, and low osmolarity caused AI-2 signal deterioration. In most situations, LuxR and LuxI homologues, as well as a homoserine lactone (HSL) counterpart, were required as regulators and AI in quorum-sensing systems. The *luxI* gene was normally expressed at a basal level, which resulted in the production of small quantities of N-(3-oxohexanoyl)-L-homoserine lactone (OHHL). The internal and extracellular concentrations of AHL increased as the cell population expanded. The cytoplasmic LuxR protein bound to OHHL when it reached a local threshold concentration, and the resulting complex acted as a transcriptional activator, which stimulated the production of *luxI* and the *lux* operon (*luxCDABEG*). An increase in concentration of the ligand that bound LuxR caused an increase in more LuxI production. Therefore, LuxI generated a positive feedback loop and facilitated the QS process. Moreover, the expression of the *lux* operon in neighbouring cells was likewise triggered by an increase in OHHL levels (106).

In *E. coli* K-12, *sdiA* was the sole known *luxR* homologue (107). It was first discovered as a transcriptional activator of the *ftsQAZ* operon, which encoded cell division proteins (108, 109). Although there are numerous promoters in this operon (P1 and -2 promoters upstream of *ftsQ*, and internal P1, -2, -3, and -4 promoters between *ftsA* and *ftsZ*), SdiA controls just the P2 promoter upstream of *ftsQ*. (15). HSL analogues or conditioned media also activated the P2 promoter in a SdiA-dependent pathway (90). SdiA was considered to regulate the P2 promoter in *E. coli* K-12 using a quorum-sensing system. The chromosomal region encoding SdiA was conserved in *E. coli* O157:H7, and the gene sequence was nearly identical to that of *E. coli* K-12 *sdiA*. However, the encoded proteins differ by only two amino acids (4th and 101st position; Lys and Ser in *E. coli* K-12, Thr and Asn in *E. coli* O157:H7, respectively) at the N terminus. Another recent study reported that SdiA was a negative regulator to the production of virulence proteins EspD and intimin in *E. coli* O157:H7 (110).

Although *E. coli* possessed three major quorum-sensing systems, significant genetic research over the past two decades failed to discover the autoinducer-3 biosynthesis pathway or chemical structure (AI-3) (49, 111, 112). Early research demonstrated that the AI-3 (87, 113) QseBC two-component signal transduction pathway responded to the host-derived adrenergic signals epinephrine and norepinephrine, which implied an interspecies sensing mechanism. (88) The locus of enterocyte effacement (LEE), which encoded a type III secretion system (T3SS) and effectors responsible for haemorrhagic colitis and the formation of lesions or eroded regions in the intestinal tract, was up-regulated by stimulation of this pathway in the enterohaemorrhagic *E. coli* (EHEC) pathotype (114). However, elucidating AI-3 quorum sensing at the molecular level across *E. coli* strains was difficult due to a lack of accessible small molecule signals. However, adequate quantities of chosen AI-3 analogues could be isolated using cellular stress-induced metabolite stimulation techniques (115), which allowed the structural, biochemical, and functional characterization of this chemical signalling family.

To activate its virulence genes, EHEC detected two types of signals: the bacterial aromatic autoinducer (AI-3) produced by the normal GI flora, and the catecholamine class of hormones, specifically epinephrine/norepinephrine produced by the host (87). AI-3 and epinephrine/norepinephrine were agonists; however, alpha and beta adrenergic hormones were antagonists (87). Norepinephrine had been shown to stimulate bacterial growth in the past (116), and there were findings that suggested norepinephrine might also operate as a siderophore (116). Moreover, it was also reported that the AI-3/epinephrine/norepinephrine signalling pathway activated the LEE genes and the flagella regulon and played an important role in pathogenesis (91) (see Figure 3.2).

3.7 QUORUM SENSING IN *HELICOBACTER PYLORI*

H. pylori was recognized to cause gastrointestinal problems such as gastritis, ulcers and gastric cancers, and stomach ulcers. It was also found in human stomach, dental plaque, oral lesions, saliva, tonsils, and adenoid tissues (117). Other disorders, such as otitis, sinusitis, pharyngitis, laryngitis, and glossitis, might be triggered (118). To survive under harsh environmental circumstances such as those created by antibiotics and disinfectants, the host's immunological defences, temperature variations, pH alterations, and the like, *H. pylori* develops biofilms rather than growing in planktonic mode.

Quorum sensing (QS) was an effective communication technique used by the cells in the biofilm. Autoinducers (AIs) were signalling molecules and part of the communication system. In *H. pylori*, QS molecules, AI-2 (cyclic furanones), and DSF (diffusible signal factors) had been identified and described (119). LuxS cleaves S-ribosylhomocysteine to create homocysteine and 4,5-dihydroxy-2, 3-pentanedione (DPD). DPD undergoes rapid dehydration and cyclization, and existed in equilibrium as a group of molecules known as AI-2 (120). DSF aided communication between two microorganisms in a biofilm, whereas AI-2 stimulated biofilm development. LuxS made molecules that were similar to AI-2, which detected signal through the chemoreceptor TlpB and initiated quorum sensing. Four chemoreceptors, TlpA, TlpB, TlpC, and TlpD, were encoded by the *H. pylori* genome. Response of chemoreceptor TlpA to arginine and biocarbonate had been discovered in strain 26695 (121), and TlpB response to acid in strain SS1 (122). LuxS was involved in the methyl cycle and cell-to-cell communication, which regulated *H. pylori* movement by altered flagella transcription and biosynthesis, as well as biofilm development (119, 123–126) (see Figure 3.1). A process of quorum quenching was reported in *H. pylori* where the quorum signalling molecules AI-2 and DSF of the quorum-sensing system (QSS) were inhibited by specific molecules called quorum quenchers or quorum-sensing inhibitors that actively regulated the QS process in the bacteria (127).

3.8 QUORUM SENSING IN *HAEMOPHILUS INFLUENZAE*

Non-typeable *H. influenzae* (NTHi), a Gram-negative bacterium, successfully colonized human tissue without causing inflammation. It was reported to be a common cause of otitis media in infants and exacerbations in people with chronic obstructive pulmonary disease (128). The most common types of invasive disease caused by *H. influenzae* are pneumonia, bloodstream infection, meningitis, epiglottitis, cellulitis, and infectious arthritis, among others.

The presence of the *luxS* genetic determinant of dihydroxypentanedione (DPD) (also referred to as autoinducer-2, or AI-2) synthesis initiated quorum signalling in *H. influenzae*. Daines and colleagues later demonstrated that a variety of *Haemophilus* strains produced quorum-sensing signals as measured by the *V. harveyi* bioluminescence assay (129). In the chinchilla infection model, *luxS* transcript levels were shown to be raised during infection of epithelial cells, and mutants lacking *luxS* retained the ability to produce biofilms, had enhanced epithelial cell invasion, and produced more severe otitis media illness (see Figure 3.1). The RbsB protein was recently discovered to be involved in the uptake of DPD quorum signals (130). RbsB, a periplasmic receptor protein, was involved in AI-2 uptake in different bacteria (131) and served as a part of an ABC transporter for ribose sugars (132). Isogenic *rbsB* mutants, much like *luxS* mutants, produced biofilms with much lower thickness and density as compared to the ancestral *Haemophilus* strain (130).

Research had shown that *Haemophilus* quorum-sensing signals might have an impact on *Moraxella catarrhalis*, an opportunistic pathogen living in many of the same host settings as *Haemophilus*. In the chinchilla infection model, growth of *M. catarrhalis* inside a polymicrobial biofilm that contained non-typeable *H. infuenzae* was demonstrated to enhance antibiotic resistance and persistence (133).

Swords (2012) (134) showed that though isogenic *Haemophilus* mutants lacking *luxS* retain the ability to form biofilms, confocal microscopy revealed significant reduction in thickness and density, which was further restored by complementation or co-culture with the parental strains (135).

3.9 QUORUM SENSING IN *NEISSERIA MENINGITIDIS*

N. meningitidis is an imperative human commensal that resides in the nasopharynx without generating symptoms. This bacteria occasionally overcome the host immunological system and make way to the bloodstream, causing meningitis and/or septicaemia, both of which are linked to high rates of morbidity and mortality (136).

This Gram-negative bacteria synthesized and secreted chemical molecules that involved S-adenosyl-l-methionine (SAM) to initiate quorum sensing. The LuxS enzyme was an established part of the activated methyl cycle (AMC) that generated SAM, the methyl donor for methylation of RNA, DNA, proteins, and certain metabolites. SAM levels were critical, because shortage of SAM could disrupt a wide range of cellular functions as SAM was the most common methyl group donor in transmethylation reactions. Through utilization of SAM, methylases produced a toxic product called S-adenosyl-l-homocysteine (SAH) which was detoxified by Pfs (5'-methylthioadenosine/S-adenosylhomocysteine nucleosidase) to S-ribosylhomocysteine (SRH) and converted to homocysteine (HCY) by the enzyme S-rybosylhomocysteinase (LuxS). Moreover, the LuxS-catalyzed reaction produced 4,5-dihydroxy-2,3-pentane dione (DPD) in addition to HCY, which was the precursor of a diffusible signal molecule autoinducer 2 (AI-2) (137). In *N. meningitidis*, the Pfs reaction was the sole intracellular source of the AI-2 signal. Inactivation of Pfs resulted in the entire loss of detectable SRH formation as well as AI-2 production (137). Furthermore, inactivation of Pfs and LuxS led to growth abnormality which could not be reversed by additional supply of exogenous AI-2. Winzer et al. (2004) showed that, during the late exponential growth phase, *N. meningitidis* produced AI-2 in a LuxS-dependent pathway and had a role in quorum sensing (138). According to studies which used an infant rat model, LuxSNM added to the capacity of the bacteria to induce bacteremia, which was essential for meningococcal pathogenicity (see Figure 3.1). Therefore, in *N. meningitidis* SAM was recycled using the two-step transformation of SAH to HCY with simultaneous production of AI-2 as it possessed both *pfs* and *luxS* genes (137).

3.10 QUORUM SENSING IN *PSEUDOMONAS AERUGINOSA*

P. aeruginosa has been the focus of several quorum-sensing (QS) investigations in the past few years, emphasizing its significance as a prevalent opportunistic pathogen. It was frequently linked to infections in immunocompromised individuals, particularly those suffering from AIDS, cancer, cystic fibrosis (CF), or serious burn wounds, leading to all possible causes of death (139). Production and secretion of extracellular virulence factors, namely proteases, elastase, exo-enzyme S, exotoxin A, haemolysins, and pyocyanin regulated the pathogenicity of *P. aeruginosa*, which led to significant tissue damage in humans and other mammals (140) (see Figure 3.1). The QS here was a multiple-level interaction where cross-talk across multiple pathways was used to build coordination among a huge number of genes (141).

P. aeruginosa harboured three QS systems: two LuxI/LuxR-type QS circuits and a third, the *Pseudomonas* quinolone signal (PQS) (1, 142). LuxI/LuxR-type QS primarily involved the synthesis of autoinducer 3-oxo-C12-homoserine lactone (3OC12HSL) from LuxI homologue LasI, which was detected by the cytoplasmic LuxR homologue LasR (143–147). Finally the LasR-3OC12HSL complex activated the expression of virulence factor genes and several other genes to complete the QS circuit (148–151).

LasR-3OC12HSL complex had several targets: *lasI*, which formed an autoinducing feed-forward loop (152); *luxI* homologue *rhlI* (153, 154), which produced another AI, named butanoyl homoserine lactone (C4HSL) (155, 156), which in high concentration bound to another LuxR homologue called RhlR that activated genes encoding elastase, proteases, pyocyanin, and siderophores (157, 158), which in turn led to the autoinduction of the second type QS circuit; and LuxI-type synthetase, which consisted of a signal molecule binding region at N-terminal and helix-turn-helix motif (HTH) at the C-terminal, acted as regulator protein in AHL synthesis (159, 160). A third related receptor, QscR, worked with 3OC12HSL and regulated gene expression in a particular regulon that overlapped with *las* and *rhl* regulons (161). Apart from stimulating the expression of virulence factors, several AIs were demonstrated to interact directly with host cells. 3OC12HSL stimulated the release of IL-8 from human bronchial epithelial cells (162) and produced cyclooxygenase-2 and prostaglandin E2 in human lung fibroblasts, which implied its role in inflammation (163). Additionally, 3OC12HSL inhibited lymphocyte proliferation, suppressed TNFα and initiated production of IL12, which in turn caused activation of T-cells to release IFNγ (163, 164), and induced apoptosis in macrophages and neutrophils which altogether implied its role in expression of virulence genes and in immunological response to infection (163, 165). This AI also was shown to affect particularly the NF-κB pathway, a key regulator in host immunological response (166).

Moreover, to further modulate virulence gene expression, *P. aeruginosa* employed a non-LuxI/LuxR QS system named *Pseudomonas* quinolone signalling (PQS) system (167). The activation of *rhl* system was regulated by the *Pseudomonas* quinolone signalling molecule 2-heptyl-3hydroxi-4-quinolone (PQS) (168). PQS controlled its own synthesis by induction of *pqsABCDE* genes through a direct interaction with PqsR (MvfR). *pqsABCDE* also produced 2-heptyl-4-quinolone (HHQ), an immediate PQS precursor (169). LasR-3OC12HSL promotes the expression of *pqsH* and *pqsR*, which plays an important role in the transition from HHQ to PQS, and RhlR-C4HSL inhibits the expression of *pqsABCD* and *pqsR* (170–172). Thus, PQS acted as a mediator between *las* and *rhl* systems and played an active role in the *Pseudomonas* QS circuit.

3.11 QUORUM SENSING IN *SALMONELLA* SPECIES

Ingestion of contaminated food and water with *S. enterica* serotype *typhi* and *paratyphi*, Gram-negative, rod-shaped, flagellated bacteria, was the primary etiologic agent of typhoid and paratyphoid fever, respectively, with high mortality and morbidity which generated major public health issues worldwide.

Quorum sensing was a common phenomenon utilized by the bacteria to survive inside host. Evidence suggested that all three types of autoinducers (AI-1, AI-2, AI-3) (see Figures 3.1 and 3.2) were involved. These bacteria do not synthetize N-acylhomoserine lactones (AHLs AI-1) due to the absence of LuxI in their genome. However, *Salmonella* produced a LuxR homologue AI-1 receptor called SdiA which was classified as a single or orphan LuxR homologue (173, 174) following a parallel evolution. Therefore, absence of LuxI and presence of SdiA indicated an interspecies communication mechanism, which might be crucial in *Salmonella* host gut colonization (175). In this way, the *Salmonella* SdiA protein could detect a wide spectrum of AHLs generated by other bacteria, with a preference for the oxoC8 (3-oxo modified unsubstituted AHL). *Salmonella*, on the other hand, might detect AHLs with various structures, based on acyl length and/or with 3-oxo modification, such as C4 and oxoC12 of *P. aeruginosa*, C4 of *A. hydrophila* and C6 and oxoC6 of *Yersinia enterocolitica* (175, 176). Hence, the regulation of the varied genes could affect the response of different serotypes to the quorum-sensing signals. The *sdiA* gene controlled two particular loci in *S. typhimurium*, the *rck* (resistance complement killing) and the *srgE* gene. The *rck* gene was located in a 90kb virulence plasmid with five other genes (*pefI*, *srgD*, *srgA*, *srgB*, *srgC*). Pef fimbriae helped the bacteria attach to crypt epithelial cells, activated pro-inflammatory cytokines, and initiated biofilm formation (177). Similarly, *srgA* assisted the folding of the PefA fimbrial subunit and the SsaC protein, a component of the type III secretion system (178, 179). *rck* encoded an outer membrane protein implicated in complement killing resistance and adhesion to fibronectin and laminin. The relevance of the *rck* operon in *S. typhimurium* virulence and the intimate link between quorum sensing and virulence control were highlighted by the fact that it was not expressed at temperatures below 37°C (180). Moreover, expression of *rck* by heterologous promoter in *Salmonella* suggested an interspecies communication (175).

The synthesis of AI-2 in *S. typhimurium* was connected to the exponential growth phase, with a peak concentration at the mid-exponential phase. Along with depletion of glucose, this signal begins to fade from the media at the onset of the stationary phase (181). AI-2-dependent signalling necessitated a low pH and high osmolarity, as low osmolarity caused deterioration in the signal (182). A variety of enzymatic processes were required for the synthesis of AI-2 in *S. typhimurium* very similar to several other Gram-negative species. First, methyltranferases converted SAM to SAH, which was subsequently transformed to DPD by methylthioadenosine/S-adenosylhomocystiene nucleosidase (Pfs) and LuxS. DPD was cyclized into (2R, 4S)–2– methyl—2,3,3,4—tetrahydroxytetrahydrofuran, known as AI-2, in the last step. The carriage protein YdgG excretes AI-2 outside the cell after its synthesis and AI-2 then binds to the autoinducer-binding protein LsrB, which was transported into the bacterial cell by an ATP transporter expressed on the *lsr* operon. AI-2 was phosphorylated by LsrK and inactivated the transcription repressor protein LsrR in a dose-dependent manner within the bacteria, which resulted in the transcription of the operon *lsrACDBFGE*. In addition, Lsr suppressed flagella production and virulence that inhibited the expression of certain genes involved in oxidative stress responses (*sodA*, *sodCI*, *sodCII*) that were critical for macrophage survival. As a result, it appeared that AI-2 inactivating LsrR was required for SPI-1 transcription and flagella expression; however, breakdown of phosphorylated AI-2 by LsrG and LsrF ceased the transcription of *lsr* operon and subsequent downstream processes (181, 183–185).

In *S. Typhimurium* the two-component regulatory systems QseC/B and QseF/E were found to be orthologues of the *E. coli* regulatory systems QseC/B and QseF/E. QseE detected sulphate and phosphate in addition to AI-3, epinephrine, and norepinephrine (186). QseC was autophosphorylated after binding to AI-3/epinephrine/norepinephrine, which resulted in the dephosphorylation of the response regulator QseB and transcription of SPI-2 genes. In macrophages, the two-component regulatory system QseC/QseB also induced some changes in the transcription of motility, invasion, and survival related genes (187). QseC/QseB linked to systemic illness and macrophage replication, and a second two-component regulatory system, QseF/E, was found to be involved in epithelial cell invasion (186). Catchecolamines were shown to stimulate growth on iron-restricted media, induced the expression of virulence genes in SPI-1 and SPI-3, and regulated *S. typhimurium* pathogenicity *in vitro* and *in vivo*. It was also implied that QseC/E might play a regulatory role in *S. typhimurium* pathogenicity (187).

3.12 QUORUM SENSING IN *BURKHOLDERIA CEPACIA*

B. cepacia (previously known as *P. cepacia*), like *P. aeruginosa*, usually was associated with CF lung infections (188). CepI (which regulates the synthesis of OHL) and CepR were two QS-related proteins expressed by one strain of this bacteria, isolated from the respiratory tract of a patient with CF). The *cepI* and *cepR* genes were transcribed differently, separated by a 727-base-pair intergenic region; the *cepI* promoter had a putative 20-bp *lux*-box-like sequence that partially overlapped the *cepR* promoter's proposed 35 region. CepR–HSL could recognize this box because *cepI*, positively regulated by CepR at high cell densities, provided a signal amplification mechanism via positive feedback control (189). *cciR* was discovered to be a worldwide regulator of quorum-sensing gene expression. CepR was predominantly responsible for positive gene regulation, whereas CciR was mostly responsible for negative gene regulation. Many of the genes controlled by both quorum-sensing systems were also regulated by CepR and CciR in a reciprocal manner (190). The system, which consisted of CepIR proteins and produced C8- and C6HSL, was similar to that found in *B. cenocepacia* K56–2 (191, 192). C8HSL and C6HSL were produced at a molar ratio of roughly 10:1 in this *B. cenocepacia* strain, H111, in which CepR reacted most efficiently to C8HSL (191, 192). CciI predominantly produced C6-HSL, with C8-HSL being formed in smaller amounts (193). CepI and cepR were transcribed differently at the genomic level, whereas *cciI* and *cciR* form a transcriptional unit (193). Since CepR was necessary for the transcription of the *cciIR* operon, the QS systems were ordered hierarchically (193). The quorum-sensing system of *B. cepacia* type strain ATCC 25416T, which was originally isolated from onion rot (194), was recently identified and characterized, and showed nearly identical characteristics to the other two CepIR systems of *B. cenocepacia* clinical isolates K56–2 and H111 (195).

In this organism, LuxRI homologues played a role in protease and siderophore expression regulation (196). *B. cepacia* strain grew in cell-free supernatant from a *Pseudomonas* species, compounds generated by *P. aeruginosa* (probably acyl HSLs) served to up-regulate production of exoproducts from another *B. cepacia* strain (197). As a result, it is reasonable to believe that acyl HSL-mediated cross-communication I between *B. cepacia* and *P. aeruginosa* populations during coinfection. In fact, an analysis of acyl HSL production by isolates of *P. aeruginosa* collected from a patient with a chronic lung infection revealed that isolates sampled during times of *B. cepacia* coinfection had a lower capacity to produce acyl HSLs than did isolates taken when the co-resident population was absent (198). The method and rationale for this decrease in signalling molecule synthesis were unknown, but it was possible that acyl HSL-mediated cross-communication input was involved.

3.13 QUORUM SENSING IN *YERSINIA* SPECIES

Bacteria that are oxidase-negative, catalase-positive, and whose cells were mostly Gram-negative rods are known as *Yersinia*. *Y. pseudotuberculosis* was a rodent parasite that infrequently infected humans. Intestinal infections and mesenteric lymphadenitis were linked to *Y. pseudotuberculosis*, while food-borne intestinal infections were linked to *Y. enterocolitica*. The most common *Yersinia* species linked to water transmission was *Y. enterocolitica*, which caused gastroenteritis. *Y. enterocolitica* was mostly responsible for acute enteritis, but it could also cause systemic infections including bacteremia, joint discomfort, and rashes. Fever, diarrhoea, and abdominal pain are the most prevalent symptoms. The creatures were discovered in surface and groundwater sources, but nothing was known about their environmental occurrence (199, 200).

Y. enterocolitica was the first *Yersinia* species to be discovered to use a quorum-sensing system. *Y. pseudotuberculosis* synthesized at least eight distinct AHLs and has two LuxI homologues for AHL synthases and two LuxR homologues for AHL-dependent response regulators (201). The genes coding for LuxR and LuxI homologues were typically conserved and organized in all *Yersinia* species studied (202). The protein encoded by the *luxI* homologue, YenI, was used to synthesize both N-hexanoyl-L-homoserine lactone (HHL) and N-(3-oxohexanoyl)-L-homoserine lactone (OHHL) by *Y. enterocolitica* (203). A second open reading frame, known as *yenR*, was located downstream of *yenI* and encoded a LuxR homologue. YenI disruption ceased acyl HSL synthesis with unaltered phenotypic changes (203). Proteomic studies revealed that a functional YenI was essential for the expression of a variety of unidentified proteins (203). An early analysis on the *Y. pseudotuberculosis* quorum-sensing system indicated an involvement of a hierarchy of at least two LuxRI homologues (YpsRI and YtbRI), expression of which were temperature dependent (202). YpsRI was reported to control motility and cellular aggregation by the flagellar regulatory cascade; however, the function of the YtbRI regulon has yet to be determined (202).

Furthermore, AHL production was investigated in stationary phase cultures for 24 hours at 28°C and 22°C in *Y. enterocolitica* (204). Similar to the

Y. pseudotuberculosis quorum-sensing system, in QS in *Y. enterocolitica*, AHL levels were found to be significantly dependent on growth temperature when AHL profiles of cultures grown at three different temperatures were compared. The higher levels were observed to be higher following growth at 22°C, significantly lower at 28°C, and nearly undetectable at 37°C. This for the first time suggested that either the expression or function of the relevant AHL synthase(s) or substrate availability would be temperature dependent, which the QS system was as well. However, it was later discovered that there was a straightforward physicochemical reason for this observation: the AHL ring's vulnerability to pH-dependent lactonolysis. Compounds that have their rings opened were unable to activate LuxR-type proteins, and so were unable to trigger AHL biosensors. Lactonolysis could be reversed and AHLs recovered from cultures maintained at 37°C by simple acidification of the spent culture supernatants to below pH 2.0 (204). Moreover, the metabolism of amino acids and the release of ammonia as a result of growth of *Y. pseudotuberculosis* in unbuffered LB quickly raised the pH to alkaline and a rise in the incubation temperature finally increased the rate of pH-dependent lactonolysis (204).

3.14 QUORUM SENSING IN *BRUCELLA* SPECIES

The zoonotic disease brucellosis is caused by *Brucella* species, facultative intracellular proteobacteria. Infection with *Brucella* species causes an acute undulating fever in humans before progressing to a chronic condition with endocarditis, meningitis, and osteomyelitis (205). Additionally, *B. abortus* is an intracellular Gram-negative bacterium that caused abortions in cattle (206).

Brucella quorum sensing, like that of other Gram-negative bacteria, relied on the generation and detection of N-acyl homoserine lactone (AHL) as a signal. Different AHLs, N-dodecanoyl-homoserine lactone (C12-HSL) and N-3-oxo-dodecanoyl-homoserine lactone (3-oxo-C12-HSL), have been reported to date (207). AHL-based QS relied on three primary components: (i) AHL, the signal molecule (AI); (ii) AHL synthase; and (iii) a transcriptional regulator that detected AI production and controlled target gene expression. In other circumstances, a fourth component, an AHL-destroying enzyme, added to the complexity and prevented the accumulation of signal molecules. The QS regulator detected AHL generated by its cognate AHL synthase to the best of its ability, ensuring specificity. While most cognate pairs of AHL synthase and QS regulator were genetically related, there were exceptions where the pairs were encoded on the chromosome at different locations (207).

Only two QS regulators had been reported in all *Brucella* genomes that had both the C-terminal HTH domain, GerE, and a classical AHL-binding domain in their N-terminus (208). Three more ORFs (BME I1583, BMEI1607, and BMEII0051) contained a C-terminal GerE domain but were members of the two-component regulator family. Although the genes encoding the synthase and their analogous regulators were frequently found together, this does not appear to be the case with

Brucella QS regulators. As a result, Patankar and González (2009) typed them as orphan or single QS regulators called VjbR and BabR (209):

VjbR (BMEII1116):

> This regulator was given the name VjbR, which stands for 'Vacuolar Jacking Brucella Regulator. The chromosomal position of *vjbR* was located downstream of a TetR family gene and close to the flagellar genes flhB and fliG. In all accessible *Brucella* genomes, the amino acid sequence of VjbR was highly conserved (>99% identity); nevertheless, only the first four 'consensus amino acids' (W57, Y61, D70, and P71) reported as important in AHL binding were present in VjbR. VjbR was essential for *virB* expression, involved in type IV secretion system and intracellular replication in this bacteria (210). C12-HSL reduced VirB8 production as it inhibited *virB* transcription from the promoter (PvirB) (211); additionally, exogenous C12-HSL reduced the capacity of the bacteria to reach its replicative niche (210). These findings implied that the repressor activity of C12-HSL's on the PvirB promoter was linked to its inhibitory effect on the VjbR regulator. The fact that two *vjbR* alleles, *vjbR* (D82A) and *vjbR* (1–180), which were mutated in the AHL binding domain and were unresponsive to signal molecules, constitutively activated PvirB (212) supported this notion.

The *virB* operon was reported as the first target of the *Brucella* QS system. Cellular infection using a *B. melitensis* strain that harboured a mutant allele of *vjbR* failed to bind the AI AHL. Furthermore, investigations had also shown that VjbR regulates *virB* expression *in vitro* (213, 214) and *in vivo* (214) through direct binding of the regulator to the *virB* promoter region (215).

BabR (BMEI1758):

> A functional screen designed to identify the AHL synthase led to the discovery of BabR, a second QS regulator of *Brucella abortus* (210). Surprisingly, this regulator was identified again and given the name BlxR (216). Two genes, an omega amino acid pyruvate amino transferase and a 5-methyltetrahydrofolatehomocysteine methyltransferase, surrounded the *babR* gene, both of which were involved in amino acid metabolism. Only *Brucella* species and *Ochrobactrum* species had this genomic architecture. The amino acid sequence of BabR was also largely conserved across all identified *Brucella* genomes. BabR successfully detected C12-HSL and was demonstrated to play an important role in QS in this bacteria through *vjbR* (215).

3.15 QUORUM SENSING IN *CAMPYLOBACTER JEJUNI*

C. jejuni belongs to the phylum Epsilonproteobacteria, a Gram-negative, motile, microaerophilic food-borne human pathogen with a wide global distribution. *C. jejuni* was identified as the primary etiologic agent that caused Campylobacterosis is an alimentary disorder of the GI tract (217). Abdominal discomfort, watery or bloody diarrhoea, fever, headache, and nausea are all frequent symptoms of campylobacteriosis. The development of Guillain-Barré or Miller-Fisher syndrome post-infection has also been documented in rare cases (218).

The LuxS system was initially described as utilizing AI-2 as the signal molecule in studies of *C. jejuni* QS (219). The counterpart of HSL was later isolated and designated cjA from *C. jejuni* 81–176 and *C. jejuni* c11 (chicken parts isolate). The existence of an HSL ring, a characteristic structure for AI-1, was verified by mass spectrometry, although without an N-acylated residue. When isolated from the cell-free supernatant of the previously mentioned strains of *C. jejuni* and cultivated in *Campylobacter*-defined medium, the molecule cjA had a significant impact on biofilm formation, virulence factor expression, and the like (220).

The amino acid sequences of LuxS from *C. jejuni* NCTC11168 were compared to those of *V. harveyi*, *E. coli*, *H. influenzae*, *P. multocida*, and *H. pylori*. It was observed that the sequence of LuxS had 74% similarity to *V. harveyi*, 71% to *E. coli*, 70% to *H. influenzae*, 73% to *Pasteurella multocida*, and 40% to *H. pylori* (219). Mutagenesis of the *luxS* gene, a useful tool for studying QS, might cause disruption of SAM metabolism together with loss of AI-2 synthesis. Both of these changes had the potential to influence the critical bacterial cell functions significantly (221). A study of the LuxS sequence in *C. jejuni* 81116 reported a naturally integrated mutation, G92D. In comparison to the wild type, this mutation resulted in the loss of AI-2 synthesis and a 100-fold catalytic reduction in the conversion of SRH to homocysteine (222). Another study revealed that *C. jejuni* NCTC11168 was unable to internalize synthetic AI-2 into the cell, which might be attributed to the fact that *C. jejuni* had a different receptor or a different signal molecule structure to manifest a different AI-2 QS mechanism (223).

Moreover, studies with *C. jejuni* had established the effect of *luxS* dysfunction that resulted in lack of AI-2 synthesis on biofilm formation, maturation, cell differentiation, and dispersion (224, 225). The *luxS* deletion mutant of *C. jejuni* showed a reduction in biofilm biomass. However, the biofilm development was reestablished on addition of a cell-free supernatant of the wild-type cells. Therefore, this finding revealed that QS played an important role in the formation of *C. jejuni* biofilms (224).

3.16 CONCLUSION

QS regulation allowed bacteria to control their gene expression in response to their population density. QS was particularly important during infection as it was the mechanism by which population density was detected. In this chapter, we have summarized the molecular perspectives of QS systems and the similarities and differences embedded among the autoinducers, their binding to corresponding receptors, and their downstream regulators in some pathogenic Gram-negative bacteria. While the list of bacteria that utilize QS systems is large, it is clearly unlikely to be complete. Usually processes that were regulated by QS were beneficial for the bacteria as it was a strategy employed to invade hosts successfully. The host's immune system could successfully detect and combat a single pathogen, but a coordinated attack by a population of bacteria might overcome host defence to manifest infection.

3.17 ACKNOWLEDGEMENTS

The authors would like to express their deepest gratitude to the director, Prof. (Dr.) Subhasish Kamal Guha, Calcutta School of Tropical Medicine, Kolkata, for his kind support. The authors also want to thank Dr. Moupriya Nag and Prof. Dibyajit Lahiri for the invitation to write this chapter.

REFERENCES

1. Sifri, C. D. (2008). Healthcare epidemiology: Quorum sensing: Bacteria talk sense. *Clinical Infectious Diseases: An Official Publication of the Infectious Diseases Society of America*, *47*(8), 1070–1076. https://doi.org/10.1086/592072
2. Li, Z., & Nair, S. K. (2012). Quorum sensing: How bacteria can coordinate activity and synchronize their response to external signals? *Protein Science: A Publication of the Protein Society*, *21*(10), 1403–1417. https://doi.org/10.1002/pro.2132
3. Kim, C. S., Gatsios, A., Cuesta, S., Lam, Y. C., Wei, Z., Chen, H., Russell, R. M., Shine, E. E., Wang, R., Wyche, T. P., Piizzi, G., Flavell, R. A., Palm, N. W., Sperandio, V., & Crawford, J. M. (2020). Characterization of autoinducer-3 structure and biosynthesis in *E. coli*. *ACS Central Science*, *6*(2), 197–206. https://doi.org/10.1021/acscentsci.9b01076
4. McNab, R., & Lamont, R. J. (2003). Microbial dinner-party conversations: The role of LuxS in interspecies communication. *Journal of Medical Microbiology*, *52*(Pt 7), 541–545. https://doi.org/10.1099/jmm.0.05128-0
5. Ryan, R. P., & Dow, J. M. (2008). Diffusible signals and interspecies communication in bacteria. *Microbiology (Reading, England)*, *154*(Pt 7), 1845–1858. https://doi.org/10.1099/mic.0.2008/017871-0
6. Pereira, C. S., Thompson, J. A., & Xavier, K. B. (2013). AI-2-mediated signalling in bacteria. *FEMS Microbiology Reviews*, *37*(2), 156–181. https://doi.org/10.1111/j.1574-6976.2012.00345.x
7. Kumar, S., Kolodkin-Gal, I., & Engelberg-Kulka, H. (2013). Novel quorum-sensing peptides mediating interspecies bacterial cell death. *mBio*, *4*(3), e00314–13. https://doi.org/10.1128/mBio.00314-13
8. Perombelon, M. C. M., & Kelman, A. (1980). Ecology of the soft rot Erwinias. *Annual Review of Phytopathology*, *18*(1), 361–387. https://doi.org/10.1146/annurev.py.18.090180.002045
9. Py, B., Barras, F., Harris, S., Robson, N., & Salmond, G. P. C. (1998). Extracellular enzymes and their role in Erwinia virulence. In *Methods in microbiology* (Vol. 27, pp. 157–168). Elsevier. https://doi.org/10.1016/S0580-9517(08)70279-4
10. Thomson, N. R., Thomas, J. D., & Salmond, G. P. C. (1999). Virulence determinants in the bacterial phytopathogen Erwinia. In *Methods in microbiology* (Vol. 29, pp. 347–426). Elsevier. https://doi.org/10.1016/S0580-9517(08)70123-5
11. Mäe, A., Heikinheimo, R., & Palva, E. T. (1995). Structure and regulation of the Erwinia carotovora subspecies carotovora SCC3193 cellulase gene celV1 and the role of cellulase in phytopathogenicity. *Molecular & General Genetics: MGG*, *247*(1), 17–26. https://doi.org/10.1007/BF00425817
12. Marits, R., Köiv, V., Laasik, E., & Mäe, A. (1999). Isolation of an extracellular protease gene of Erwinia carotovora subsp. carotovora strain SCC3193 by transposon mutagenesis and the role of protease in phytopathogenicity. *Microbiology (Reading, England)*, *145(Pt 8)*, 1959–1966. https://doi.org/10.1099/13500872-145-8-1959

13. Salmond, G. P. C. (1994). Secretion of extracellular virulence factors by plant pathogenic bacteria. *Annual Review of Phytopathology*, *32*(1), 181–200. https://doi.org/10.1146/annurev.py.32.090194.001145
14. Vieira, F., Nadal-Jimenez, P., Teixeira, L., & Xavier, K. B. (2020). Erwinia carotovora quorum sensing system regulates host-specific virulence factors and development delay in drosophila melanogaster. *mBio*, *11*(3), e01292–20. https://doi.org/10.1128/mBio.01292-20
15. Sjöblom, S., Brader, G., Koch, G., & Palva, E. T. (2006). Cooperation of two distinct ExpR regulators controls quorum sensing specificity and virulence in the plant pathogen Erwinia carotovora. *Molecular Microbiology*, *60*(6), 1474–1489. https://doi.org/10.1111/j.1365-2958.2006.05210.x
16. Klapwijk, P. M., Scheulderman, T., & Schilperoort, R. A. (1978). Coordinated regulation of octopine degradation and conjugative transfer of Ti plasmids in Agrobacterium tumefaciens: Evidence for a common regulatory gene and separate operons. *Journal of Bacteriology*, *136*(2), 775–785. https://doi.org/10.1128/jb.136.2.775-785.1978
17. Ellis, J. G., Kerr, A., Petit, A., & Tempe, J. (1982). Conjugal transfer of nopaline and agropine Ti-plasmids—the role of agrocinopines. *Molecular and General Genetics: MGG*, *186*(2), 269–274. https://doi.org/10.1007/BF00331861
18. Habeeb, L. F., Wang, L., & Winans, S. C. (1991). Transcription of the octopine catabolism operon of the Agrobacterium tumor-inducing plasmid pTiA6 is activated by a LysR-type regulatory protein. *Molecular Plant-Microbe Interactions: MPMI*, *4*(4), 379–385. https://doi.org/10.1094/mpmi-4-379
19. Wang, L., Helmann, J. D., & Winans, S. C. (1992). The A. tumefaciens transcriptional activator OccR causes a bend at a target promoter, which is partially relaxed by a plant tumor metabolite. *Cell*, *69*(4), 659–667. https://doi.org/10.1016/0092-8674(92)90229-6
20. Fuqua, W. C., & Winans, S. C. (1994). A LuxR-LuxI type regulatory system activates Agrobacterium Ti plasmid conjugal transfer in the presence of a plant tumor metabolite. *Journal of Bacteriology*, *176*(10), 2796–2806. https://doi.org/10.1128/jb.176.10.2796-2806.1994
21. Fuqua, C., & Winans, S. C. (1996a). Localization of OccR-activated and TraR-activated promoters that express two ABC-type permeases and the traR gene of Ti plasmid pTiR10. *Molecular Microbiology*, *20*(6), 1199–1210. https://doi.org/10.1111/j.1365-2958.1996.tb02640.x
22. Piper, K. R., Beck von Bodman, S., & Farrand, S. K. (1993). Conjugation factor of Agrobacterium tumefaciens regulates Ti plasmid transfer by autoinduction. *Nature*, *362*(6419), 448–450. https://doi.org/10.1038/362448a0
23. Zhang, L., Murphy, P. J., Kerr, A., & Tate, M. E. (1993). Agrobacterium conjugation and gene regulation by N-acyl-L-homoserine lactones. *Nature*, *362*(6419), 446–448. https://doi.org/10.1038/362446a0
24. Fuqua, C., & Winans, S. C. (1996b). Conserved cis-acting promoter elements are required for density-dependent transcription of Agrobacterium tumefaciens conjugal transfer genes. *Journal of Bacteriology*, *178*(2), 435–440. https://doi.org/10.1128/jb.178.2.435-440.1996
25. Farrand, S. K., Hwang, I., & Cook, D. M. (1996). The tra region of the nopaline-type Ti plasmid is a chimera with elements related to the transfer systems of RSF1010, RP4, and F. *Journal of Bacteriology*, *178*(14), 4233–4247. https://doi.org/10.1128/jb.178.14.4233-4247.1996
26. Cho, H., & Winans, S. C. (2007). TraA, TraC and TraD autorepress two divergent quorum-regulated promoters near the transfer origin of the Ti plasmid of Agrobacterium tumefaciens. *Molecular Microbiology*, *63*(6), 1769–1782. https://doi.org/10.1111/j.1365-2958.2007.05624.x

27. Li, P. L., Everhart, D. M., & Farrand, S. K. (1998). Genetic and sequence analysis of the pTiC58 trb locus, encoding a mating-pair formation system related to members of the type IV secretion family. *Journal of Bacteriology*, *180*(23), 6164–6172. https://doi.org/10.1128/JB.180.23.6164-6172.1998
28. Cho, H., Pinto, U. M., & Winans, S. C. (2009). Transsexuality in the rhizosphere: Quorum sensing reversibly converts Agrobacterium tumefaciens from phenotypically female to male. *Journal of Bacteriology*, *191*(10), 3375–3383. https://doi.org/10.1128/JB.01608-08
29. Hwang, I., Li, P. L., Zhang, L., Piper, K. R., Cook, D. M., Tate, M. E., & Farrand, S. K. (1994). TraI, a LuxI homologue, is responsible for production of conjugation factor, the Ti plasmid N-acylhomoserine lactone autoinducer. *Proceedings of the National Academy of Sciences of the United States of America*, *91*(11), 4639–4643. https://doi.org/10.1073/pnas.91.11.4639
30. Oger, P., Kim, K. S., Sackett, R. L., Piper, K. R., & Farrand, S. K. (1998). Octopine-type Ti plasmids code for a mannopine-inducible dominant-negative allele of traR, the quorum-sensing activator that regulates Ti plasmid conjugal transfer. *Molecular Microbiology*, *27*(2), 277–288. https://doi.org/10.1046/j.1365-2958.1998.00671.x
31. Zhu, J., & Winans, S. C. (1998). Activity of the quorum-sensing regulator TraR of Agrobacterium tumefaciens is inhibited by a truncated, dominant defective TraR-like protein. *Molecular Microbiology*, *27*(2), 289–297. https://doi.org/10.1046/j.1365-2958.1998.00672.x
32. Fuqua, C., Burbea, M., & Winans, S. C. (1995). Activity of the Agrobacterium Ti plasmid conjugal transfer regulator TraR is inhibited by the product of the traM gene. *Journal of Bacteriology*, *177*(5), 1367–1373. https://doi.org/10.1128/jb.177.5.1367-1373.1995
33. Hwang, I., Cook, D. M., & Farrand, S. K. (1995). A new regulatory element modulates homoserine lactone-mediated autoinduction of Ti plasmid conjugal transfer. *Journal of Bacteriology*, *177*(2), 449–458. https://doi.org/10.1128/jb.177.2.449-458.1995
34. Hwang, I., Smyth, A. J., Luo, Z. Q., & Farrand, S. K. (1999). Modulating quorum sensing by antiactivation: TraM interacts with TraR to inhibit activation of Ti plasmid conjugal transfer genes. *Molecular Microbiology*, *34*(2), 282–294. https://doi.org/10.1046/j.1365-2958.1999.01595.x
35. Luo, Z. Q., Qin, Y., & Farrand, S. K. (2000). The antiactivator TraM interferes with the autoinducer-dependent binding of TraR to DNA by interacting with the C-terminal region of the quorum-sensing activator. *The Journal of Biological Chemistry*, *275*(11), 7713–7722. https://doi.org/10.1074/jbc.275.11.7713
36. Su, S., Khan, S. R., & Farrand, S. K. (2008). Induction and loss of Ti plasmid conjugative competence in response to the acyl-homoserine lactone quorum-sensing signal. *Journal of Bacteriology*, *190*(13), 4398–4407. https://doi.org/10.1128/JB.01684-07
37. Piper, K. R., & Farrand, S. K. (2000). Quorum sensing but not autoinduction of Ti plasmid conjugal transfer requires control by the opine regulon and the antiactivator TraM. *Journal of Bacteriology*, *182*(4), 1080–1088. https://doi.org/10.1128/JB.182.4.1080-1088.2000
38. Goryachev, A. B., Toh, D. J., Wee, K. B., Lee, T., Zhang, H. B., & Zhang, L. H. (2005). Transition to quorum sensing in an Agrobacterium population: A stochastic model. *PLoS Computational Biology*, *1*(4), e37. https://doi.org/10.1371/journal.pcbi.0010037
39. White, C. E., & Winans, S. C. (2005). Identification of amino acid residues of the Agrobacterium tumefaciens quorum-sensing regulator TraR that are critical for positive control of transcription. *Molecular Microbiology*, *55*(5), 1473–1486. https://doi.org/10.1111/j.1365-2958.2004.04482.x

40. Pinto, U. M., Pappas, K. M., & Winans, S. C. (2012). The ABCs of plasmid replication and segregation. *Nature Reviews: Microbiology*, *10*(11), 755–765. https://doi.org/10.1038/nrmicro2882
41. Li, P. L., & Farrand, S. K. (2000). The replicator of the nopaline-type Ti plasmid pTiC58 is a member of the repABC family and is influenced by the TraR-dependent quorum-sensing regulatory system. *Journal of Bacteriology*, *182*(1), 179–188. https://doi.org/10.1128/JB.182.1.179-188.2000
42. Pappas, K. M., & Winans, S. C. (2003a). A LuxR-type regulator from Agrobacterium tumefaciens elevates Ti plasmid copy number by activating transcription of plasmid replication genes. *Molecular Microbiology*, *48*(4), 1059–1073. https://doi.org/10.1046/j.1365-2958.2003.03488.x
43. Pappas, K. M., & Winans, S. C. (2003b). The RepA and RepB autorepressors and TraR play opposing roles in the regulation of a Ti plasmid repABC operon. *Molecular Microbiology*, *49*(2), 441–455. https://doi.org/10.1046/j.1365-2958.2003.03560.x
44. Cho, H., & Winans, S. C. (2005). VirA and VirG activate the Ti plasmid repABC operon, elevating plasmid copy number in response to wound-released chemical signals. *Proceedings of the National Academy of Sciences of the United States of America*, *102*(41), 14843–14848. https://doi.org/10.1073/pnas.0503458102
45. Pappas K. M. (2008). Cell-cell signaling and the Agrobacterium tumefaciens Ti plasmid copy number fluctuations. *Plasmid*, *60*(2), 89–107. https://doi.org/10.1016/j.plasmid.2008.05.003
46. Baker-Austin, C., Oliver, J. D., Alam, M., Ali, A., Waldor, M. K., Qadri, F., & Martinez-Urtaza, J. (2018). Vibrio spp. infections. *Nature Reviews Disease Primers*, *4*(1), 1–19. https://doi.org/10.1038/s41572-018-0005-8
47. Milton D. L. (2006). Quorum sensing in vibrios: Complexity for diversification. *International Journal of Medical Microbiology: IJMM*, *296*(2–3), 61–71. https://doi.org/10.1016/j.ijmm.2006.01.044
48. Lin, B., Wang, Z., Malanoski, A. P., O'Grady, E. A., Wimpee, C. F., Vuddhakul, V., Alves, N., Jr, Thompson, F. L., Gomez-Gil, B., & Vora, G. J. (2010). Comparative genomic analyses identify the Vibrio harveyi genome sequenced strains BAA-1116 and HY01 as Vibrio campbellii. *Environmental Microbiology Reports*, *2*(1), 81–89. https://doi.org/10.1111/j.1758-2229.2009.00100.x
49. Ng, W. L., & Bassler, B. L. (2009). Bacterial quorum-sensing network architectures. *Annual Review of Genetics*, *43*, 197–222. https://doi.org/10.1146/annurev-genet-102108-134304.
50. Moré, M. I., Finger, L. D., Stryker, J. L., Fuqua, C., Eberhard, A., & Winans, S. C. (1996). Enzymatic synthesis of a quorum-sensing autoinducer through use of defined substrates. *Science (New York, N.Y.)*, *272*(5268), 1655–1658. https://doi.org/10.1126/science.272.5268.1655
51. Hanzelka, B. L., Parsek, M. R., Val, D. L., Dunlap, P. V., Cronan, J. E., Jr, & Greenberg, E. P. (1999). Acylhomoserine lactone synthase activity of the Vibrio fischeri AinS protein. *Journal of Bacteriology*, *181*(18), 5766–5770. https://doi.org/10.1128/JB.181.18.5766-5770.1999
52. Parsek, M. R., Val, D. L., Hanzelka, B. L., Cronan, J. E., Jr, & Greenberg, E. P. (1999). Acyl homoserine-lactone quorum-sensing signal generation. *Proceedings of the National Academy of Sciences of the United States of America*, *96*(8), 4360–4365. https://doi.org/10.1073/pnas.96.8.4360
53. Val, D. L., & Cronan, J. E., Jr. (1998). In vivo evidence that S-adenosylmethionine and fatty acid synthesis intermediates are the substrates for the LuxI family of autoinducer synthases. *Journal of Bacteriology*, *180*(10), 2644–2651. https://doi.org/10.1128/JB.180.10.2644-2651.1998

54. Cao, J. G., & Meighen, E. A. (1989). Purification and structural identification of an autoinducer for the luminescence system of Vibrio harveyi. *The Journal of Biological Chemistry*, *264*(36), 21670–21676.
55. Bassler, B. L., Wright, M., Showalter, R. E., & Silverman, M. R. (1993). Intercellular signalling in Vibrio harveyi: Sequence and function of genes regulating expression of luminescence. *Molecular Microbiology*, *9*(4), 773–786. https://doi.org/10.1111/j.1365-2958.1993.tb01737.x
56. Milton, D. L., Hardman, A., Camara, M., Chhabra, S. R., Bycroft, B. W., Stewart, G. S., & Williams, P. (1997). Quorum sensing in Vibrio anguillarum: Characterization of the vanI/vanR locus and identification of the autoinducer N-(3-oxodecanoyl)-L-homoserine lactone. *Journal of Bacteriology*, *179*(9), 3004–3012. https://doi.org/10.1128/jb.179.9.3004-3012.1997
57. Milton, D. L., Chalker, V. J., Kirke, D., Hardman, A., Cámara, M., & Williams, P. (2001). The LuxM homologue VanM from Vibrio anguillarum directs the synthesis of N-(3-hydroxyhexanoyl)homoserine lactone and N-hexanoylhomoserine lactone. *Journal of Bacteriology*, *183*(12), 3537–3547. https://doi.org/10.1128/JB.183.12.3537-3547.2001
58. Wang, Y., Wang, H., Liang, W., Hay, A. J., Zhong, Z., Kan, B., & Zhu, J. (2013). Quorum sensing regulatory cascades control Vibrio fluvialis pathogenesis. *Journal of Bacteriology*, *195*(16), 3583–3589. https://doi.org/10.1128/JB.00508-13
59. Lerat, E., & Moran, N. A. (2004). The evolutionary history of quorum-sensing systems in bacteria. *Molecular Biology and Evolution*, *21*(5), 903–913. https://doi.org/10.1093/molbev/msh097
60. Jobling, M. G., & Holmes, R. K. (1997). Characterization of hapR, a positive regulator of the Vibrio cholerae HA/protease gene hap, and its identification as a functional homologue of the Vibrio harveyi luxR gene. *Molecular Microbiology*, *26*(5), 1023–1034. https://doi.org/10.1046/j.1365-2958.1997.6402011.x
61. McDougald, D., Rice, S. A., & Kjelleberg, S. (2001). SmcR-dependent regulation of adaptive phenotypes in Vibrio vulnificus. *Journal of Bacteriology*, *183*(2), 758–762. https://doi.org/10.1128/JB.183.2.758-762.2001
62. Vendeville, A., Winzer, K., Heurlier, K., Tang, C. M., & Hardie, K. R. (2005). Making 'sense' of metabolism: Autoinducer-2, LuxS and pathogenic bacteria. *Nature Reviews. Microbiology*, *3*(5), 383–396. https://doi.org/10.1038/nrmicro1146
63. Doherty, N., Holden, M. T., Qazi, S. N., Williams, P., & Winzer, K. (2006). Functional analysis of luxS in Staphylococcus aureus reveals a role in metabolism but not quorum sensing. *Journal of Bacteriology*, *188*(8), 2885–2897. https://doi.org/10.1128/JB.188.8.2885-2897.2006
64. Yu, D. N., Ren, Z. M., Han, Y. Z., & Zhang, W. J. (2011). Function of LuxS gene in sulfur metabolism of Streptococcus mutans. *Chinese Journal of Stomatology*, *46*(4), 210–213.
65. Miller, M. B., Skorupski, K., Lenz, D. H., Taylor, R. K., & Bassler, B. L. (2002). Parallel quorum sensing systems converge to regulate virulence in Vibrio cholerae. *Cell*, *110*(3), 303–314. https://doi.org/10.1016/s0092-8674(02)00829-2
66. Henke, J. M., & Bassler, B. L. (2004). Three parallel quorum-sensing systems regulate gene expression in Vibrio harveyi. *Journal of Bacteriology*, *186*(20), 6902–6914. https://doi.org/10.1128/JB.186.20.6902-6914.2004
67. Ünüvar, S. (2018). Microbial foodborne diseases. In *Foodborne diseases* (pp. 1–31). Elsevier. https://doi.org/10.1016/B978-0-12-811444-5.00001-4
68. Swift, S., Karlyshev, A. V., Fish, L., Durant, E. L., Winson, M. K., Chhabra, S. R., Williams, P., Macintyre, S., & Stewart, G. S. (1997). Quorum sensing in Aeromonas hydrophila and Aeromonas salmonicida: Identification of the LuxRI homologs

AhyRI and AsaRI and their cognate N-acylhomoserine lactone signal molecules. *Journal of Bacteriology*, *179*(17), 5271–5281. https://doi.org/10.1128/jb.179.17.5271-5281.1997

69. Swift, S., Lynch, M. J., Fish, L., Kirke, D. F., Tomás, J. M., Stewart, G. S., & Williams, P. (1999). Quorum sensing-dependent regulation and blockade of exoprotease production in Aeromonas hydrophila. *Infection and Immunity*, *67*(10), 5192–5199. https://doi.org/10.1128/IAI.67.10.5192-5199.1999
70. Garde, C., Bjarnsholt, T., Givskov, M., Jakobsen, T. H., Hentzer, M., Claussen, A., Sneppen, K., Ferkinghoff-Borg, J., & Sams, T. (2010). Quorum sensing regulation in Aeromonas hydrophila. *Journal of Molecular Biology*, *396*(4), 849–857. https://doi.org/10.1016/j.jmb.2010.01.002
71. Kirke, D. F., Swift, S., Lynch, M. J., & Williams, P. (2004). The Aeromonas hydrophila LuxR homologue AhyR regulates the N-acyl homoserine lactone synthase, AhyI positively and negatively in a growth phase-dependent manner. *FEMS Microbiology Letters*, *241*(1), 109–117. https://doi.org/10.1016/j.femsle.2004.10.011
72. Kozlova, E. V., Khajanchi, B. K., Popov, V. L., Wen, J., & Chopra, A. K. (2012). Impact of QseBC system in c-di-GMP-dependent quorum sensing regulatory network in a clinical isolate SSU of Aeromonas hydrophila. *Microbial Pathogenesis*, *53*(3–4), 115–124. https://doi.org/10.1016/j.micpath.2012.05.008
73. Jangid, K., Kong, R., Patole, M. S., & Shouche, Y. S. (2007). luxRI homologs are universally present in the genus Aeromonas. *BMC Microbiology*, *7*, 93. https://doi.org/10.1186/1471-2180-7-93
74. Lynch, M. J., Swift, S., Kirke, D. F., Keevil, C. W., Dodd, C. E., & Williams, P. (2002). The regulation of biofilm development by quorum sensing in Aeromonas hydrophila. *Environmental Microbiology*, *4*(1), 18–28. https://doi.org/10.1046/j.1462-2920.2002.00264.x
75. Khajanchi, B. K., Sha, J., Kozlova, E. V., Erova, T. E., Suarez, G., Sierra, J. C., Popov, V. L., Horneman, A. J., & Chopra, A. K. (2009). N-acylhomoserine lactones involved in quorum sensing control the type VI secretion system, biofilm formation, protease production, and in vivo virulence in a clinical isolate of Aeromonas hydrophila. *Microbiology (Reading, England)*, *155*(Pt 11), 3518–3531. https://doi.org/10.1099/mic.0.031575-0
76. Bi, Z. X., Liu, Y. J., & Lu, C. P. (2007). Contribution of AhyR to virulence of Aeromonas hydrophila J-1. *Research in Veterinary Science*, *83*(2), 150–156. https://doi.org/10.1016/j.rvsc.2007.01.003
77. Schwenteit, J., Gram, L., Nielsen, K. F., Fridjonsson, O. H., Bornscheuer, U. T., Givskov, M., & Gudmundsdottir, B. K. (2011). Quorum sensing in Aeromonas salmonicida subsp. achromogenes and the effect of the autoinducer synthase AsaI on bacterial virulence. *Veterinary Microbiology*, *147*(3–4), 389–397. https://doi.org/10.1016/j.vetmic.2010.07.020
78. Van Houdt, R., Aertsen, A., & Michiels, C. W. (2007). Quorum-sensing-dependent switch to butanediol fermentation prevents lethal medium acidification in Aeromonas hydrophila AH-1N. *Research in Microbiology*, *158*(4), 379–385. https://doi.org/10.1016/j.resmic.2006.11.015
79. Chan, K. G., Puthucheary, S. D., Chan, X. Y., Yin, W. F., Wong, C. S., Too, W. S., & Chua, K. H. (2011). Quorum sensing in Aeromonas species isolated from patients in Malaysia. *Current Microbiology*, *62*(1), 167–172. https://doi.org/10.1007/s00284-010-9689-z
80. Bruhn, J. B., Dalsgaard, I., Nielsen, K. F., Buchholtz, C., Larsen, J. L., & Gram, L. (2005). Quorum sensing signal molecules (acylated homoserine lactones) in gram-negative fish pathogenic bacteria. *Diseases of Aquatic Organisms*, *65*(1), 43–52. https://doi.org/10.3354/dao065043

81. Huang, Y., Zhang, J., Yu, Z., Zeng, Y., & Chen, Y. (2012). Isolation and characterization of acyl homoserine lactone-producing bacteria during an urban river biofilm formation. *Archives of Microbiology, 194*(12), 1043–1048. https://doi.org/10.1007/s00203-012-0849-3
82. Ochiai, S., Morohoshi, T., Kurabeishi, A., Shinozaki, M., Fujita, H., Sawada, I., & Ikeda, T. (2013). Production and degradation of N-acylhomoserine lactone quorum sensing signal molecules in bacteria isolated from activated sludge. *Bioscience, Biotechnology, and Biochemistry, 77*(12), 2436–2440. https://doi.org/10.1271/bbb.130553
83. Zeng, Y., Yu, Z., & Huang, Y. (2014). Combination of culture-dependent and -independent methods reveals diverse acyl homoserine lactone-producers from rhizosphere of wetland plants. *Current Microbiology, 68*(5), 587–593. https://doi.org/10.1007/s00284-013-0513-4
84. Rasch, M., Kastbjerg, V. G., Bruhn, J. B., Dalsgaard, I., Givskov, M., & Gram, L. (2007). Quorum sensing signals are produced by Aeromonas salmonicida and quorum sensing inhibitors can reduce production of a potential virulence factor. *Diseases of Aquatic Organisms, 78*(2), 105–113. https://doi.org/10.3354/dao01865
85. Ponnusamy, K., Paul, D., Kim, Y. S., & Kweon, J. H. (2010). 2(5H)-Furanone: A prospective strategy for biofouling-control in membrane biofilm bacteria by quorum sensing inhibition. *Brazilian Journal of Microbiology, 41*(1), 227–234. https://doi.org/10.1590/S1517-83822010000100032
86. Kozlova, E. V., Khajanchi, B. K., Sha, J., & Chopra, A. K. (2011). Quorum sensing and c-di-GMP-dependent alterations in gene transcripts and virulence-associated phenotypes in a clinical isolate of Aeromonas hydrophila. *Microbial Pathogenesis, 50*(5), 213–223. https://doi.org/10.1016/j.micpath.2011.01.007
87. Sperandio, V., Torres, A. G., Jarvis, B., Nataro, J. P., & Kaper, J. B. (2003). Bacteria-host communication: The language of hormones. *Proceedings of the National Academy of Sciences of the United States of America, 100*(15), 8951–8956. https://doi.org/10.1073/pnas.1537100100
88. Clarke, M. B., Hughes, D. T., Zhu, C., Boedeker, E. C., & Sperandio, V. (2006). The QseC sensor kinase: A bacterial adrenergic receptor. *Proceedings of the National Academy of Sciences of the United States of America, 103*(27), 10420–10425. https://doi.org/10.1073/pnas.0604343103
89. Allocati, N., Masulli, M., Alexeyev, M. F., & Di Ilio, C. (2013). Escherichia coli in Europe: An overview. *International Journal of Environmental Research and Public Health, 10*(12), 6235–6254. https://doi.org/10.3390/ijerph10126235
90. Sitnikov, D. M., Schineller, J. B., & Baldwin, T. O. (1996). Control of cell division in Escherichia coli: Regulation of transcription of ftsQA involves both rpoS and SdiA-mediated autoinduction. *Proceedings of the National Academy of Sciences of the United States of America, 93*(1), 336–341. https://doi.org/10.1073/pnas.93.1.336
91. Sperandio, V., Mellies, J. L., Nguyen, W., Shin, S., & Kaper, J. B. (1999). Quorum sensing controls expression of the type III secretion gene transcription and protein secretion in enterohemorrhagic and enteropathogenic Escherichia coli. *Proceedings of the National Academy of Sciences of the United States of America, 96*(26), 15196–15201. https://doi.org/10.1073/pnas.96.26.15196
92. Surette, M. G., & Bassler, B. L. (1999). Regulation of autoinducer production in Salmonella typhimurium. *Molecular Microbiology, 31*(2), 585–595. https://doi.org/10.1046/j.1365-2958.1999.01199.x
93. Bassler, B. L., & Silverman, M. R. (2014). Intercellular communication in marine *Vibrio* species: Density-dependent regulation of the expression of bioluminescence. *Two-component signal transduction* (pp. 431–445). ASM Press. https://doi.org/10.1128/9781555818319.ch28

94. Bassler, B. L., Wright, M., Showalter, R. E., & Silverman, M. R. (1993). Intercellular signalling in Vibrio harveyi: Sequence and function of genes regulating expression of luminescence. *Molecular Microbiology*, *9*(4), 773–786. https://doi.org/10.1111/j.1365-2958.1993.tb01737.x
95. Bassler, B. L., Greenberg, E. P., & Stevens, A. M. (1997). Cross-species induction of luminescence in the quorum-sensing bacterium Vibrio harveyi. *Journal of Bacteriology*, *179*(12), 4043–4045. https://doi.org/10.1128/jb.179.12.4043-4045.1997
96. Schauder, S., Shokat, K., Surette, M. G., & Bassler, B. L. (2001). The LuxS family of bacterial autoinducers: Biosynthesis of a novel quorum-sensing signal molecule. *Molecular Microbiology*, *41*(2), 463–476. https://doi.org/10.1046/j.1365-2958.2001.02532.x
97. Chen, X., Schauder, S., Potier, N., Van Dorsselaer, A., Pelczer, I., Bassler, B. L., & Hughson, F. M. (2002). Structural identification of a bacterial quorum-sensing signal containing boron. *Nature*, *415*(6871), 545–549. https://doi.org/10.1038/415545a
98. Quiocho, F. A., & Ledvina, P. S. (1996). Atomic structure and specificity of bacterial periplasmic receptors for active transport and chemotaxis: Variation of common themes. *Molecular Microbiology*, *20*(1), 17–25. https://doi.org/10.1111/j.1365-2958.1996.tb02484.x
99. Dwyer, M. A., & Hellinga, H. W. (2004). Periplasmic binding proteins: A versatile superfamily for protein engineering. *Current Opinion in Structural Biology*, *14*(4), 495–504. https://doi.org/10.1016/j.sbi.2004.07.004
100. Bassler, B. L., Wright, M., & Silverman, M. R. (1994). Multiple signalling systems controlling expression of luminescence in Vibrio harveyi: Sequence and function of genes encoding a second sensory pathway. *Molecular Microbiology*, *13*(2), 273–286. https://doi.org/10.1111/j.1365-2958.1994.tb00422.x
101. Freeman, J. A., & Bassler, B. L. (1999). A genetic analysis of the function of LuxO, a two-component response regulator involved in quorum sensing in Vibrio harveyi. *Molecular Microbiology*, *31*(2), 665–677. https://doi.org/10.1046/j.1365-2958.1999.01208.x
102. Lenz, D. H., Mok, K. C., Lilley, B. N., Kulkarni, R. V., Wingreen, N. S., & Bassler, B. L. (2004). The small RNA chaperone Hfq and multiple small RNAs control quorum sensing in Vibrio harveyi and Vibrio cholerae. *Cell*, *118*(1), 69–82. https://doi.org/10.1016/j.cell.2004.06.009
103. Martin, M., Showalter, R., & Silverman, M. (1989). Identification of a locus controlling expression of luminescence genes in Vibrio harveyi. *Journal of Bacteriology*, *171*(5), 2406–2414. https://doi.org/10.1128/jb.171.5.2406-2414.1989
104. Xavier, K. B., & Bassler, B. L. (2005). Regulation of uptake and processing of the quorum-sensing autoinducer AI-2 in Escherichia coli. *Journal of Bacteriology*, *187*(1), 238–248. https://doi.org/10.1128/JB.187.1.238-248.2005
105. Surette, M. G., & Bassler, B. L. (1998). Quorum sensing in Escherichia coli and Salmonella typhimurium. *Proceedings of the National Academy of Sciences of the United States of America*, *95*(12), 7046–7050. https://doi.org/10.1073/pnas.95.12.7046
106. Whitehead, N. A., Barnard, A. M., Slater, H., Simpson, N. J., & Salmond, G. P. (2001). Quorum-sensing in Gram-negative bacteria. *FEMS Microbiology Reviews*, *25*(4), 365–404. https://doi.org/10.1111/j.1574-6976.2001.tb00583.x
107. Henikoff, S., Wallace, J. C., & Brown, J. P. (1990). Finding protein similarities with nucleotide sequence databases. *Methods in Enzymology*, *183*, 111–132. https://doi.org/10.1016/0076-6879(90)83009-x
108. Aldea, M., Garrido, T., Pla, J., & Vicente, M. (1990). Division genes in Escherichia coli are expressed coordinately to cell septum requirements by gearbox promoters. *The EMBO Journal*, *9*(11), 3787–3794.

109. Wang, X. D., de Boer, P. A., & Rothfield, L. I. (1991). A factor that positively regulates cell division by activating transcription of the major cluster of essential cell division genes of Escherichia coli. *The EMBO Journal*, *10*(11), 3363–3372.
110. Kanamaru, K., Kanamaru, K., Tatsuno, I., Tobe, T., & Sasakawa, C. (2000). SdiA, an Escherichia coli homologue of quorum-sensing regulators, controls the expression of virulence factors in enterohaemorrhagic Escherichia coli O157:H7. *Molecular Microbiology*, *38*(4), 805–816. https://doi.org/10.1046/j.1365-2958.2000.02171.x
111. Rutherford, S. T., & Bassler, B. L. (2012). Bacterial quorum sensing: Its role in virulence and possibilities for its control. *Cold Spring Harbor Perspectives in Medicine*, *2*(11), a012427. https://doi.org/10.1101/cshperspect.a012427
112. Papenfort, K., & Bassler, B. L. (2016). Quorum sensing signal–response systems in Gram-negative bacteria. *Nature Reviews Microbiology*, *14*(9), 576–588. https://doi.org/10.1038/nrmicro.2016.89
113. Sperandio, V., Torres, A. G., & Kaper, J. B. (2002). Quorum sensing Escherichia coli regulators B and C (QseBC): A novel two-component regulatory system involved in the regulation of flagella and motility by quorum sensing in E. coli. *Molecular Microbiology*, *43*(3), 809–821. https://doi.org/10.1046/j.1365-2958.2002.02803.x
114. Carlson-Banning, K. M., & Sperandio, V. (2018). Enterohemorrhagic Escherichia coli outwits hosts through sensing small molecules. *Current Opinion in Microbiology*, *41*, 83–88. https://doi.org/10.1016/j.mib.2017.12.002
115. Crawford, J. M., Kontnik, R., & Clardy, J. (2010). Regulating alternative lifestyles in entomopathogenic bacteria. *Current Biology: CB*, *20*(1), 69–74. https://doi.org/10.1016/j.cub.2009.10.059
116. Freestone, P. P., Lyte, M., Neal, C. P., Maggs, A. F., Haigh, R. D., & Williams, P. H. (2000). The mammalian neuroendocrine hormone norepinephrine supplies iron for bacterial growth in the presence of transferrin or lactoferrin. *Journal of Bacteriology*, *182*(21), 6091–6098. https://doi.org/10.1128/JB.182.21.6091-6098.2000
117. Challa, S., & Neelapu, N. R. R. (2018). Quorum sensing in Helicobacter pylori: Role of biofilm and its implications for antibiotic resistance and immune evasion. In *Implication of quorum sensing system in biofilm formation and virulence* (pp. 277–286). Springer. https://doi.org/10.1007/978-981-13-2429-1_18
118. Kurtaran, H., Uyar, M. E., Kasapoglu, B., Turkay, C., Yilmaz, T., Akcay, A., & Kanbay, M. (2008). Role of Helicobacter pylori in pathogenesis of upper respiratory system diseases. *Journal of the National Medical Association*, *100*(10), 1224–1230. https://doi.org/10.1016/s0027-9684(15)31471-1
119. Zhou, L., Yu, Y., Chen, X., Diab, A. A., Ruan, L., He, J., Wang, H., & He, Y. W. (2015). The multiple DSF-family QS signals are synthesized from carbohydrate and branched-chain Amino Acids via the FAS elongation cycle. *Scientific Reports*, *5*, 13294. https://doi.org/10.1038/srep13294
120. Schauder, S., Shokat, K., Surette, M. G., & Bassler, B. L. (2001). The LuxS family of bacterial autoinducers: Biosynthesis of a novel quorum-sensing signal molecule. *Molecular Microbiology*, *41*(2), 463–476. https://doi.org/10.1046/j.1365-2958.2001.02532.x
121. Cerda, O., Rivas, A., & Toledo, H. (2003). Helicobacter pylori strain ATCC700392 encodes a methyl-accepting chemotaxis receptor protein (MCP) for arginine and sodium bicarbonate. *FEMS Microbiology Letters*, *224*(2), 175–181. https://doi.org/10.1016/S0378-1097(03)00423-3
122. Croxen, M. A., Sisson, G., Melano, R., & Hoffman, P. S. (2006). The Helicobacter pylori chemotaxis receptor TlpB (HP0103) is required for pH taxis and for colonization of the gastric mucosa. *Journal of Bacteriology*, *188*(7), 2656–2665. https://doi.org/10.1128/JB.188.7.2656-2665.2006

123. Forsyth, M. H., & Cover, T. L. (2000). Intercellular communication in Helicobacter pylori: LuxS is essential for the production of an extracellular signaling molecule. *Infection and Immunity*, *68*(6), 3193–3199. https://doi.org/10.1128/IAI.68.6.3193-3199.2000
124. Joyce, E. A., Bassler, B. L., & Wright, A. (2000). Evidence for a signaling system in Helicobacter pylori: Detection of a luxS-encoded autoinducer. *Journal of Bacteriology*, *182*(13), 3638–3643. https://doi.org/10.1128/JB.182.13.3638-3643.2000
125. Lee, W. K., Ogura, K., Loh, J. T., Cover, T. L., & Berg, D. E. (2006). Quantitative effect of luxS gene inactivation on the fitness of Helicobacter pylori. *Applied and Environmental Microbiology*, *72*(10), 6615–6622. https://doi.org/10.1128/AEM.01291-06
126. Doherty, N. C., Shen, F., Halliday, N. M., Barrett, D. A., Hardie, K. R., Winzer, K., & Atherton, J. C. (2010). In Helicobacter pylori, LuxS is a key enzyme in cysteine provision through a reverse transsulfuration pathway. *Journal of Bacteriology*, *192*(5), 1184–1192. https://doi.org/10.1128/JB.01372-09
127. Gohil, N., Ramírez-García, R., Panchasara, H., Patel, S., Bhattacharjee, G., & Singh, V. (2018). Book review: Quorum sensing vs. quorum quenching: A battle with no end in sight. *Frontiers in Cellular and Infection Microbiology*, *8*, 106. https://doi.org/10.3389/fcimb.2018.00106
128. Langereis, J. D., & Hermans, P. W. M. (2013). Novel concepts in nontypeable *Haemophilus influenza* biofilm formation. *FEMS Microbiology Letters*, *346*(2), 81–89. https://doi.org/10.1111/1574-6968.12203
129. Daines, D. A., Bothwell, M., Furrer, J., Unrath, W., Nelson, K., Jarisch, J., Melrose, N., Greiner, L., Apicella, M., & Smith, A. L. (2005). Haemophilus influenzae luxS mutants form a biofilm and have increased virulence. *Microbial Pathogenesis*, *39*(3), 87–96. https://doi.org/10.1016/j.micpath.2005.06.003
130. Armbruster, C. E., Pang, B., Murrah, K., Juneau, R. A., Perez, A. C., Weimer, K. E., & Swords, W. E. (2011). RbsB (NTHI_0632) mediates quorum signal uptake in nontypeable Haemophilus influenzae strain 86–028NP. *Molecular Microbiology*, *82*(4), 836–850. https://doi.org/10.1111/j.1365-2958.2011.07831.x
131. Shao, H., James, D., Lamont, R. J., & Demuth, D. R. (2007). Differential interaction of Aggregatibacter (Actinobacillus) actinomycetemcomitans LsrB and RbsB proteins with autoinducer 2. *Journal of Bacteriology*, *189*(15), 5559–5565. https://doi.org/10.1128/JB.00387-07
132. Park, Y., Cho, Y. J., Ahn, T., & Park, C. (1999). Molecular interactions in ribose transport: The binding protein module symmetrically associates with the homodimeric membrane transporter. *The EMBO Journal*, *18*(15), 4149–4156. https://doi.org/10.1093/emboj/18.15.4149
133. Armbruster, C. E., Hong, W., Pang, B., Weimer, K. E., Juneau, R. A., Turner, J., & Swords, W. E. (2010). Indirect pathogenicity of Haemophilus influenzae and Moraxella catarrhalis in polymicrobial otitis media occurs via interspecies quorum signaling. *mBio*, *1*(3), e00102–10. https://doi.org/10.1128/mBio.00102-10
134. Swords W. E. (2012). Quorum signaling and sensing by nontypeable Haemophilus influenzae. *Frontiers in Cellular and Infection Microbiology*, *2*, 100. https://doi.org/10.3389/fcimb.2012.00100
135. Armbruster, C. E., Hong, W., Pang, B., Weimer, K. E., Juneau, R. A., Turner, J., & Swords, W. E. (2010). Indirect pathogenicity of Haemophilus influenzae and Moraxella catarrhalis in polymicrobial otitis media occurs via interspecies quorum signaling. *mBio*, *1*(3), e00102–10. https://doi.org/10.1128/mBio.00102-10
136. Yazdankhah, S. P., & Caugant, D. A. (2004). Neisseria meningitidis: An overview of the carriage state. *Journal of Medical Microbiology*, *53*(Pt 9), 821–832. https://doi.org/10.1099/jmm.0.45529-0

137. Heurlier, K., Vendeville, A., Halliday, N., Green, A., Winzer, K., Tang, C. M., & Hardie, K. R. (2009). Growth deficiencies of Neisseria meningitidis pfs and luxS mutants are not due to inactivation of quorum sensing. *Journal of Bacteriology*, *191*(4), 1293–1302. https://doi.org/10.1128/JB.01170-08
138. Winzer, K., Sun, Y. H., Green, A., Delory, M., Blackley, D., Hardie, K. R., Baldwin, T. J., & Tang, C. M. (2002). Role of Neisseria meningitidis luxS in cell-to-cell signaling and bacteremic infection. *Infection and Immunity*, *70*(4), 2245–2248. https://doi.org/10.1128/IAI.70.4.2245-2248.2002
139. Koch, C., & Høiby, N. (1993). Pathogenesis of cystic fibrosis. *Lancet (London, England)*, *341*(8852), 1065–1069. https://doi.org/10.1016/0140-6736(93)92422-p
140. D'Agata, E. (2015). Pseudomonas aeruginosa and other Pseudomonas species. In *Mandell, Douglas, and Bennett's principles and practice of infectious diseases* (pp. 2518–2531.e3). Elsevier. https://doi.org/10.1016/B978-1-4557-4801-3.00221-6
141. Lee, B., Haagensen, J. A., Ciofu, O., Andersen, J. B., Høiby, N., & Molin, S. (2005). Heterogeneity of biofilms formed by nonmucoid Pseudomonas aeruginosa isolates from patients with cystic fibrosis. *Journal of Clinical Microbiology*, *43*(10), 5247–5255. https://doi.org/10.1128/JCM.43.10.5247-5255.2005
142. Rutherford, S. T., & Bassler, B. L. (2012). Bacterial quorum sensing: Its role in virulence and possibilities for its control. *Cold Spring Harbor Perspectives in Medicine*, *2*(11), a012427. https://doi.org/10.1101/cshperspect.a012427
143. Moré, M. I., Finger, L. D., Stryker, J. L., Fuqua, C., Eberhard, A., & Winans, S. C. (1996). Enzymatic synthesis of a quorum-sensing autoinducer through use of defined substrates. *Science (New York, N.Y.)*, *272*(5268), 1655–1658. https://doi.org/10.1126/science.272.5268.1655
144. Val, D. L., & Cronan, J. E., Jr (1998). In vivo evidence that S-adenosylmethionine and fatty acid synthesis intermediates are the substrates for the LuxI family of autoinducer synthases. *Journal of Bacteriology*, *180*(10), 2644–2651. https://doi.org/10.1128/JB.180.10.2644-2651.1998
145. Parsek, M. R., Val, D. L., Hanzelka, B. L., Cronan, J. E., Jr, & Greenberg, E. P. (1999). Acyl homoserine-lactone quorum-sensing signal generation. *Proceedings of the National Academy of Sciences of the United States of America*, *96*(8), 4360–4365. https://doi.org/10.1073/pnas.96.8.4360
146. Gould, T. A., Schweizer, H. P., & Churchill, M. E. (2004). Structure of the Pseudomonas aeruginosa acyl-homoserinelactone synthase LasI. *Molecular Microbiology*, *53*(4), 1135–1146. https://doi.org/10.1111/j.1365-2958.2004.04211.x
147. Bottomley, M. J., Muraglia, E., Bazzo, R., & Carfì, A. (2007). Molecular insights into quorum sensing in the human pathogen Pseudomonas aeruginosa from the structure of the virulence regulator LasR bound to its autoinducer. *The Journal of Biological Chemistry*, *282*(18), 13592–13600. https://doi.org/10.1074/jbc.M700556200
148. Gambello, M. J., & Iglewski, B. H. (1991). Cloning and characterization of the Pseudomonas aeruginosa lasR gene, a transcriptional activator of elastase expression. *Journal of Bacteriology*, *173*(9), 3000–3009. https://doi.org/10.1128/jb.173.9.3000-3009.1991
149. Gambello, M. J., Kaye, S., & Iglewski, B. H. (1993). LasR of Pseudomonas aeruginosa is a transcriptional activator of the alkaline protease gene (apr) and an enhancer of exotoxin A expression. *Infection and Immunity*, *61*(4), 1180–1184. https://doi.org/10.1128/iai.61.4.1180-1184.1993
150. Schuster, M., Lostroh, C. P., Ogi, T., & Greenberg, E. P. (2003). Identification, timing, and signal specificity of Pseudomonas aeruginosa quorum-controlled genes: A transcriptome analysis. *Journal of Bacteriology*, *185*(7), 2066–2079. https://doi.org/10.1128/JB.185.7.2066-2079.2003

151. Schuster, M., Urbanowski, M. L., & Greenberg, E. P. (2004). Promoter specificity in Pseudomonas aeruginosa quorum sensing revealed by DNA binding of purified LasR. *Proceedings of the National Academy of Sciences of the United States of America, 101*(45), 15833–15839. https://doi.org/10.1073/pnas.0407229101
152. Seed, P. C., Passador, L., & Iglewski, B. H. (1995). Activation of the Pseudomonas aeruginosa lasI gene by LasR and the Pseudomonas autoinducer PAI: An autoinduction regulatory hierarchy. *Journal of Bacteriology, 177*(3), 654–659. https://doi.org/10.1128/jb.177.3.654-659.1995
153. Latifi, A., Foglino, M., Tanaka, K., Williams, P., & Lazdunski, A. (1996). A hierarchical quorum-sensing cascade in Pseudomonas aeruginosa links the transcriptional activators LasR and RhIR (VsmR) to expression of the stationary-phase sigma factor RpoS. *Molecular Microbiology, 21*(6), 1137–1146. https://doi.org/10.1046/j.1365-2958.1996.00063.x
154. Pesci, E. C., Pearson, J. P., Seed, P. C., & Iglewski, B. H. (1997). Regulation of las and rhl quorum sensing in Pseudomonas aeruginosa. *Journal of Bacteriology, 179*(10), 3127–3132. https://doi.org/10.1128/jb.179.10.3127-3132.1997
155. Ochsner, U. A., Koch, A. K., Fiechter, A., & Reiser, J. (1994). Isolation and characterization of a regulatory gene affecting rhamnolipid biosurfactant synthesis in Pseudomonas aeruginosa. *Journal of Bacteriology, 176*(7), 2044–2054. https://doi.org/10.1128/jb.176.7.2044-2054.1994
156. Pearson, J. P., Passador, L., Iglewski, B. H., & Greenberg, E. P. (1995). A second N-acylhomoserine lactone signal produced by Pseudomonas aeruginosa. *Proceedings of the National Academy of Sciences of the United States of America, 92*(5), 1490–1494. https://doi.org/10.1073/pnas.92.5.1490
157. Schuster, M., Lostroh, C. P., Ogi, T., & Greenberg, E. P. (2003). Identification, timing, and signal specificity of Pseudomonas aeruginosa quorum-controlled genes: A transcriptome analysis. *Journal of Bacteriology, 185*(7), 2066–2079. https://doi.org/10.1128/JB.185.7.2066-2079.2003
158. Schuster, M., & Greenberg, E. P. (2007). Early activation of quorum sensing in Pseudomonas aeruginosa reveals the architecture of a complex regulon. *BMC Genomics, 8*, 287. https://doi.org/10.1186/1471-2164-8-287
159. Hanzelka, B. L., & Greenberg, E. P. (1995). Evidence that the N-terminal region of the Vibrio fischeri LuxR protein constitutes an autoinducer-binding domain. *Journal of Bacteriology, 177*(3), 815–817. https://doi.org/10.1128/jb.177.3.815-817.1995
160. Haque, S., Yadav, D. K., Bisht, S. C., Yadav, N., Singh, V., Dubey, K. K., Jawed, A., Wahid, M., & Dar, S. A. (2019). Quorum sensing pathways in Gram-positive and -negative bacteria: Potential of their interruption in abating drug resistance. *Journal of Chemotherapy (Florence, Italy), 31*(4), 161–187. https://doi.org/10.1080/1120009X.2019.1599175
161. Lequette, Y., Lee, J. H., Ledgham, F., Lazdunski, A., & Greenberg, E. P. (2006). A distinct QscR regulon in the Pseudomonas aeruginosa quorum-sensing circuit. *Journal of Bacteriology, 188*(9), 3365–3370. https://doi.org/10.1128/JB.188.9.3365-3370.2006
162. DiMango, E., Zar, H. J., Bryan, R., & Prince, A. (1995). Diverse Pseudomonas aeruginosa gene products stimulate respiratory epithelial cells to produce interleukin-8. *The Journal of Clinical Investigation, 96*(5), 2204–2210. https://doi.org/10.1172/JCI118275
163. Smith, R. S., Harris, S. G., Phipps, R., & Iglewski, B. (2002). The Pseudomonas aeruginosa quorum-sensing molecule N-(3-oxododecanoyl)homoserine lactone contributes to virulence and induces inflammation in vivo. *Journal of Bacteriology, 184*(4), 1132–1139. https://doi.org/10.1128/jb.184.4.1132-1139.2002

164. Skindersoe, M. E., Zeuthen, L. H., Brix, S., Fink, L. N., Lazenby, J., Whittall, C., Williams, P., Diggle, S. P., Froekiaer, H., Cooley, M., & Givskov, M. (2009). Pseudomonas aeruginosa quorum-sensing signal molecules interfere with dendritic cell-induced T-cell proliferation. *FEMS Immunology and Medical Microbiology*, *55*(3), 335–345. https://doi.org/10.1111/j.1574-695X.2008.00533.x
165. Smith, R. S., & Iglewski, B. H. (2003). P. aeruginosa quorum-sensing systems and virulence. *Current Opinion in Microbiology*, *6*(1), 56–60. https://doi.org/10.1016/s1369-5274(03)00008-0
166. Kravchenko, V. V., Kaufmann, G. F., Mathison, J. C., Scott, D. A., Katz, A. Z., Grauer, D. C., Lehmann, M., Meijler, M. M., Janda, K. D., & Ulevitch, R. J. (2008). Modulation of gene expression via disruption of NF-kappaB signaling by a bacterial small molecule. *Science (New York, N.Y.)*, *321*(5886), 259–263. https://doi.org/10.1126/science.1156499
167. Pesci, E. C., Milbank, J. B., Pearson, J. P., McKnight, S., Kende, A. S., Greenberg, E. P., & Iglewski, B. H. (1999). Quinolone signaling in the cell-to-cell communication system of Pseudomonas aeruginosa. *Proceedings of the National Academy of Sciences of the United States of America*, *96*(20), 11229–11234. https://doi.org/10.1073/pnas.96.20.11229
168. Diggle, S. P., Winzer, K., Chhabra, S. R., Worrall, K. E., Cámara, M., & Williams, P. (2003). The Pseudomonas aeruginosa quinolone signal molecule overcomes the cell density-dependency of the quorum sensing hierarchy, regulates rhl-dependent genes at the onset of stationary phase and can be produced in the absence of LasR. *Molecular Microbiology*, *50*(1), 29–43. https://doi.org/10.1046/j.1365-2958.2003.03672.x
169. Diggle, S. P., Cornelis, P., Williams, P., & Cámara, M. (2006). 4-quinolone signalling in Pseudomonas aeruginosa: Old molecules, new perspectives. *International Journal of Medical Microbiology: IJMM*, *296*(2–3), 83–91. https://doi.org/10.1016/j.ijmm.2006.01.038
170. Gallagher, L. A., McKnight, S. L., Kuznetsova, M. S., Pesci, E. C., & Manoil, C. (2002). Functions required for extracellular quinolone signaling by Pseudomonas aeruginosa. *Journal of Bacteriology*, *184*(23), 6472–6480. https://doi.org/10.1128/JB.184.23.6472-6480.2002
171. Déziel, E., Lépine, F., Milot, S., He, J., Mindrinos, M. N., Tompkins, R. G., & Rahme, L. G. (2004). Analysis of Pseudomonas aeruginosa 4-hydroxy-2-alkylquinolines (HAQs) reveals a role for 4-hydroxy-2-heptylquinoline in cell-to-cell communication. *Proceedings of the National Academy of Sciences of the United States of America*, *101*(5), 1339–1344. https://doi.org/10.1073/pnas.0307694100
172. Xiao, G., He, J., & Rahme, L. G. (2006). Mutation analysis of the Pseudomonas aeruginosa mvfR and pqsABCDE gene promoters demonstrates complex quorum-sensing circuitry. *Microbiology (Reading, England)*, *152*(Pt 6), 1679–1686. https://doi.org/10.1099/mic.0.28605-0
173. Ahmer, B. M., van Reeuwijk, J., Timmers, C. D., Valentine, P. J., & Heffron, F. (1998). Salmonella typhimurium encodes an SdiA homolog, a putative quorum sensor of the LuxR family, that regulates genes on the virulence plasmid. *Journal of Bacteriology*, *180*(5), 1185–1193. https://doi.org/10.1128/JB.180.5.1185-1193.1998
174. Patankar, A. V., & González, J. E. (2009). Orphan LuxR regulators of quorum sensing. *FEMS Microbiology Reviews*, *33*(4), 739–756. https://doi.org/10.1111/j.1574-6976.2009.00163.x
175. Michael, B., Smith, J. N., Swift, S., Heffron, F., & Ahmer, B. M. (2001). SdiA of Salmonella enterica is a LuxR homolog that detects mixed microbial communities. *Journal of Bacteriology*, *183*(19), 5733–5742. https://doi.org/10.1128/JB.183.19.5733-5742.2001

176. Dyszel, J. L., Smith, J. N., Lucas, D. E., Soares, J. A., Swearingen, M. C., Vross, M. A., Young, G. M., & Ahmer, B. M. (2010). Salmonella enterica serovar Typhimurium can detect acyl homoserine lactone production by Yersinia enterocolitica in mice. *Journal of Bacteriology*, *192*(1), 29–37. https://doi.org/10.1128/JB.01139-09
177. Nicholson, B., & Low, D. (2000). DNA methylation-dependent regulation of pef expression in Salmonella typhimurium. *Molecular Microbiology*, *35*(4), 728–742. https://doi.org/10.1046/j.1365-2958.2000.01743.x
178. Miki, T., Okada, N., Shimada, Y., & Danbara, H. (2004). Characterization of Salmonella pathogenicity island 1 type III secretion-dependent hemolytic activity in Salmonella enterica serovar Typhimurium. *Microbial Pathogenesis*, *37*(2), 65–72. https://doi.org/10.1016/j.micpath.2004.04.006
179. Samudrala, R., Heffron, F., & McDermott, J. E. (2009). Accurate prediction of secreted substrates and identification of a conserved putative secretion signal for type III secretion systems. *PLoS Pathogens*, *5*(4), e1000375. https://doi.org/10.1371/journal.ppat.1000375
180. Smith, J. N., & Ahmer, B. M. (2003). Detection of other microbial species by Salmonella: Expression of the SdiA regulon. *Journal of Bacteriology*, *185*(4), 1357–1366. https://doi.org/10.1128/JB.185.4.1357-1366.2003
181. Taga, M. E., & Bassler, B. L. (2003). Chemical communication among bacteria. *Proceedings of the National Academy of Sciences of the United States of America*, *100*(Suppl 2), 14549–14554. https://doi.org/10.1073/pnas.1934514100
182. Surette, M. G., Miller, M. B., & Bassler, B. L. (1999). Quorum sensing in Escherichia coli, Salmonella typhimurium, and Vibrio harveyi: A new family of genes responsible for autoinducer production. *Proceedings of the National Academy of Sciences of the United States of America*, *96*(4), 1639–1644. https://doi.org/10.1073/pnas.96.4.1639
183. Herzberg, M., Kaye, I. K., Peti, W., & Wood, T. K. (2006). YdgG (TqsA) controls biofilm formation in Escherichia coli K-12 through autoinducer 2 transport. *Journal of Bacteriology*, *188*(2), 587–598. https://doi.org/10.1128/JB.188.2.587-598.2006
184. Xue, T., Zhao, L., Sun, H., Zhou, X., & Sun, B. (2009). LsrR-binding site recognition and regulatory characteristics in Escherichia coli AI-2 quorum sensing. *Cell Research*, *19*(11), 1258–1268. https://doi.org/10.1038/cr.2009.91
185. Pereira, C. S., Santos, A. J., Bejerano-Sagie, M., Correia, P. B., Marques, J. C., & Xavier, K. B. (2012). Phosphoenolpyruvate phosphotransferase system regulates detection and processing of the quorum sensing signal autoinducer-2. *Molecular Microbiology*, *84*(1), 93–104. https://doi.org/10.1111/j.1365-2958.2012.08010.x
186. Moreira, C. G., & Sperandio, V. (2012). Interplay between the QseC and QseE bacterial adrenergic sensor kinases in Salmonella enterica serovar Typhimurium pathogenesis. *Infection and Immunity*, *80*(12), 4344–4353. https://doi.org/10.1128/IAI.00803-12
187. Moreira, C. G., Weinshenker, D., & Sperandio, V. (2010). QseC mediates Salmonella enterica serovar typhimurium virulence in vitro and in vivo. *Infection and Immunity*, *78*(3), 914–926. https://doi.org/10.1128/IAI.01038-09
188. Govan, J. R., Hughes, J. E., & Vandamme, P. (1996). Burkholderia cepacia: Medical, taxonomic and ecological issues. *Journal of Medical Microbiology*, *45*(6), 395–407. https://doi.org/10.1099/00222615-45-6-395
189. Lewenza, S., & Sokol, P. A. (2001). Regulation of ornibactin biosynthesis and N-acyl-L-homoserine lactone production by CepR in Burkholderia cepacia. *Journal of Bacteriology*, *183*(7), 2212–2218. https://doi.org/10.1128/JB.183.7.2212-2218.2001
190. O'Grady, E. P., Viteri, D. F., Malott, R. J., & Sokol, P. A. (2009). Reciprocal regulation by the CepIR and CciIR quorum sensing systems in Burkholderia cenocepacia. *BMC Genomics*, *10*, 441. https://doi.org/10.1186/1471-2164-10-441

191. Gotschlich, A., Huber, B., Geisenberger, O., Tögl, A., Steidle, A., Riedel, K., Hill, P., Tümmler, B., Vandamme, P., Middleton, B., Camara, M., Williams, P., Hardman, A., & Eberl, L. (2001). Synthesis of multiple N-acylhomoserine lactones is widespread among the members of the Burkholderia cepacia complex. *Systematic and Applied Microbiology*, *24*(1), 1–14. https://doi.org/10.1078/0723-2020-00013
192. Huber, B., Riedel, K., Hentzer, M., Heydorn, A., Gotschlich, A., Givskov, M., Molin, S., & Eberl, L. (2001). The cep quorum-sensing system of Burkholderia cepacia H111 controls biofilm formation and swarming motility. *Microbiology (Reading, England)*, *147*(Pt 9), 2517–2528. https://doi.org/10.1099/00221287-147-9-2517
193. Malott, R. J., Baldwin, A., Mahenthiralingam, E., & Sokol, P. A. (2005). Characterization of the cciIR quorum-sensing system in Burkholderia cenocepacia. *Infection and Immunity*, *73*(8), 4982–4992. https://doi.org/10.1128/IAI.73.8.4982-4992.2005
194. Burkholder, W. H. (1950). Sour skin, a bacterial rot. *Phytopathology*, *40*, 115–117.
195. Aguilar, C., Bertani, I., & Venturi, V. (2003). Quorum-sensing system and stationary-phase sigma factor (rpoS) of the onion pathogen Burkholderia cepacia genomovar I type strain, ATCC 25416. *Applied and Environmental Microbiology*, *69*(3), 1739–1747. https://doi.org/10.1128/AEM.69.3.1739-1747.2003
196. Lewenza, S., Conway, B., Greenberg, E. P., & Sokol, P. A. (1999). Quorum sensing in Burkholderia cepacia: Identification of the LuxRI homologs CepRI. *Journal of Bacteriology*, *181*(3), 748–756. https://doi.org/10.1128/JB.181.3.748-756.1999
197. McKenney, D., Brown, K. E., & Allison, D. G. (1995). Influence of Pseudomonas aeruginosa exoproducts on virulence factor production in Burkholderia cepacia: Evidence of interspecies communication. *Journal of Bacteriology*, *177*(23), 6989–6992. https://doi.org/10.1128/jb.177.23.6989-6992.1995
198. Geisenberger, O., Givskov, M., Riedel, K., Høiby, N., Tümmler, B., & Eberl, L. (2000). Production of N-acyl-L-homoserine lactones by P. aeruginosa isolates from chronic lung infections associated with cystic fibrosis. *FEMS Microbiology Letters*, *184*(2), 273–278. https://doi.org/10.1111/j.1574-6968.2000.tb09026.x
199. Percival, S. L., & Williams, D. W. (2014). Yersinia. In *Microbiology of waterborne diseases* (pp. 249–259). Elsevier. https://doi.org/10.1016/B978-0-12-415846-7.00013-5
200. Barton, M. D. (2016). Yersinia enterocolitica. In *Reference module in food science* (p. B9780081005965010000). Elsevier. https://doi.org/10.1016/B978-0-08-100596-5.00996-3
201. Atkinson, S., Throup, J. P., Stewart, G. S., & Williams, P. (1999). A hierarchical quorum-sensing system in Yersinia pseudotuberculosis is involved in the regulation of motility and clumping. *Molecular Microbiology*, *33*(6), 1267–1277. https://doi.org/10.1046/j.1365-2958.1999.01578.x
202. Atkinson, S., Sockett, R. E., Cámara, M., & Williams, P. (2006). Quorum sensing and the lifestyle of Yersinia. *Current Issues in Molecular Biology*, *8*(1), 1–10.
203. Throup, J. P., Camara, M., Briggs, G. S., Winson, M. K., Chhabra, S. R., Bycroft, B. W., Williams, P., & Stewart, G. S. (1995). Characterisation of the yenI/yenR locus from Yersinia enterocolitica mediating the synthesis of two N-acylhomoserine lactone signal molecules. *Molecular Microbiology*, *17*(2), 345–356. https://doi.org/10.1111/j.1365-2958.1995.mmi_17020345.x
204. Yates, E. A., Philipp, B., Buckley, C., Atkinson, S., Chhabra, S. R., Sockett, R. E., Goldner, M., Dessaux, Y., Cámara, M., Smith, H., & Williams, P. (2002). N-acylhomoserine lactones undergo lactonolysis in a pH-, temperature-, and acyl chain length-dependent manner during growth of Yersinia pseudotuberculosis and Pseudomonas aeruginosa. *Infection and Immunity*, *70*(10), 5635–5646. https://doi.org/10.1128/IAI.70.10.5635-5646.2002

205. Starr, T., Child, R., Wehrly, T. D., Hansen, B., Hwang, S., López-Otin, C., Virgin, H. W., & Celli, J. (2012). Selective subversion of autophagy complexes facilitates completion of the Brucella intracellular cycle. *Cell Host & Microbe*, *11*(1), 33–45. https://doi.org/10.1016/j.chom.2011.12.002
206. Golding, B., Scott, D. E., Scharf, O., Huang, L. Y., Zaitseva, M., Lapham, C., Eller, N., & Golding, H. (2001). Immunity and protection against Brucella abortus. *Microbes and Infection*, *3*(1), 43–48. https://doi.org/10.1016/s1286-4579(00)01350-2
207. Terwagne, M., Uzureau, S., & Jean-Jacques, L. (2012). Brucella quorum sensing: Much more than sensing quorum. In *Brucella: Molecular microbiology and genomics*. Caister Academic Press.
208. Zhang, R. G., Pappas, K. M., Brace, J. L., Miller, P. C., Oulmassov, T., Molyneaux, J. M., Anderson, J. C., Bashkin, J. K., Winans, S. C., & Joachimiak, A. (2002). Structure of a bacterial quorum-sensing transcription factor complexed with pheromone and DNA. *Nature*, *417*(6892), 971–974. https://doi.org/10.1038/nature00833
209. Patankar, A. V., & González, J. E. (2009). An orphan LuxR homolog of Sinorhizobium meliloti affects stress adaptation and competition for nodulation. *Applied and Environmental Microbiology*, *75*(4), 946–955. https://doi.org/10.1128/AEM.01692-08
210. Taminiau, B., Daykin, M., Swift, S., Boschiroli, M. L., Tibor, A., Lestrate, P., De Bolle, X., O'Callaghan, D., Williams, P., & Letesson, J. J. (2002). Identification of a quorum-sensing signal molecule in the facultative intracellular pathogen Brucella melitensis. *Infection and Immunity*, *70*(6), 3004–3011. https://doi.org/10.1128/IAI.70.6.3004-3011.2002
211. Weeks, J. N., Galindo, C. L., Drake, K. L., Adams, G. L., Garner, H. R., & Ficht, T. A. (2010). Brucella melitensis VjbR and C12-HSL regulons: Contributions of the N-dodecanoyl homoserine lactone signaling molecule and LuxR homologue VjbR to gene expression. *BMC Microbiology*, *10*, 167. *https://doi.org/10.1186/1471-2180-10-167*
212. Uzureau, S., Godefroid, M., Deschamps, C., Lemaire, J., De Bolle, X., & Letesson, J. J. (2007). Mutations of the quorum sensing-dependent regulator VjbR lead to drastic surface modifications in Brucella melitensis. *Journal of Bacteriology*, *189*(16), 6035–6047. https://doi.org/10.1128/JB.00265-07
213. de Jong, M. F., Sun, Y. H., den Hartigh, A. B., van Dijl, J. M., & Tsolis, R. M. (2008). Identification of VceA and VceC, two members of the VjbR regulon that are translocated into macrophages by the Brucella type IV secretion system. *Molecular Microbiology*, *70*(6), 1378–1396. https://doi.org/10.1111/j.1365-2958.2008.06487.x
214. Arocena, G. M., Sieira, R., Comerci, D. J., & Ugalde, R. A. (2010). Identification of the quorum-sensing target DNA sequence and N-Acyl homoserine lactone responsiveness of the Brucella abortus virB promoter. *Journal of Bacteriology*, *192*(13), 3434–3440. https://doi.org/10.1128/JB.00232-10
215. Uzureau, S., Lemaire, J., Delaive, E., Dieu, M., Gaigneaux, A., Raes, M., De Bolle, X., & Letesson, J. J. (2010). Global analysis of quorum sensing targets in the intracellular pathogen Brucella melitensis 16 M. *Journal of Proteome Research*, *9*(6), 3200–3217. https://doi.org/10.1021/pr100068p
216. Rambow-Larsen, A. A., Rajashekara, G., Petersen, E., & Splitter, G. (2008). Putative quorum-sensing regulator BlxR of Brucella melitensis regulates virulence factors including the type IV secretion system and flagella. *Journal of Bacteriology*, *190*(9), 3274–3282. https://doi.org/10.1128/JB.01915-07
217. Backert, S., & Hofreuter, D. (2013). Molecular methods to investigate adhesion, transmigration, invasion and intracellular survival of the foodborne pathogen Campylobacter jejuni. *Journal of Microbiological Methods*, *95*(1), 8–23. https://doi.org/10.1016/j.mimet.2013.06.031

218. Horrocks, S. M., Anderson, R. C., Nisbet, D. J., & Ricke, S. C. (2009). Incidence and ecology of Campylobacter jejuni and coli in animals. *Anaerobe*, *15*(1–2), 18–25. https://doi.org/10.1016/j.anaerobe.2008.09.001
219. Elvers, K. T., & Park, S. F. (2002). Quorum sensing in Campylobacter jejuni: Detection of a luxS encoded signalling molecule. *Microbiology (Reading, England)*, *148*(Pt 5), 1475–1481. https://doi.org/10.1099/00221287-148-5-1475
220. Moorhead, S. M., & Griffiths, M. W. (2011). Expression and characterization of cell-signalling molecules in Campylobacter jejuni. *Journal of Applied Microbiology*, *110*(3), 786–800. https://doi.org/10.1111/j.1365-2672.2010.04934.x
221. Plummer, P., Sahin, O., Burrough, E., Sippy, R., Mou, K., Rabenold, J., Yaeger, M., & Zhang, Q. (2012). Critical role of LuxS in the virulence of Campylobacter jejuni in a guinea pig model of abortion. *Infection and Immunity*, *80*(2), 585–593. https://doi.org/10.1128/IAI.05766-11
222. Plummer, P., Zhu, J., Akiba, M., Pei, D., & Zhang, Q. (2011). Identification of a key amino acid of LuxS involved in AI-2 production in Campylobacter jejuni. *PLoS One*, *6*(1), e15876. https://doi.org/10.1371/journal.pone.0015876
223. Adler, L., Alter, T., Sharbati, S., & Gölz, G. (2015). The signalling molecule autoinducer-2 is not internalised in Campylobacter jejuni. *Berliner und Munchener tierarztliche Wochenschrift*, *128*(3–4), 111–116.
224. Reeser, R. J., Medler, R. T., Billington, S. J., Jost, B. H., & Joens, L. A. (2007). Characterization of Campylobacter jejuni biofilms under defined growth conditions. *Applied and Environmental Microbiology*, *73*(6), 1908–1913. https://doi.org/10.1128/AEM.00740-06
225. Bezek, K., Kurinčič, M., Knauder, E., Klančnik, A., Raspor, P., Bucar, F., & Smole Možina, S. (2016). Attenuation of adhesion, biofilm formation and quorum sensing of Campylobacter jejuni by Euodia ruticarpa. *Phytotherapy Research: PTR*, *30*(9), 1527–1532. https://doi.org/10.1002/ptr.5658

4 Fundamentals of Bacterial Quorum Sensing and Antimicrobial Resistance

Tahseena Naaz, Vandana Singh, Subhasree Ray, Kalpana Sharma, Barun Kumar, and Soumya Pandit

4.1 INTRODUCTION

Quorum sensing is a mechanism of detecting neighbouring cell population density by bacterial cells to survive adverse conditions such as extremes of pH, temperature, pressure, nutrient scarcity, and so on. This mechanism governs various biological behaviours of the bacteria. It involves a cascade of genes regulating the population density (Zhao et al., 2020). As the cell population density increases, the bacterial cell begins to produce small signalling molecules inducing various cellular mechanisms like virulence factors, antibiotic tolerance, formation of biofilm mobility, and mechanism of drug resistance. This allows the bacteria to express its mechanisms only when there is a maximum effect on the environment. Quorum sensing controls the gene that controls how bacteria act when a bunch of bacteria acts at the same time. The total bacterial population behaviour is controlled by the timely synthesis of small signalling molecules known as autoinducers (AIs). For instance, Gram-positive bacteria utilize as a signalling molecule auto-inducing peptides that activate the receptor's kinase activity. In contrast, Gram-negative bacteria use small molecules such as acyl-homoserine lactones AHLs (Papenfort and Bassler, 2016) (see Figure 4.1).

The particular gene regulates the habituation of the bacterial population as the autoinducer concentration, and the density of the population reaches the threshold. The basic principle of quorum sensing consists of three steps. First is the secretion of the signalling molecule. If the population density is high, it permits the detection of the signalling molecule and response to it (Zhao et al., 2020). The second step involves detection of the signalling molecule by the receptor present in the cytoplasm of the bacterial cell and formation of a QS signalling molecule receptor complex, followed by activation of the specific gene that regulates the QS system. This process finally triggers the synchronizing activity of the bacterial population (Zhao et al., 2020).

Any substance derived naturally or synthetically that can kill, destroy, or impede the growth of microbes is defined as an antimicrobial agent (Abdula

DOI: 10.1201/9781003300557-4

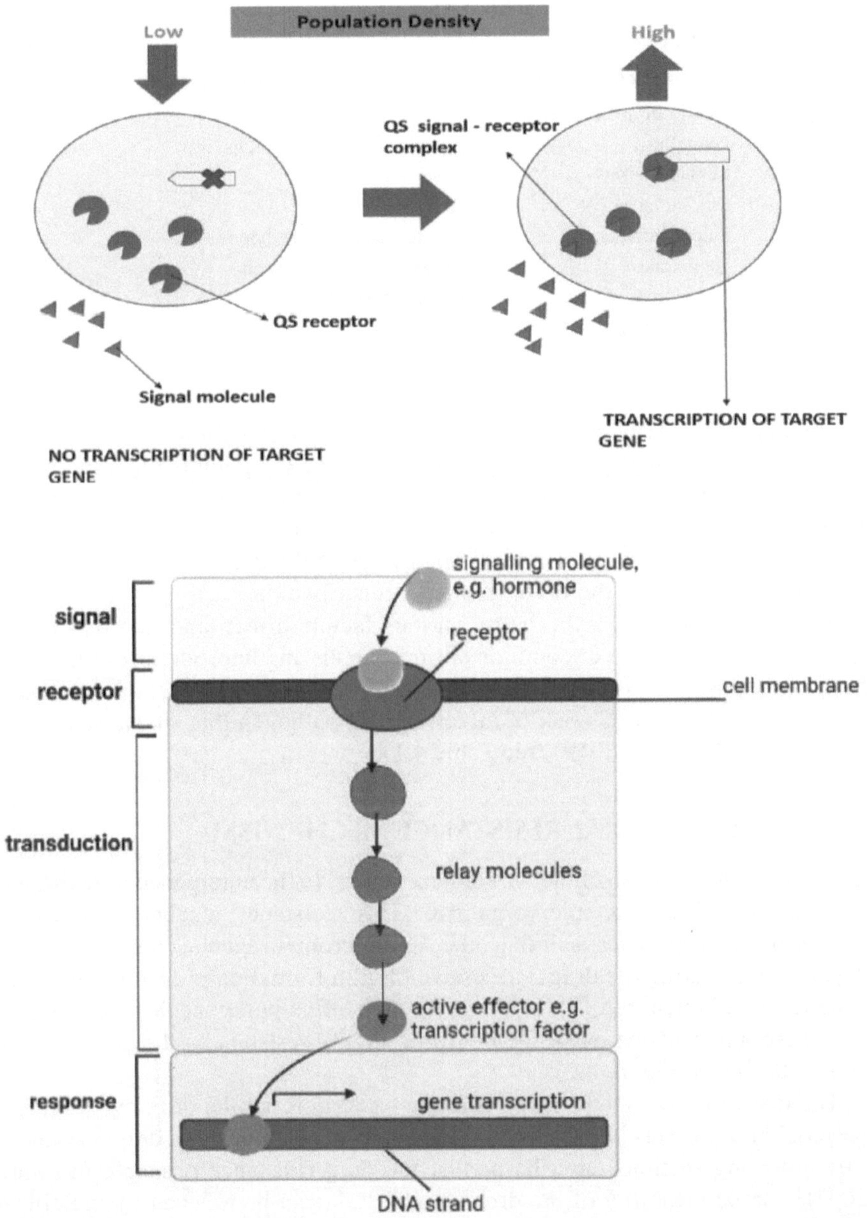

FIGURE 4.1 Effect of signalling molecule and population density on transcription of target gene.

et al., 2016). Antimicrobial resistance is the condition wherein the microbes develop resistance toward the drug. The microbes can show resistance toward the drug through various processes, such as inactivating the drug or preventing

TABLE 4.1
Antimicrobial Agent and Its Corresponding Group

Group of antibiotics	Antimicrobial agent
Penicillin	Augmentin, amoxicillin, ampicillin
Cephalosporins	Ceftriaxone
Chloramphenicol	Chloramphenicol
Sulphonamides	Sulphamethazine, trimethoprim
Quinolones	Nalidixic acid, ciprofloxacin
Oxazolidinones	Linezolid, fosfomycin
Carbapenems	Imipenem

the uptake of the drug by going through certain modifications in its cell wall. Further, a special membrane pump is used after the drug enters the microbes. It can be classified according to the antimicrobial activity mechanism. The majority of antimicrobial substances typically hinder the production of cell walls, proteins, nucleic acids, or metabolic pathways (Nikaido, 2009). The ability of the microbes to resist the antimicrobial agents is either acquired naturally or from other microbes. Furthermore, various factors affect antimicrobial resistance, such as overusage of common antimicrobials and improper prescription of broad-spectrum drugs (Liu et al., 2019). Overdosage of an antimicrobial agent can increase the patient's risk of infection by a pathogen that shows sensitivity to the drug (Liu et al., 2019) (see Table 4.1).

4.2 ANTIMICROBIAL RESISTANCE MECHANISM

Excessive drug usage in clinical treatment has led to the emergence of resistance mechanisms within the microorganism. This resistance mechanism involves chemical changes encapsulating the antimicrobial agents, regulated gene expression targeting the drug, successive biofilm formation preventing the drug penetration, efficient removal of drugs through efflux pump mechanisms, and so on. These mechanisms make the microbes highly resistant to the drug (Rajput et al., 2018) (see Figure 4.2).

The latest discovery has shown that the bacteria resist the drug by forming a bacterial biofilm. This is an organized and complex mechanism of drug resistance. The following sections describe part of the drug resistance principle (Stewart, 2002). The permeability of the drug into bacteria can be reduced by the efficacious barrier formed by the bacterial biofilm. Bacteria in the biofilm are linked together by various protein DNA and the extracellular polysaccharide, which form an impenetrable barrier that significantly reduces antibiotic drug permeability and increases bacteria viability in the biofilm (Rajput et al., 2018).

Furthermore, different areas of the biofilm had different concentrations of bacterial secretion and nutrients. Consequently, bacterial growth in the different areas of biofilm is inconsistent. This led to the formation of a microenvironment of

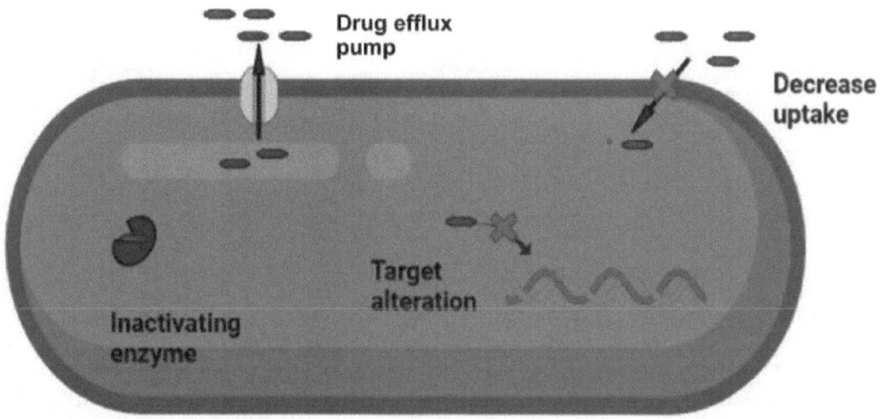

FIGURE 4.2 Various mechanisms of antimicrobial resistance.

the biofilm that allows the bacteria in the membrane to show resistivity toward the antibiotic and generate heterogeneity (Zhao et al., 2020). Depending upon the biofilm, the bacteria have various levels of resistivity. Drug resistance can be seen within the membrane in certain extreme surroundings. But in drastic conditions, the outside surroundings are altered, including the temperature, pH, and certain chemical concentrations, which reduce the growth of the bacteria and resistance toward the drug by affecting the biofilm functions that govern the biochemical and physiological function (Freilich et al., 2011).

In addition, the chemical modification inactivates the mechanism of an antimicrobial agent by producing the altered enzyme which modified the chemical structure of the antimicrobial agent (Liu et al., 2019). This led the antimicrobial agent to lose its activity. Producing enzymes is one of the best ways to deal with antibiotics, since it degrades the antimicrobial agent and converts the chemical group in the antimicrobial agent (Rajput et al., 2018). This conversion of the chemical compound impedes the antimicrobial agent from interacting with the target substance. An efflux pump plays a crucial role in antimicrobial resistance by pumping out the drug that has entered the microbes through the membrane (Wasaznik et al., 2009). The invasion of the antimicrobial agent is an important condition for interacting with or attacking the specific target. The microorganism uses the efflux pump to expel the invading antimicrobial agent. by assembly of the efflux pump protein on the bacterial cell's membrane. The drug excretion is faster than drug penetration, boosting the cell's antimicrobial agent concentration to an insensitive level (Zhao et al., 2020). Various efflux pumps, such as lipophilic and hydrophilic efflux systems, are found in the microbes that attack drugs of various chemical compositions.

The interaction of the antimicrobial agent with its specific target gene is an important condition for its activity. Thus, the modification of the target gene inhibits antimicrobial activity. The bacterial cell modifies the target gene in order

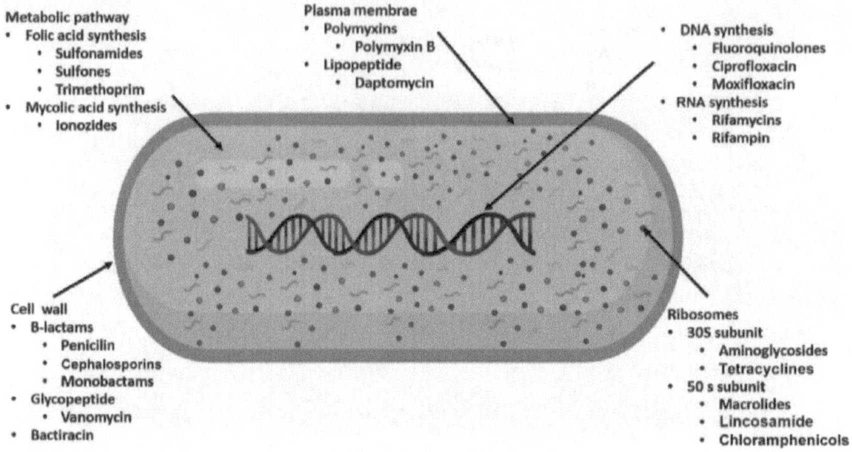

FIGURE 4.3 Classes of antimicrobial agent and their modes of action.

to develop resistance toward the antimicrobial agent. As a result, the antimicrobial agent loses its specific target gene and becomes inactive since the affinity of the antimicrobial agent toward the binding site is reduced. β-lactam antibiotics kill bacteria by inhibiting mucopeptide synthase and penicillin-binding proteins in the bacterial cell wall trans-peptide process. It prevents the bacteria from forming a complete cell wall, causing them to die. The bacterial cell adapts to unfavourable environments and harsh conditions that cause bacteria stress. This is a type of cellular adaptation to survive in a harsh environment. Bacteria cells compete for the nutrient oxygen and withstand the substance secreted by other competing microbes (Zhao et al., 2019). Mostly bacteria get attacked by the immune system of the host. Thus, to survive in these environments, bacteria must cope with the harsh conditions and adapt to them (see Figure 4.3).

Bacteria undergo various mechanisms to prevent disturbance to their cellular mechanism by unfavourable conditions such as membrane homeostasis and cell wall synthesis (see Table 4.2) – for instance, the development of resistance by *Staphylococcus aureus* toward vancomycin and daptomycin. Superbacteria evolved due to the increasing multiple resistance mechanism in bacteria. Examples of super-resistant bacteria are *S. aureus* and *Pseudomonas aeruginosa*. The ability to prevent loss of the antimicrobial agent through resistance genes is achieved by horizontal gene transfer changes in the body shape of the bacteria and production of various efflux pumps. Besides that, degradation of penicillin via β-lactamase and the PBP2 protein production that prevent the ability of the methicillin cell wall mucin synthase can observed in *S. aureus*. Therefore, in order to decrease antimicrobial resistance, research on microbial resistance by various mechanisms is needed (Ma et al., 2019).

TABLE 4.2
Antimicrobial Groups Based on Their Modes of Action

Antimicrobial class	Antimicrobial resistance mechanism
Target modification	
Aminoglycosides	Ribosome protein alteration
β-lactam antibiotics	New/altered penicillin-binding protein
erythromycin and streptogramin	Methylation of the bacteria ribosome
Trimethoprim	New drug-insensitive dihydropteroate
Tetracycline	Ribosomal protection
Vancomycin	Affinity decreases by alteration of cell wall precursors
Detoxifying enzymes	Acetyltransferase
Chloramphenicol	Inhibiting enzymes in the folate synthesis pathway
Trimethoprim-sulphamethoxazole	
Decreased drug uptake	Outer membrane protein alteration
Decline in permeability: tetracycline/chloramphenicol/quinolone/β-lactam antibiotics	New membrane transport system
Active efflux: tetracycline/trimethoprim	

4.3 MICROBIAL QUORUM-SENSING MECHANISM AND ITS REGULATION

Nowadays, the research results of quorum-sensing systems in microbes (mainly bacteria) provide new ideas for understanding the regulatory mechanism and its limitation factors of antibiotic resistance. These observations make the clear explanation that quorum-sensing systems regulate the cellular mechanism of different microorganisms, including gene expression of pathogen, toxic substance production, and extracellular polysaccharide synthesis, and all these help in regulation of drug efflux pumps along with microbial biofilm formation (Zhang and Li, 2015). There are three different types of bacterial quorum-sensing systems in accordance with the self-induced molecules: 1) In Gram-negative bacteria, the quorum-sensing system with acyl-homoserine lactone (AHL) is the self-induced molecule; 2) In Gram-positive bacteria, the oligopeptides are quorum-sensing systems that are self-induced molecules; and 3) In others the furan borate diesters are used as autoinducing molecules in both positive and negative strains of bacteria (see Figure 4.4).

In signal molecules of quorum sensing mainly Gram-positive bacteria are represented as small molecules of autoinducing oligopeptides (AIP). The precursor molecules of AIP are formed by modifying the main controlling peptide, and this can be transported throughout the body for outer cellular secretion with the help of ATP-binding cassette (Monnet and Gardan, 2015). The best example is the AGR system in Gram-positive bacteria (*S. aureus*). The AGR system is the coordination

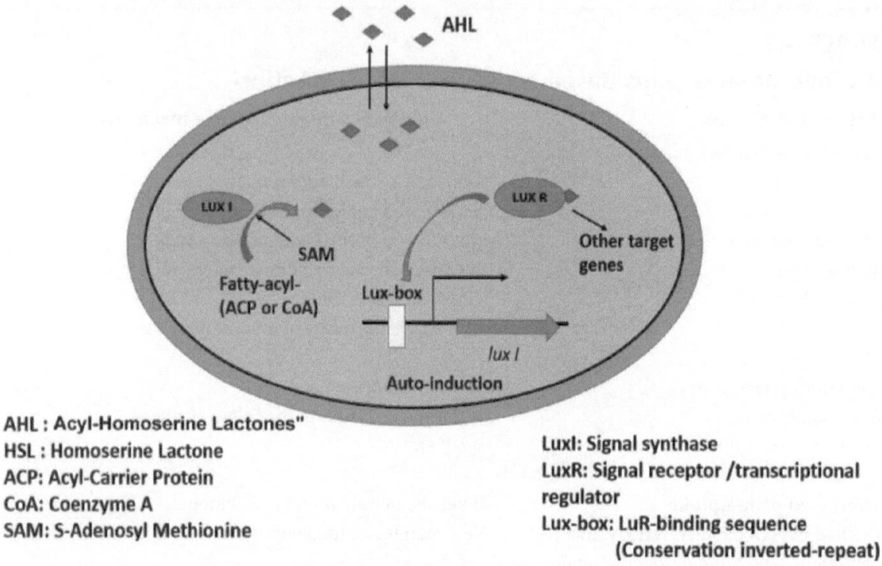

FIGURE 4.4 Quorum-sensing system of AHL.

of gene expression in response to bacterial cell population (Peacock and Paterson, 2015). Bacteria initiate the synthesis of a greater number of virulence factors with more pathogenic activity by the increase in cell density. This entire process is the reflex made by the oligopeptide signal molecules that activates the cell and regulates the expression of the gene. When the concentration of the extracellular oligopeptide signal molecules is just right, they bind to the receptor protein on the cell membrane and transfer the oligopeptide to the intracellular promoter, which binds via phosphorylation to start the transcriptional mechanism and modifies post-translational modifications that either activate (or inhibit) gene expression (Zhao et al., 2020).

Mainly in Gram-negative bacteria the signal molecules of quorum sensing are represented by autoinducers (AI-1) and AHLs (Zhang and Li, 2015). The LuxR–LuxI system has AHLs as the formed product in negative stained bacteria which helps it to diffuse easily outside and inside bacteria and helps it to grow in the surrounding environment (Ng et al., 2018). Lux-R binding proteins are encoded by Lux-R, which is the activator for transcription that produces the response known as transcription regulator. This regulation needs the AHL's activation molecule, which helps to encode Lux-I protein and synthetase of AHL's (Chen and Xiang, 2016). Substrates that are required to function are SAM and acyl carrier protein linked to acyl which forms the binding of acyl-acyl carrier protein. Later, through the binding of Lux-R to the AHL, multimerization occurs and the product binds to the region of upstream regulation of the targeted gene which activates or stops the targeted gene expression. The AHL is specified by different acyl side chain groups and the binding capacity of AHL with Lux-R is very strong. Thus, Gram-negative

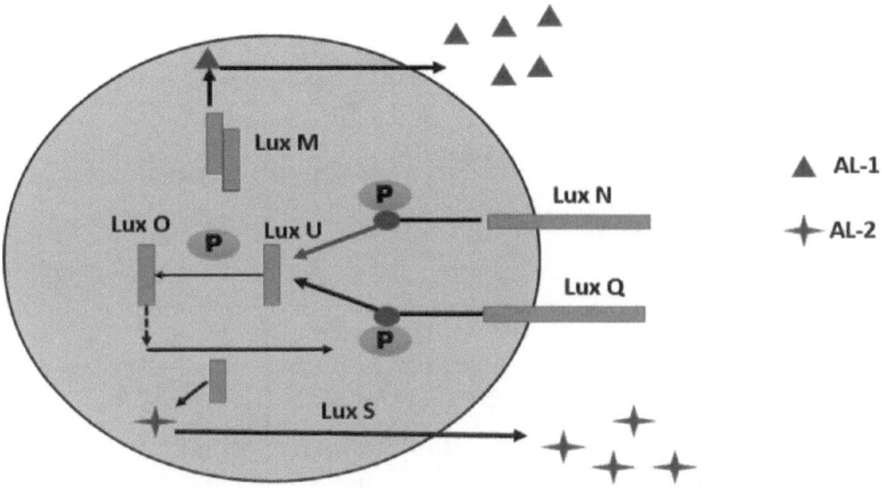

FIGURE 4.5 Quorum-sensing system of AI2.

bacteria use this process to receive information from the same species of bacteria and avoid the disturbance caused by other groups of bacteria.

The signal molecules of quorum sensing in both positive and negative stained bacteria by furan borate diester are represented by the AI-2 that mediates the quorum sensing signals between two different bacteria species and regulates the gene expression. AI-2 is formed from SAM in activated methyl cycle (AMC) with three step enzymatic reactions. In that reaction the methyl donor is SAM and the mediator product is SAH, which is hydrolyzed by 5'-methylthioadenosine to S-ribosyl homocysteine (SRH) and adenine. Cleavage of SRH to 4,5-dihydroxy-2,3-pentanedione (DPD) and homocysteine is catalyzed by LuxS and the main substance for AI-2 synthesis. DPD molecular structure is not stable where the other reaction changes over time, but it can form different molecules of AI-2 with differences in structure and composition which can be identified by other bacterial species. AI-2 is simply the mixture of many of the same small molecules with different compositions. Scientists discovered 2 different AI-2 structures: 1) a furanosyl borate diester of AI-2 in *V. harveyi* was found by Li et al. (2019); and 2) a furanosyl borate diester lacking boron in *S. typhimurium* was found by Miller et al. (2004) (see Figure 4.5).

4.4 MICROBIAL COMPETITION AND QUORUM SENSING

In general, "ecological (or biological) competition" is the term which refers to the superiority complex by which one organism reduces the survival rate or reproductive rate of another organism (Cornforth and Foster, 2013). *Exploitative* and *interference* are the two different types of competition. Exploitative competition can be defined as indirect competition of one organism by utilizing another organism's

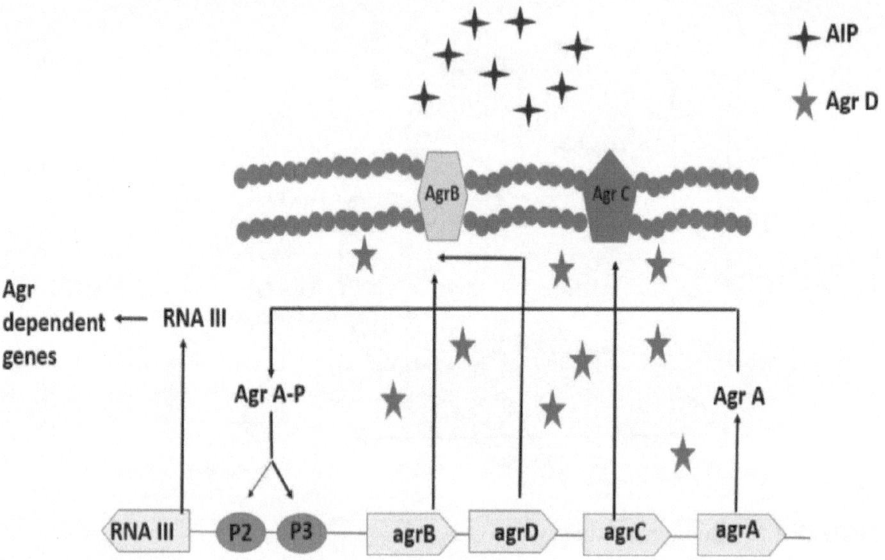

FIGURE 4.6 Quorum sensing system in *S. aureus*.

resources (nutrients) as their source of living. This exploitative competition also occurs in microorganisms, when these microbes form dense communities on surfaces. In such conditions, restriction of nutrients is very common and it results in high exploitative competition between cells of same genotype and different genotype. Interference competition is when one organism directly competes with another organism. In microorganisms this competition occurs via the release of antibiotics and synthetic polymers. Many experiments show that these metabolites released from the organism can determine the genotype that can survive in a mixed community. The diversity of bacteria and its survival is strongly affected by the exploitative and interference competitions (Freilich et al., 2011). By evolution, the bacteria are able to detect the ecological or biological competition, and can respond to regulate the favourable conditions in various bacterial stress responses. This is known as bacterial quorum sensing.

In biological competition the bacterial stress response is related to toxin release because the bacterial community competition has both evolved and foreign genotypes (Li and Tian, 2016). Toxin release in the bacteria population describes the quorum-sensing role in microbial diversity regulation. Toxin metabolites released by the bacteria damage the genotype of other bacterial cells, killing them. Toxins mainly help in finding the functional characteristics of bacteria that effect the other bacterial activity and kill them. This includes the ribosomal synthesized antimicrobial peptide produced by bacteria that targets other bacteria and secondary metabolites which affects both nutrition and metabolism (Doekes et al., 2019). The toxin-regulated bacterial diversity is as follows: Quorum sensing mediates the information related to microbial communities that sense the biological competition. The density of the same number of cells releases toxins. Quorum-sensing regulation provides the bacteria with the same genotype as in the bacterial community to release toxins and to increase growth products. Microbial competition intensity can be detailed with quorum-sensing information. The diversity potential of microbial communities rests not only in community density information but also the specific signal genotype that is produced by other cells and signal molecules. Due to the identification of certain strains or species as competitors, such as *Proteus mirabilis* (Gram-negative bacteria), it gives population density information. This also allows it to find the other genotype strains in species and identify future competitors. The variety of genotype strains in a microbial community can surely affect the order of various genotypes in the community, thereby managing biodiversity.

4.5 MICROBIAL REGULATION SYSTEM BY QUORUM SENSING

4.5.1 Bacterial Efflux Pump Regulation by Quorum Sensing

Bacterial active efflux pumps are important in multidrug resistance because they can efficaciously deliver antibiotics from the bacteria. Bacterial active efflux pumps are made up of three parts: cytoplasmic membrane efflux protein, fusion protein, and outer membrane channel protein, which are arranged from inside to outside the participation of energy. the fusion protein attaches to the outer membrane channel protein and the cytoplasmic membrane efflux (12). Protein, foreign substances, such as antibiotics, and their metabolites are all eliminated from bacteria, either selectively or nonselectively (Subhadra et al., 2019). The regulating effect of the QS system on the bacterial efflux pump has been affirmed. The regulating effect of the QS system on the expression of multidrug-resistant pumps is that the expression of multidrug-resistant pumps can be controlled, and the expression level of multidrug-resistant pumps affects the QS system as well (Li et al., 2015). On the one hand, the QS system can control efflux pump gene expression. *Bacteroides fragilis* ATCC25285 was cultivated in vitro with or without the self-inducible molecules C6-HSL and C8-HSL, for example. The biofilm structure and growth-resistant drug-sensitive efflux pump gene bmeB expression were investigated, and it was discovered that LuxR, a self-inducing molecule receptor of the QS system, can interact with exogenous AHL, up-regulate the expression of the bmeB efflux pump, and create antibiotic resistance (Pumbwe et al., 2008).

Similarly, autoinducers have been found to increase the multidrug resistance pump MexAB-OprM, leading to multidrug resistance in bacteria (Li et al., 2015). The efflux pump's expression level, on the other hand, has an impact on the QS system. Some researchers have discovered that overexpression of the MexCD-OprJ multidrug efflux pump inhibits *P. aeruginosa*'s QS response. While some efflux pumps (such as RND) expel drugs from cells, QS system self-inducible molecules can also be expelled from cells, enhancing the concentration of extracellular self-inducible molecules, which can lead to an increase in bacterial infections.

This suggests that high efflux pump expression may enhance further QS system activation, promote QS system regulation of toxin infection factor synthesis and efflux pump expression, and increase bacterial infectivity and invasiveness. Simultaneously, researchers discovered that the QS system controls biofilm-forming genes as well as bacterial resistance genes. In *B. fragilis*, the LuxR receptor of the QS system has been shown to regulate biofilm formation and bmeB expression in the efflux pump. In biofilms, the resistance gene ampC can be highly expressed in the presence of imipenem. The presence of high biofilm-forming bacteria without free or low-expression genes suggests that the QS system regulates the expression of these resistance-related genes while governing biofilm formation. This has piqued people's interest, and it could be another way for the QS system to directly regulate drug resistance-related genes.

4.5.2 Quorum Sensing in Regulation of Bacterial Biofilm Formation

In order to adapt to the living environment, bacteria trapped on the surface of inert or active materials form a biofilm. It is made up of polysaccharides and proteins, as well as an extracellular matrix. It has a distinct spatial structure and exhibits new biological characteristics as well as greater environmental adaptability. The majority of human bacterial infections are linked to biofilm, according to recent research, and biofilm formation is one of the main reasons why clinical bacterial infections are challenging to medicate and often relapse. The development of drug resistance is linked to the development of bacterial biofilms. Bacterial biofilms are thought to contribute to bacterial drug resistance via penetration restriction mechanisms, nutrition restriction mechanisms, and phenotypic drug resistance mechanisms (Saxena et al., 2019). The main mechanism of limiting permeability is the formation of a molecular barrier and a charge barrier (mostly negatively charged) by the polysaccharides in the biofilm, which can prevent or delay the penetration of certain antibiotics (Roy et al., 2018). The permeation limitation mechanism is linked to the nutrient limitation mechanism. Nutrients cannot conveniently cross through the biofilm due to its permeation limitation, leaving the nutrient in the biofilm deficient in nutrients and slowing the development of the inner bacteria. Slow growth is also known as starvation, and bacteria that are starved are less susceptible to antibiotics (Stubbendieck et al., 2016; Chebotar et al., 2012). Furthermore, the study discovered that drug resistance is linked to differential expression of some biofilm-related and phenotype-related genes, implying that biofilm resistance may also be linked to a drug-resistant phenotype mechanism.

The QS system plays a critical regulatory role in the formation of biofilms. Gram-negative and Gram-positive bacteria's biofilm formation is regulated by the QS system. The QS system, which is made up of signal molecules and corresponding receptors, regulates the formation of Gram-negative bacteria biofilms by using AHL as a signal molecule. The QS system in *P. aeruginosa*, for example, has two signalling systems: lasl/lasR and rhlI/rhlI. Signal molecule synthetases and receptors are encoded by the lasl/rhlI and lasR and rhlR genes, respectively.

As the number of bacteria grows, so does the secretion of signal molecules. When a signal molecule reaches a certain threshold, it binds to the receptor for the corresponding signal molecule and activates it. The stimulated receptor then induces bacteria to form biofilms by activating transcriptional regulators that synthesize extracellular polysaccharides, toxic factors, alginates, and other compounds. The QS system regulates Gram-positive bacteria biofilms by using oligopeptides as signal molecules that can be recognized by the 2-sensing protein after alteration, and it regulates AIM gene expression by phosphorylation and dephosphorylation of the protein. This regulates the emergence of biofilms even more. Different oligopeptide signalling molecules and QS system regulation pathways are found in different types of bacteria. The histidine protein kinase and response regulatory protein (McCluskey et al., 2004) are two components of the two-component system in *Streptococcus*, while the QS system in *S. aureus* is highly conserved (see Figure 4.6).

The YycG/YycF 2 system, which directly controls the formation of biofilms, is also known as WalK/WalR (Dubrac et al., 2007). There are recognized regulatory factors that affect biofilm formation in addition to two-component systems. RNA nucleic acid polymerase III, for example, can enhance the formation of *S. aureus* biofilms while inhibiting protein peptides and significantly inhibiting the output of its biofilms (Tan et al., 2018). Competent stimulating factors in *Streptococcus* can help the QS system regulate the biofilm. The QS system–related comprehensive regulator sarA is linked to and is a positive regulator of biofilm formation in *S. epidermidis*.

4.5.3 Bacterial Secretion System Regulation by Quorum Sensing

Proteins are secreted by pathogenic bacteria through the cell membrane. They can use this basic process to attack other microorganisms, evade the immune system of the host, cause tissue damage, and invade host cells. Secreted proteins have the potential to be toxic, produce toxic substances in host cells, and promote cell adhesion. Several secretion systems transport proteins across phospholipid membranes (Green and Mecsas, 2016). The secretion system plays a critical role in bacterial transmission. The structure, composition, and activity of eight secretion systems (T1, T2, T3, T4, T5, T6, T7, T9) have been determined so far. The variations between Gram-positive and Gram-negative bacteria are the cause of these differences.

In Gram-negative bacteria, the type I secretion system (T1SS) is extensively distributed. ABC transporter, membrane fusion protein, and outer membrane factor are the three structural elements (Thomas et al., 2014). So far, the Has system of *Serratia marcescens* and *P. aeruginosa* and the haemolysin system of *V. cholerae*, *N. meningitidis*, and *E. coli* are the only two systems that can regulate the expression and secretion of T1SS substrates. T1SS is positively regulated by QS in the transcriptional study of *P. aeruginosa* because the expression of its effector alkaline protease AprA is dependent on QS. T2SS is primarily responsible for obtaining nutrition. It is in charge of secreting a large number of exoproteins.

The QS-regulated virulence factors elastase and exotoxin are secreted by the Xcp system in *P. aeruginosa*, which is also positively regulated by QS (Maura et al., 2016). The QS system is directly responsible for biofilm production in *V. cholerae* T2SS (Teschler et al., 2015). Gram-negative and Gram-positive bacteria both have the type IV secretion system. T4SS is the most widely used secretion system on the planet. It can transfer not only proteins but also DNA (1). In *Brucella abortus*, the QS system is linked to T4SS. LuxR-type QS is responsible for the virulence characteristics of *B. abortus* for the virB operon encoding

4.6 NEW STRATEGY FOR PREVENTIVE TREATMENT OF MICROBIAL RESISTANCE

By controlling the creation of biofilms and the direct control of drug efflux pumps, the QS system plays a crucial role in the development of bacterial drug-resistant pathways (see Figure 4.7). The challenge of disease prevention has been heightened by the increase in bacterial resistance, and the side effects of excessive antibiotic usage may potentially endanger human health. In recent years, the identification of control techniques for quorum quenching disorders has opened up new possibilities for fighting and resolving the issue of microbial resistance. Quorum quenching is the phenomena that occurs when the QS system of certain bacteria is disturbed

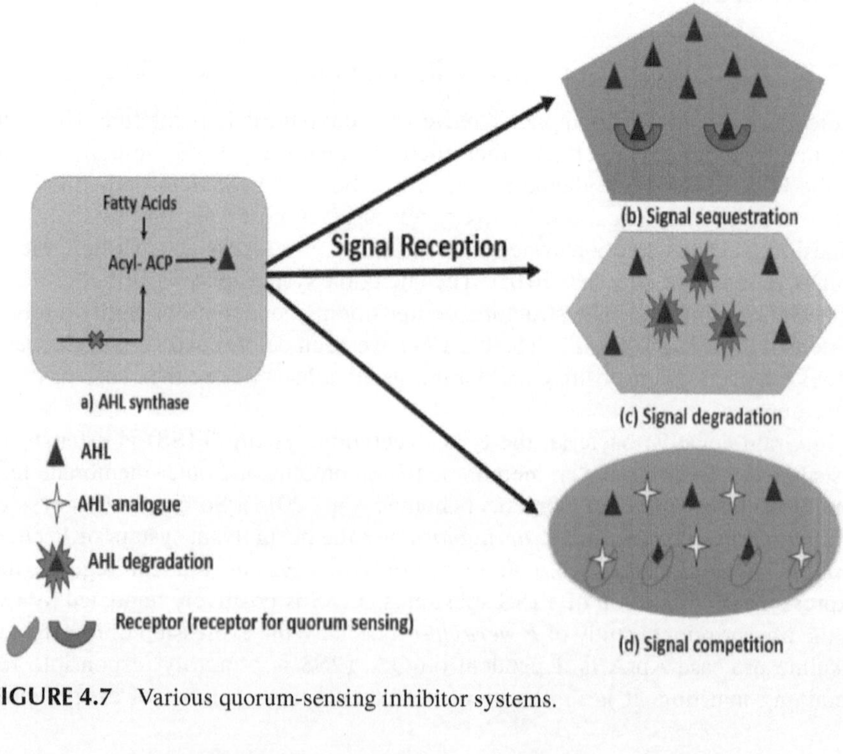

FIGURE 4.7 Various quorum-sensing inhibitor systems.

and interchange of signals between microorganisms is inhibited, where the expression level of risk factors is decreased (Roy et al., 2010). Quorum quenching properties include the role of molecules (i.e., enzymes, chemical compounds) and a mode of action that includes competition, inhibition, and quorum sensing signal interdict. The targets include several stages in the quorum-sensing pathway, which include the synthesis, diffusion, grouping, and detection of the quorum sensing signal. In general, the enzymes which are responsible for signal inactivation in quorum sensing are known as quorum quenching enzymes and the chemicals that inhibit the pathway of quorum sensing are called quorum sensing inhibitors (Stotani et al., 2019). Combined with quenching, the most effective chemical is referred to as quorum sensing inhibitors. At present, the antibiotics and quorum quenching agents decrease the microbial infection by inhibiting the microbial quorum induction, and this does not affect the growth of microbes. The three main pathways of quorum quenching are: (1) production of signal molecule inhibition; (2) signal molecule degradation; (3) conduction of signal molecule inhibition or receptor binding (see Figure 4.7). These quorum-sensing systems focus on stopping the growth of microorganisms or killing microorganisms; otherwise the microbial resistance development will be inhibited (Blöcher et al., 2018). Thus, the finding that quorum quenching depends on quorum sensing inhibition makes researchers interested in developing new strategies for control of harmful microorganisms.

4.7 FUTURE OUTLOOK

Quorum sensing came to the forefront in microbial research during the last decade of the 20th century. But even now, in the third decade of the 21st century, many studies are yet incomplete and antibiotic resistance is not well understood. With that in mind, opportunities and research in quorum sensing and microbial resistance should be very high in future. Of great interest in quorum sensing and microbial resistance a few areas in particular should be taken into consideration:

1) Molecular biology should strive for better understanding of the quorum-sensing mechanism.
2) Microbial quorum sensing research must focus on complex microbial antibiotic resistance.
3) Present quorum-sensing inhibition screening is inefficient, so we must improve the technology.

The different research done on microbial resistance mechanisms helps to give a better idea of the problem of resistance in bacteria. The generation and control of bacterial resistance in microorganisms are researched for the prevention and management of illnesses brought on by germs. However, research into the mechanisms that control microbial resistance is now quite few. Numerous studies demonstrate that *Pseudomonas aeruginosa* and other pathogenic bacteria, as well as quorum sensing, play important roles in the biofilm formation, drug efflux gene expression, and antibiotic resistance mechanisms of other medical diseases (Williams and Cámara, 2009). With the advancement in technology and the research in the field of quorum sensing and mechanism systems gives the idea of antimicrobial resistance in bacteria.

4.8 CHALLENGES

Antimicrobial resistance has a more complicated impact on patients' health because of things like lack of understanding of how microorganisms become resistant to drugs, their adverse effects, and the incorrect regulatory mechanisms of drugs in use. The main issue is the misuse of antimicrobial drugs and the quality of the drugs, both of which affect antimicrobial activity. Furthermore, temperature and humidity can decrease drug quality. Missed drug doses in the patient can result in antimicrobial resistance. Antimicrobials are also used in prevention and treatment of diseases in animals and to promote growth in the animals. Antimicrobial-resistant bacteria have the ability to transfer through food and direct contact. For instance, multidrug-resistant bacteria are found in meat and in the humans who are in contact with the livestock, potentially causing cause serious health issues. During vaccine preparation, antimicrobial resistance can lead to failure vaccines that act on disease-causing bacteria.

4.9 CONCLUSION

It is very clear that communication between cells is significant in the microbial world, and knowing these processes is very basic to microbiology as a whole. To date we have learned that quorum sensing helps in promoting communication between different species, and it can help us to find the cell density and population. Microbial quorum sensing signal detection is complex and organized in different configurations. LuxS/AI-2 quorum sensing has an important role in antimicrobial resistance in bacteria. The LuxS/AI-2 system controls the different gene expressions and helps in regulation of cellular activities to adapt to all environments.

REFERENCES

Abdula N, Macharia J, Motsoaledi A, Swaminathan S, VijayRaghavan K. National action for global gains in antimicrobial resistance. Lancet Lond Engl. 2016 Jan 9;387(10014):e3–5.

Blöcher R, Rodarte Ramírez A, Castro-Escarpulli G, Curiel-Quesada E, Reyes-Arellano A. Design, synthesis, and evaluation of alkyl-quinoxalin-2(1H)-one derivatives as anti-quorum sensing molecules, inhibiting biofilm formation in Aeromonas caviae Sch3. Molecules. 2018 Nov 24;23(12):3075.

Chebotar IV, Mayansky ANKE, Lazareva AVCV. Antimicrobial resistance of bacteria in biofilms. Clin Microbiol Antimicrob Chemother. 2012;14(1):51–8.

Chen Z, Xiang J. Advances in the research of LuxR family protein in quorum-sensing system of gram-negative bacteria. Zhonghua Shao Shang Za Zhi Zhonghua Shaoshang Zazhi Chin J Burns. 2016 Sep 20;32(9):536–8.

Cornforth DM, Foster KR. Competition sensing: the social side of bacterial stress responses. Nat Rev Microbiol. 2013 Apr;11(4):285–93.

Doekes HM, De Boer RJ, Hermsen R. Toxin production spontaneously becomes regulated by local cell density in evolving bacterial populations. PLoS Comput Biol. 2019;15(8):e1007333.

Dubrac S, Boneca IG, Poupel O, Msadek T. New insights into the WalK/WalR (YycG/YycF) essential signal transduction pathway reveal a major role in controlling cell wall metabolism and biofilm formation in Staphylococcus aureus. J Bacteriol. 2007 Nov;189(22):8257–69.

Freilich S, Zarecki R, Eilam O, Segal ES, Henry CS, Kupiec M, Gophna U, Sharan R, Ruppin E. Competitive and cooperative metabolic interactions in bacterial communities. Nat Commun. 2011 Dec 13;2:589.

Green ER, Mecsas J. Bacterial secretion systems: an overview. Microbiol Spectr. 2016;4. doi:10.1128/microbiolspec. VMBF-0012-2015.

Li X, Zhang G, Zhu Y, Bi J, Hao H, Hou H. Effect of the luxI/R gene on AHL-signaling molecules and QS regulatory mechanism in Hafnia alvei H4. AMB Express. 2019 Dec 5;9(1):197.

Li XZ, Plésiat P, Nikaido H. The challenge of efflux-mediated antibiotic resistance in Gram-negative bacteria. Clin Microbiol Rev. 2015;28(2):337–418.

Li Y-H, Tian X-L. Quorum sensing and bacterial social interactions in biofilms: bacterial cooperation and competition. In: Stress and Environmental Regulation of Gene Expression and Adaptation in Bacteria [Internet]. John Wiley & Sons, Ltd; 2016, pp. 1195–205.

Liu L, Ye C, Soteyome T, Zhao X, Xia J, Xu W, Mao Y, Peng R, Chen J, Xu Z, Shirtliff ME, Harro JM. Inhibitory effects of two types of food additives on biofilm formation by foodborne pathogens. Microbiologyopen. 2019 Sep;8(9):e00853.

Ma Y, Lan G, Li C, Cambaza EM, Liu D, Ye X, Chen S, Ding T. Stress tolerance of Staphylococcus aureus with different antibiotic resistance profiles. Microb Pathog. 2019 Aug;133:103549.

Maura D, Hazan R, Kitao T, Ballok AE, Rahme LG. Evidence for direct control of virulence and defense gene circuits by the Pseudomonas aeruginosa quorum sensing regulator, MvfR. Sci Rep. 2016 Sep 28;6:34083.

McCluskey J, Hinds J, Husain S, Witney A, Mitchell TJ. A two-component system that controls the expression of pneumococcal surface antigen A (PsaA) and regulates virulence and resistance to oxidative stress in Streptococcus pneumoniae. Mol Microbiol. 2004 Mar;51(6):1661–75.

Miller ST, Xavier KB, Campagna SR, Taga ME, Semmelhack MF, Bassler BL, Hughson FM. Salmonella typhimurium recognizes a chemically distinct form of the bacterial quorum-sensing signal AI-2. Mol Cell. 2004 Sep 10;15(5):677–87.

Monnet V, Gardan R. Quorum-sensing regulators in Gram-positive bacteria: "cherchez le peptide." Mol Microbiol. 2015 Jul;97(2):181–4.

Ng Y-K, Grasso M, Wright V, Garcia V, Williams P, Atkinson S. The quorum sensing system of yersinia enterocolitica 8081 regulates swimming motility, host cell attachment, and virulence plasmid maintenance. Genes. 2018 Jun 20;9(6):E307.

Nikaido H. Multidrug resistance in bacteria. Annu Rev Biochem. 2009;78:119–46.

Papenfort K, Bassler BL. Quorum sensing signal-response systems in gram-negative bacteria. Nat Rev Microbiol. 2016 Aug 11;14(9):576–88.

Peacock SJ, Paterson GK. Mechanisms of methicillin resistance in Staphylococcus aureus. Annu Rev Biochem. 2015;84:577–601.

Pumbwe L, Skilbeck CA, Wexler HM. Presence of quorum-sensing systems associated with multidrug resistance and biofilm formation in Bacteroides fragilis. Microb Ecol. 2008 Oct;56(3):412–9.

Rajput A, Thakur A, Sharma S, Kumar M. aBiofilm: A resource of anti-biofilm agents and their potential implications in targeting antibiotic drug resistance. Nucleic Acids Res. 2018 Jan 4;46(D1):D894–900.

Roy R, Tiwari M, Donelli G, Tiwari V. Strategies for combating bacterial biofilms: a focus on anti-biofilm agents and their mechanisms of action. Virulence. 2018;9(1):522–54.

Roy V, Fernandes R, Tsao C-Y, Bentley WE. Cross species quorum quenching using a native AI-2 processing enzyme. ACS Chem Biol. 2010 Feb 19;5(2):223–32.

Saxena P, Joshi Y, Rawat K, Bisht R. Biofilms: architecture, resistance, quorum sensing and control mechanisms. Indian J Microbiol. 2019;59:3–12.

Stewart PS. Mechanisms of antibiotic resistance in bacterial biofilms. Int J Med Microbiol IJMM. 2002 Jul;292(2):107–13.

Stotani S, Gatta V, Medarametla P, Padmanaban M, Karawajczyk A, Giordanetto F, Tammela P, Laitinen T, Poso A, Tzalis D, Collina S. DPD-inspired discovery of novel LsrK kinase inhibitors: an opportunity to fight antimicrobial resistance. J Med Chem. 2019 Mar 14;62(5):2720–37.

Stubbendieck RM, Vargas-Bautista C, Straight PD. Bacterial communities: interactions to scale. Front Microbiol. 2016;7:1234.

Subhadra B, Oh MH, Choi CH. RND efflux pump systems in Acinetobacter, with special emphasis on their role in quorum sensing. J Bacteriol Virol. 2019 Mar 9;49(1):1–11.

Tan L, Li SR, Jiang B, Hu XM, Li S. Therapeutic targeting of the Staphylococcus aureus accessory gene regulator (agr) system. Front Microbiol. 2018;9:55.

Teschler JK, Zamorano-Sánchez D, Utada AS, Warner CJ, Wong GC, Linington RG, Yildiz FH. Living in the matrix: assembly and control of Vibrio cholerae biofilms. Nat Rev Microbiol. 2015;13(5):255–68.

Thomas S, Holland IB, Schmitt L. The Type 1 secretion pathway—the hemolysin system and beyond. Biochim Biophys Acta. 2014 Aug;1843(8):1629–41.

Wasaznik A, Grinholc M, Bielawski KP. Active efflux as the multidrug resistance mechanism. Postepy Hig Med Doswiadczalnej. 2009 Apr 9;63:123–33.

Williams P, Cámara M. Quorum sensing and environmental adaptation in Pseudomonas aeruginosa: a tale of regulatory networks and multifunctional signal molecules. Curr Opin Microbiol. 2009 Apr;12(2):182–91.

Zhang W, Li C. Exploiting quorum sensing interfering strategies in gram-negative bacteria for the enhancement of environmental applications. Front Microbiol. 2015;6:1535.

Zhao X, Xia J, Liu Y. Contrast of real-time fluorescent PCR methods for detection of Escherichia coli O157:H7 and of introducing an internal amplification control. Microorganisms. 2019; 7(8):230.

Zhao X, Yu Z, Ding T. Quorum-sensing regulation of antimicrobial resistance in bacteria. Microorganisms. 2020 Mar 17;8(3):425.

5 Phytocompound-Mediated Inhibition of Quorum-Sensing Cascade in Pathogenic Bacteria

M.P. Roy, M.D. Adhikari, and B.K. Tiwary

5.1 INTRODUCTION

Quorum sensing (QS) is essentially a regulatory system that regulates gene expression in response to cell density by sensing the concentration of certain signal molecules released by the bacterial cells themselves, and thereby controls the behavioural pattern of the population. The whole process is orchestrated by the secretion of chemical signal molecules known as autoinducers. Autoinducers' concentration rises as population size increases, and when the minimal threshold stimulatory concentration of an autoinducer is reached, that leads to a signalling cascade culminating in alteration in collective target gene expression in the group. As reaching the sufficient cell density – the quorum – is a prerequisite for such autoinduction, the process has been named quorum sensing (Bettenworth et al. 2019). This name stands for any form of activity in which bacteria create, secrete, and then respond to diffusible signal molecules (Heilmann et al. 2015). QS is prevalent in the bacterial world, ranging from at least one to several such systems being detected in every species. This inevitably established a vast variety among autoinducer molecules and their corresponding regulatory networks (Hagen and Son 2017). Studies have categorized the identified autoinducers into three major classes: N-acyl-homoserine lactones (AHLs) used by Gram-negative bacteria; autoinducing peptides (AIPs) used by Gram-positive bacteria; and autoinducer-2 (AI-2), a type of universal signalling molecules used by both Gram-negative and Gram-positive bacteria (Li et al. 2021).

QS communication circuits enable bacteria to synchronize their physiological activities as a function of population changes and adapt collectively towards a diverse array of cooperative behaviours, such as symbiosis, virulence, competence, nutrient acquisition, conjugation, antibiotic production, motility, bioluminescence,

DOI: 10.1201/9781003300557-5

sporulation, biofilm formation, and others (Papenfort and Bassler 2016). To perform a variety of collective, multicellular behaviours, bacteria secrete exquisite public goods to be shared by all members of the population. The secretion of these extracellular public goods is in fact often under QS regulation, as the overall benefit of the population as a whole often lies in strategically ramping up the production at a precise population size (Heilmann et al. 2015). These secreted molecules are usually helpful for various purposes like virulence, nutrient accession, inter- and/or intraspecies competition, among others. (Goo et al. 2012). Virulence, being a collective trait, often involves QS in a number of mechanisms that contribute to the effective development of bacterial infections, especially in production of virulence factors, which are responsible for defying the host's defences or for exploiting the host's resources. The collective production of these molecules at a higher population size helps in maximizing the impact on the host. Hence, the strategy of blocking the QS system and inhibiting the production of virulence factors is being considered as a new target for antimicrobial therapy, which attenuates virulence without killing the pathogens. This essentially weakens the selective pressure on pathogens, thereby thwarting the possibility of rapid evolution of drug resistance (Zhang et al. 2019).

Recently, there has been a significant increase in research interest in developing and formulating a new generation of antimicrobials. According to the World Health Organization (WHO 2020), antibiotic resistance, particularly multidrug resistance or pandrug resistance, has now emerged as a gigantic threat to life. Hence, screening and development of new, improved, and effective antimicrobials that follow different approaches for combating microbial invasion and infection, strategically thwarting bacterial adaptation towards resistance, has become the need of the hour. Plants have been used as remedies in traditional medicine from the dawn of time as they are impressively rich in antimicrobial compounds, the majority of which are effective against a wide range of species including fungi, bacteria, insects, nematodes, and others. (Abreu et al. 2013, Baldé et al. 2020). Plants also continue to be investigated as sources for novel antimicrobial agents with unique mechanisms of action. Additionally, there is already a long line of phytochemicals that have been reported to exert assistance to antibiotics' effectivity, thereby adding to their activity as resistance-modulating molecules (Abreu et al. 2012). Phytocompounds are preferred for their easy availability, nontoxicity, compatibility, biodegradability, and environmental friendliness (Adnan et al. 2021). Many of the phytochemicals that show antimicrobial effect commonly target microbial cell wall biosynthesis, membrane integrity, or signal transduction in QS (Monte et al. 2014), and may cause inhibition at different steps of the signalling pathway, like blocking enzymatic activity for autoinducer production or inhibition of signal molecules or their receptors (Khan et al. 2009). A wide range of phytocompounds, such as flavonoids, alkaloids, terpenoids, phenolics, polyacetylenes, and essential oils, effectively showing capability of inhibiting QS, has been reported by many researchers (Cetin-Karaca and Newman 2015, Nazzaro et al. 2013). In this chapter, we present an update and brief review of different kinds of phytocompounds potentially exhibiting anti-QS ability along with their underlying mechanisms of action.

5.2 OVERVIEW OF QS MECHANISM

QS is an intercellular communication system based on stimuli and responses in relation to bacterial cell population density that regulates different kinds of gene expression, including several factors that, directly or indirectly, contribute to establishment of infections such as virulence determinants, biofilm formation, and antibiotic resistance. It is essentially a complex signal transduction cascade initialized through the production of certain chemical signalling molecules called autoinducers or bacterial pheromones, which are produced at a low but constant basal level but can be accumulated up to a recognition threshold only with build-up of a distinct and adequate number of cells, referred to as a quorum (Nazzaro et al. 2013). At the threshold level of the local autoinducer concentration, it can productively bind to its cognate receptor to activate a transcriptional regulator, modulating the expression of target genes (Vasquez and Blackwell 2019). The autoinducer molecules can vary chemically and functionally among bacteria, classically falling into three categories: i) N-acylhomoserine lactone (AHL), reportedly used by Gram-negative bacteria; ii) auto-inducing oligopeptide (AIPs), used by Gram-positive bacteria (see Figure 5.1); and iii) autoinducer-2 (AI-2), a furanosyl borate diester or tetrahydroxy furan (species dependent) that can be used by both Gram-positive and Gram-negative bacteria. However, autoinducers share common links even among distantly related species, with similarities to autocrine and paracrine signalling (Doğaner et al. 2016). Typically, a QS circuit comprises a signal synthase, a signalling molecule, and a signal-responsive protein that regulates the expression of target genes (Kalia 2018).

5.2.1 QS BASED ON AHL MOLECULES

The most studied intraspecies autoinducer, the type I autoinducer (AI-1), is N-acyl homoserine lactone (AHL), widespread in Gram-negative bacteria, that constitutes a QS system often termed QS-1. It was discovered in the marine symbiotic bacterium *Vibrio fischeri,* where it is responsible for regulating bioluminescence (Rezzonico et al. 2012). In *V. fischeri,* the system consists of LuxI, a signal synthase coding for the signal molecule N-(3-oxohexanoyl)-L-homoserine lactone (3-oxo-C6-HSL or OHHL), and LuxR, a transcriptional activator encoded by the genes luxI (first gene of Lux operon) and luxR, respectively. A sufficient amount of AHL concentration leads to LuxR activation, and the LuxR/3-oxo-C6-HSL complex in turn initiates transcription of the lux operon, including luxAB genes encoding luciferase and luxCDE encoding the enzymes to produce the substrate for luciferase, thus prompting bioluminescence (Nazzaro et al. 2013). In other Gram-negative bacteria studied, similar signalling circuits utilizing similar components have been reported. A variety of AHL molecules, having a common homoserinelactone core but diverging in their acyl side chain moieties and associated receptors, has been reported (Rezzonico et al. 2012). Although there could be some crosstalk between bacteria that employ chemically similar signals, interspecies communication is largely restricted to species that share the same system (Rezzonico et al. 2012).

5.2.2 QS Based on Auto-Inducing Oligopeptides

The Gram-positive bacteria utilize alternative signalling molecules, that is, oligopeptide pheromones or auto-inducing oligopeptides (AIPs). These peptides are often modified post-translationally, then secreted extracellularly by assigned transporters, and finally recognized by target cells via membrane-bound receptor kinases that are the part of a two-component regulatory system having a sensor and response-regulator protein, which creates a phosphorylation cascade that ends in activation of particular genes leading to changes in cellular physiology in response to changes in the environment (Nazzaro et al. 2013). For example, many staphylococcal species, such as *Staphyloccoccus aureus*, evidently have the accessory gene regulator (agr) operon, which coordinates the autocatalytic signal transduction for QS by encoding four proteins (AgrBDCA), of which AgrC, a transmembrane receptor histidine kinase to which the AIP binds, and AgrA, an intracellular transcription factor, are part of the classical two-component regulatory system (Vasquez and Blackwell 2019). On the other hand, AgrB, an integral membrane endopeptidase, recognizes and modifies the conformation of AgrD, which is then transported outside the plasma membrane, getting further modification in the process and ultimately liberating as the mature AIP signal outside of the cell (Wang et al. 2015). Owing to the cellular multiplication, the AIP concentration increases in the vicinity soon reaches a quorum, becoming able to actively bind to AgrC causing up-regulation of the agr operon and thereby amplifying the QS signal, in a typical autoinduction cycle that is a hallmark of QS systems (Vasquez and Blackwell 2019). Finally, the activated AgrA persuades production of a range of virulence factors that are important for *S. aureus* pathogenesis. In an obvious way, many other Gram-positive bacteria are shown to engage a diversity of autoinducers and concerned receptors, as in the case of Gram-negative bacteria.

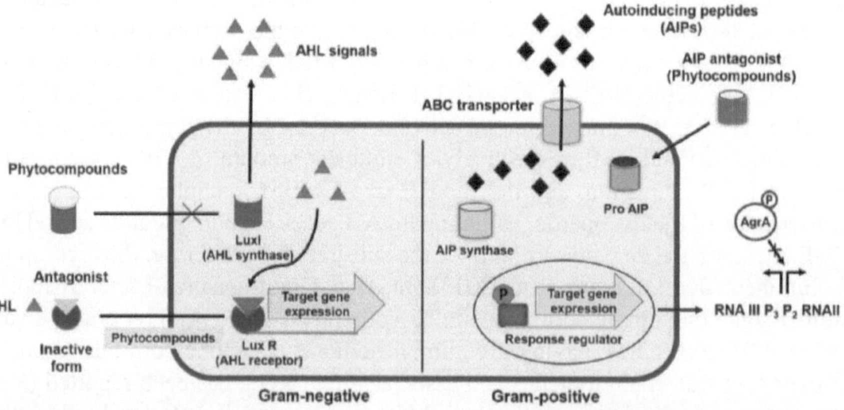

FIGURE 5.1 Inhibition of quorum sensing by phytocompounds in Gram-positive and Gram-negative bacteria.

5.2.3 QS Based on Production of AI-2 System

Apart from the QS-1 system, another QS system, QS-2, was first identified in *V. harveyi*, followed by discovery in many other Gram-negative and Gram-positive bacterial species like *Pasteurella, Photorhabdus, Haemophilus, Bacillus*, etc. This system is regulated by a different autoinducer molecule, named autoinducer-2 (AI-2), chemically known as (2S, 4S)-2-methyl-2,3,3,4-tetrahydroxytetrahydrofuryl borate (S-THMF-borate), a furanosyl borate diester. The LuxS protein is responsible for the production of (S)-4,5-dihydroxy-2,3-pentanedione (DPD), and this DPD molecule is actually the direct precursor of the AI-2 signal molecule. The signal is usually detected by a two-component sensor kinase LuxPQ leading to a similar signal transduction pathway as that of the QS-1 system and a third system, based on signal molecule (S)-3-hydroxytridecan-4-one (CAI-1) in *Vibrio* species. All these signals are channelled through LuxU, a relay protein, and LuxO, the transcription regulator, culminating in modulation of gene expression (Rezzonico et al. 2012).

5.3 PHYTOCOMPOUNDS WITH INHIBITORY EFFECTS AGAINST QUORUM SENSING IN PATHOGENIC BACTERIA

With antibiotic resistance alarmingly on the rise, there is an urgent need for innovative and effective medicines that target and destroy microorganisms without increasing the risk of selection pressure. Hence, an ideal therapeutic strategy should target the processes underlying virulence to assist the host's immunity to win the battle, but at the same time not allowing the bacteria to evolve fast under the survival pressure. QS inhibitors (QSIs) fits the description well, as QS is a regulatory system that not only promotes pathogenesis but also enables bacteria to turn recalcitrant against immune attacks as well as antimicrobials (Borges et al. 2016). Basically, those molecules that can bring about inhibition of quorum sensing cascade, thereby suppressing formation of biofilms, antibiotic resistance, and virulence factors, are said to be QSIs (Subramanian et al. 2022). Disruption of QS networks has been gaining immense worldwide research attention as a potential solution to this therapeutic riddle, especially in the fight against multiresistant bacteria as QSI impart no such toxicity on target bacteria to trigger the selection pressure (LaSarre and Federle 2013). A prospective anti-QS agent might employ any of three possible mechanisms: (i) targeting the signal generation; (ii) degradation of signal molecule; and/or by (iii) blocking the signal receptors (Khan et al. 2018). Moreover, for a model QSI to act as a successful 'antivirulence' therapeutic drug, it should show high target specificity towards the target bacterium along with negative toxicity towards the host (Ahmad et al. 2011). Given the widespread usage of plants to fulfill dietary as well as medicinal needs in humans from time immemorial, plants and their derived compounds have long been a natural choice to investigate possible therapeutic principles; yet they still remain huge, undiscovered, and relatively safe pools of various bioactive chemicals that also attract the favour of the pharmaceutical sector for their efficacy, affordability, and reduced side effects (Silva et al. 2016). Phytocompounds or phytochemicals are generally defined as chemicals of plant origin, produced through primary or secondary metabolism, showing a variety of biological activity in the host and prominently including a role in plant defence against competitors, pathogens, or predators (Breslin 2017). In fact, having potent antimicrobial effects on display, phytocompounds are now being considered as the richest reservoir available for therapeutics and many of them have already been reported to interfere with QS-mediated molecular communication. Studies suggest that phytocompounds, as a part of plant defence system, often interrupt inter- and intraspecies communication signals of pathogens via multiple approaches, such as enzymatic degradation of QS signals, inhibition of autoinducer synthesis, receptor antagonism, or disruption of signal secretion (Khan 2018, Díaz-Nuñez et al. 2021). Recent breakthroughs in plant-based research, on the other hand, are more particularly focused on gene expression variation as well as addressing global regulatory variables governing QS mechanism (Kalia 2018).

The constant interactions between plants and microorganisms have led to evolution of unconventional defence mechanisms against pathogenicity; one such mechanism is the disruption of QS pathways. Phytochemicals inhibit QS pathways often by blocking autoinducers like AHL (see Figure 5.1) and AI-2 (Ćirić et al. 2019). They may also play an important role in inhibition of biofilm formation through disruption of the QS network leading to suppression of genes, reduced accessibility to nutrients, and interference in bacterial adhesion and growth (Adnan et al. 2020). According to many researchers, quorum quenchers (QQ), when applied in combination with antibiotics, are actually the best alternative as antibiofilm agents (Paluch et al. 2020). The first quorum quenchers, or anti-QS chemicals, discovered were halogenated furanones generated by the benthic marine macroalga *Delisea pulchra* for their involvement in biofilm inhibition (Olivero-Verbel et al. 2014). Medicinal and nonmedicinal plants and their extracts (e.g., *Phyllanthus emblica, Terminalia bellirica, T. chebula, Punica granatum, Syzygium cumini, Mangifera indica, Laurus nobilis, Amphipterygium adstringensi, Sonchus oleraceus, Rosmarinus officinalis, Tecoma capensis, Jasminum sambac, Populus alba*, and *Pinus nigra,*), essential oils, extracts and isolated molecules from fruits (e.g., apple, pineapple, plantain, and sapodilla), herbs and spices (e.g., *Centella asiatica* L.), plant beverages (e.g., tea and coconut water) are naturally rich sources of several bioactive phytocompounds, many of which have anti-QS potential (see Table 5.1) and thereby are being explored extensively in foods and medicines (Nazzaro et al. 2013, Borges et al. 2016). Plant extracts may contain QSIs that are chemically analogous to those of QS signals (e.g., AHL) or having ability to block signal receptors (e.g., LuxR/LasR). For example, GABA (γ-aminobutyric acid) found in plants has been shown to degrade the OHC8HSL AHL signal by lactonase (AttM) in *Agrobacterium tumefaciens*, thus controlling the QS-dependent infection process; in the case of *Emblica officinalis*, pyrogallol and its analogues exhibit antagonism against AI-2; *Medicago truncatula* modulates QS by interfering with AhyR, CviR, and LuxR receptors in *Sinorhizobium meliloti* and *P. aeruginosa*; and curcumin in *Curcuma longa* showed inhibition of the expression of virulence genes in *P. aeruginosa* PA01 (Nazzaro et al. 2013). Also demonstrating QSI activities are plant extracts from *L. nobilis, S. oleraceus, R. officinalis, T. capensis, J. sambac, P. alba*, and *P. nigra, Allium sativum, Zingiber officinale* Roscoe, vanilla, and varieties of apple (e.g., Annurca) and apple derivatives (e.g., cider); and pineapple, plantain, and sapodilla. Leaf extracts from *Psidium guajava* L. (Ghosh et al. 2014), *Fragaria nubicola, Osbeckia nepalensis*, and *Astilbe rivularis* (Tiwary et al. 2017) are reported for their anti-QS activity in *Chromobacterium violaceum* and *P. aeruginosa*. Plant extracts and plant-derived chemicals often exhibited suppression of several gene expressions to block QS, such as *rhlI, pqsA, lasB*, as their gene products are either key QS components or the important virulence factor regulated by QS; rhlI catalyzes the synthesis of the synthase enzyme for producing N-butyryl-L-homoserine lactone, whereas *pqsA* and *lasB* encodes quinolone synthase and elastase, respectively (Wang et al. 2016). Miscellaneous phytocompounds encompassing a diverse range of chemical groups including polyphenols, flavonoids, fatty acid derivatives, alkaloids, coumarins, lignans, terpenoids, and phytosterols have been reported to show potent anti-QS effect (Silva et al. 2016). A schematic representation of different strategies of QS inhibition by phytocompounds is shown in Figure 5.2.

5.3.1 Flavonoids

Flavonoids are a type of plant secondary metabolite with a polyphenolic structure that is abundant in fruits, vegetables, and certain beverages. They have antioxidative, antibacterial, anti-inflammatory, antimutagenic, and anticarcinogenic properties, as well as the ability to modulate key cellular enzyme functions (Panche et al. 2016, Cushnie and Lamb 2011). These molecules are derived from phenylpropanoids, all having the same basic C15 skeleton, diversifying with degree of

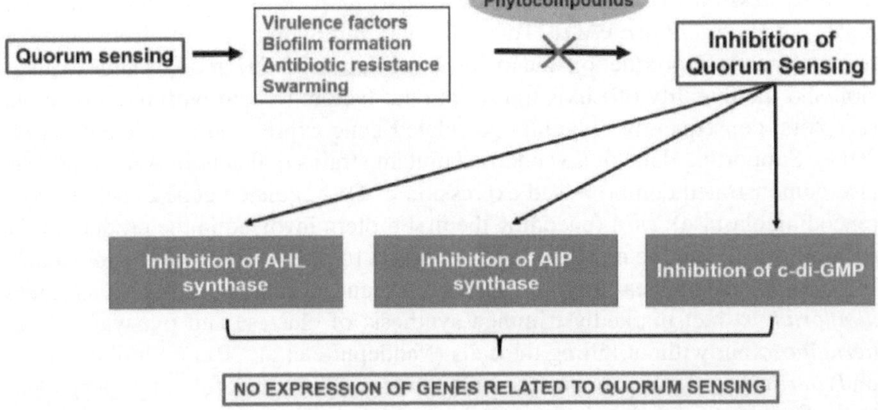

FIGURE 5.2 A schematic representation of different strategies of QS inhibition by phytocompounds.

unsaturation and oxidation of C-ring and carbon skeleton substitutions through hydroxylation, glycosylation, methylation, acylation, and prenylation (Bouyahya et al. 2017, Ghosh et al. 2022). The most important subgroups of flavonoids are flavones, flavonols, flavanones, flavanonols, flavanols or catechins, anthocyanins, and chalcones (Panche et al. 2016). In plants, flavonoids are important not only in different levels of germination, growth, and development but they also play multiple defensive roles, such as protection from stresses both biotic and abiotic, and from UV damage. They promote activity in cell signalling as signal molecules, function as allopathic compounds, phytoalexins, detoxifying agents and antimicrobial compounds, and aid in frost hardiness and freeze tolerance, drought resistance, and heat acclimatization (Samanta et al. 2011, Panche et al. 2016).

Flavonoids are abundantly found in citrus plants and they have shown to interrupt QS, thereby blocking biofilm formation and bioluminescence (Vikram et al. 2010). For example, naringenin interfered with expression of type III secretion system genes controlled by cell-to-cell signalling and also with AI-2 and HAI-1 signals, thereby bringing down bioluminescence induction in *V. harveyi* and biofilm formation in both *V. harveyi* BB120 and *E. coli* 0157: H7 (Vikram et al. 2010). Similarly, flavonoids (naringenin, kaempferol, quercetin, and apigenin) were found to inhibit HAI-1- or AI-2-mediated bioluminescence in *V. harveyi* strains BB886 and MM32, and quercetin, sinensetin, apigenin, and naringenin displayed antibiofilm activity against *V. harveyi* BB120 and *E. coli* O157:H7 (Nazzaro et al. 2013). Baicalein, a flavone, inhibited QS-mediated biofilm formation in *P. aeruginosa* PAO1 even at micromolar concentrations, and also caused proteolytic degradation of *A. tumefaciens* signal receptor for QS in TraR of *E. coli* (Ghosh et al. 2022). Vasavi et al. showed that the highly rich concentrations of flavonoids (flavanones like naringenin, eriodictyol, and taxifolin) in the ethyl acetate extract of *Centella asiatica* reduced pyocyanin production and biofilm formation by inhibiting elastolitic and proteolytic activity in *C. violaceum* (Vasavi et al. 2016).

As per studies, naringenin and taxifolin showed anti-QS activities, thus scaling down expression of several genes, *lasR, lasI, lasB, lasA, rhlR, rhlA, rhlI,* and *phzA1,* in *P. aeruginosa* PAO1. This activity of naringenin resulted from both the reduction in AHL production due to down-regulation of *rhl1* and *lasI* gene expression and the inability of LuxR transcriptional factors to bind with the respective receptors, consequently reducing QS-related gene expression (Vandeputte et al. 2011). Supporting data on *lasI* and *rhlI* mutant strains deficient in AHL synthesis also demonstrated compromised expression of QS-regulated genes, such as *lasB* (encoding elastase), *rhlA* (encoding the first protein involved in the production of rhamnolipids), and the phz operon (responsible for the production of pyocyanin). In addition, flavonones taxifolin, naringenin, and eriodictyol from *Combretum albiflorum* extract markedly trimmed synthesis of elastase and pyocyanin in *P. aeruginosa* but without killing the cells (Vandeputte et al. 2011). Similar studies on *P. aeruginosa* with quercetin showed impairment of a handful of QS-dependent traits like biofilm formation, exopolysaccharide (EPS) production, violacein production, alginate production, motility, production of virulence factors (e.g., pyocyanins, proteases, elastases) in a concentration-dependent manner, as quercetin is structurally a competitive inhibitor for the ligands in the lasR receptor pathway and decreases transcription of lasI, lasR, rhlI, and rhlR genes (Ouyang et al. 2016). Catechin from *C. albiflorum* impeded perceptivity of the QS signal molecule N-butanoyl-l-homoserine lactone by RhlR, therefore reducing QS factor production, expression of the QS-regulated genes lasB and rhlA and QS-regulatory genes *lasI, lasR, rhlI,* and *rhlR*, and finally synthesis of pyocyanin, elastase, and biofilms (Vandeputte et al. 2010). Flavan-3-ol catechin reduced the production of QS-mediated virulence factors, such as pyocyanin and elastase, and also biofilm formation in case of *P. aeruginosa* PAO1 (Nazzaro et al. 2013).

As per the study conducted by Gopu et al. (2015), it has been affirmed that two phytochemicals, cyanidin and anthocyanin, have the ability to significantly suppress QS-mediated traits including violacein production (73.96%) and EPS formation (72.43%), although in a dose-dependent manner in the opportunistic pathogen *Klebsiella pnuemoniae*. Similarly, anthocyanin from *Syzgium cumini* also caused obstruction in EPS production (64.29%) as well as biofilm formation (79.94%) in *K. pneumoniae* (Gopu et al. 2015). Malvidin, an O-methylated anthocyanidin, is in fact the major contributing factor for the QS-suppression property of the total anthocyanin fraction of *S. cumini*. Malvidin led to decrease in production of violacein, EPS, and overall biofilm structure depending on treatment concentration. On the other hand, rosmarinic acid, obtained from sweet basil, has been shown to compete with the ligand N-butanoyl-homoserine lactone (C4-HSL) to get bound to the QS-regulator RhlR in *P. aeruginosa* (Ghosh et al. 2022). Thus, acting as a competitive inhibitor for the proper receptor function, it caused increased stimulation of RhlR-associated transcription *in vitro*, compared to the desired transcription level mediated by the actual ligand C4-HSL.

Zhang et al. (2014) reported the anti-QS and antibiofilm property of polyphenol extract of *Rosa rugosa* tea, which contained flavonoids and polyphenols as the primary active constituents responsible for the inhibitory activities. This extract showed the ability to block violacein synthesis, which is under QS regulation, in *C. violaceum* without influencing growth considerably. Furthermore, swarming motility and biofilm formation in *E. coli* and *P. aeruginosa* were also inhibited by the same extract. When leaf extracts from *Cecropia pachystachya* Trécul were tested for QS inhibition, it was discovered that these extracts with their glycosyl flavonoids (isoorientin, chlorogenic acid, isovitexin, orientin, rutin, and vitexin) disrupted QS signalling in *C. violaceum* and *E. coli* (Brango-Vanegas et al. 2014). Varieties of apples and their derivative extracts, such as cider, demonstrated QS inhibitory effect, which is most likely owing to the flavonoids and polyphenols found in them, such as hydroxycinnamic acids, epicatechin, and rutin, that work synergistically as anti-QS agents against *C. violaceum* (Subramanian et al. 2022).

5.3.2 Alkaloids

Alkaloids represent naturally occurring complex organic molecules containing a heterocyclic nitrogen ring that constitute a structurally diverse group having a wide range of biological activities. Plants are a major source of alkaloids and many have pharmaceutical uses; examples include caffeine, quinine, morphine, cocaine, and strychnine (O'Connor 2010). QSIs have chiefly been identified in the indole, 1,3,4-oxadiazole, piperidine, and steroidal alkaloid classes (Cushnie et al. 2014). In fact, a good number of alkaloids inhibit formation and/or dispersal of bacterial biofilms (e.g., piperidines, imidazoles, pyrrole-imidazoles, pyrrolidines, isoquinolines, cinchona alkaloids). This case-specific capability has been attributed to different mechanisms, like direct antibacterial activity and QS disruption, or to unknown causes, possibly inhibition of sortase or adhesins (Cushnie et al. 2014).

The biofilm formation in *Streptococcus mutans* has been found to be inhibited by piperine, an alkaloid present in black pepper (*Piper nigrum*) (minimum inhibitory concentration being 0.0407 ± 0.03 mg/mL) and embelin, a quinone derivative (minimum inhibitory concentration being 0.0620 ± 0.03 mg/mL) in microtiter plate assay (Dwivedi and Singh 2016). The antibiofilm effect of these substances was presumptively due to blocking of the receptors and signal molecules functioning in the QS system that lead to biofilm formation (Asfour 2018).

Zhou et al. (2018) reported effective inhibition of QS as well as biofilm development in the food-borne pathogen *P. aeruginosa*, using hordenine, a polyphenolic compound from barley, through down-regulation in the expression of QS regulatory genes *lasR, lasI, rhlI,* and *rhlR*, and inhibition of AHL levels, thus leading to repression of QS-related extracellular virulence factors. When a drug-herb combination of hordenine and netilmicin (an aminoglycoside antibiotic) was used, the findings were more encouraging. As conjugated with gold nanoparticles (Au NPs) to give hordenine-Au NPs, hordenine displayed greater biofilm inhibition as compared to hordenine alone, which is encouraging, as nanomedicines, in general, are paving a new way towards better treatment against multidrug-resistant pathogens (Rajkumari et al. 2017).

Berberine, an isoquinoline alkaloid that exerts both direct antibacterial and antivirulence effects, retarded violacein production in *C. violaceum*, interfering with QS, and also exhibited QS and biofilm inhibitory activities as well as anti-infective capability against the pathogenic strains *Salmonella enterica* sv Typhimurium and *P. aeruginosa* PA01 (Aswathanarayan and Vittal 2018). Such inhibitory activities of berberine come from the fact that it interacted with the QS signal receptors LasR and RhlR, according to in silico research. Studies have shown that the bioactive macrocyclic alkaloid spermine derived from *Albizia schimperiana*, a tropical decidu-ous tree, has antibacterial action against a variety of bacterial pathogens, including methicillin-resistant *S. aureus* (MRSA) and *E. coli* (Bacha et al. 2016). Wang et al. (2016) reported that spermidine suppressed expression of the QS genes *rhlI* and *pqsA* in *P. aeruginosa*. They also tested a laboratory synthesized conjugate R101SPM – a combination of rhodamine 101 and spermine – which showed substantial efficacy in inhibiting gene expression of type III secretion system (T3SS) and attenuating T3SS-mediated cytotoxicity on human cells, for T3SS in *P. aeruginosa* is a prime contributor to its pathogenicity while attacking eukaryotic hosts. Additionally, two spermine derivatives isolated from marine sponges, spermatinamine and its derivative pseudoceramine B, reportedly have the ability to block the effector secretion by the T3SS of *Yersinia pseudotuberculosis* along with strong antibacterial activity (Yin et al. 2011).

5.3.3 Terpenoids and Limonoids

Terpenoids, also known as isoprenoids, are the most abundant and structurally diversified natural compounds found in many plants, with a wide range of essential pharmacological effects (Ludwiczuk et al. 2017). Terpenoids are a modified variety of terpenes with different functional groups and oxidized methyl groups relocated or deleted at various places. They are classified as monoterpenes, diterpenes, triterpenes, sesquiterpenes, and sesterpenes, based on their carbon units (Perveen 2018). Limonoids are phytochemicals of the triterpenoid class commonly obtained from citrus fruits. Limonoids are highly oxygenated modified terpenoids with a typical structure that either contains or is derived from a common precursor having a 4,4,8-trimethyl-17-furanylsteroid skeleton (Zhang et al. 2013).

Sesquiterpenoid viridiflorol and triterpenoids betulinic and ursolic acids, isolated from the liverwort *Lepidozia chordulifera*, presented strong QSI ability to impede the formation of biofilm in both *P. aeruginosa* and *S. aureus* and to lower the elastolytic activity in the former (Gilabert et al. 2015). Methanolic preparations of the Italian medicinal plant *Castanea sativa*, rich in pentacyclic triterpenes (ursene and oleanene derivatives), have shown demonstrated anti-QS activity against *S. aureus*. These extracts were able to suppress the haemolytic activity and delta-toxin production by *S. aureus*; its biofilm formation was also hampered, though at a lower degree (Quave et al. 2015). On the other hand, in case of *P. aeruginosa*, diterpene phytol not only reduced biofilm formation, flagellar motility, and twitching of the strain PAO1, it al-so exhibited effective pyocyanin inhibitory ability (Pejin et al. 2015).

Citrus limonoids (ichangin and isolimonic acid) strongly inhibited biofilm development and attachment to Caco-2 cells, suppressed enterocyte effacement-encoded genes, and flhDC operon genes in enterohaemorrhagic *E. coli* (EHEC) serotypes; and isolimonic acid could restrain the AI-3/epinephrine activated cell-to-cell signalling pathway (Vikram et al. 2012). In other studies, ichangin, isolimonic acid, and deacetylnomilinic acid glucoside exhibited substantial inhibition of AI-mediated cell-to-cell signalling and biofilm formation (Vikram et al. 2011). The response regulator gene luxO, as well, was triggered by ichangin and isolimonic acid (Vikram et al. 2011). Sour-orange seed extracts containing limonoids, such as isolimonic acid, ichangin, and deacetyl nomilinic acid 17 β-D-glucopyranoside, demonstrated greater than 90% inhibition of AI-2 activity in *V. harveyi* at a concentration of 100 μg/mL and also showed activity against HAI- and AI-2-mediated bioluminescence (Vikram et al. 2011). Furocoumarins present in grapefruit restricted the AI-1 and AI-2 activities in *V. harveyi* as well as biofilm formation by pathogens such as *E. coli, S. typhimurium*, and *P. aeruginosa* (Nazzaro et al. 2013). Additionally, another limonoid, obacunone, isolated from grapefruit, was tested to express strong inhibitory activities against both QS-1 and QS-2 systems, biofilm development, and EHEC virulence (Asfour 2018).

5.3.4 Phenol Derivatives

Brackman et al. (2011) observed that cinnamaldehyde, baicalin hydrate, and hamamelitannin improved the susceptibility of *B Burkholderia cenocepacia, P. aeruginosa,* and *S. aureus*, and even MRSA biofilms while treating in combination with conventional antibiotics (vancomycin, clindamycin, and tobramycin) and also enhanced the chances of survival of infected *Caenorhabditis elegans* and *Galleria mellonella*. Brackman et al. (2016) also reported that in vitro and in vivo susceptibility of MRSA biofilms were increased to the combined effect of antibiotic vancomycin and QSI hamamelitannin, owing to its capability to link with the TraP receptor, influencing the release of extracellular DNA, a critical event in biofilm synthesis. Phenolic plant secondary metabolites, e.g., salicylic acid, were found to up-regulate expression of AHL-lactonase. Thus, ursolic acid, even at a low concentration of 10 μg/mL, was able to diminish biofilm formation by 57%–95% in *V. harveyi* and *P. aeruginosa* PAO1 and by 79% in *E. coli* (Nazzaro et al. 2013).

Polyphenolic compounds such as epigallocatechin gallate, ellagic acid, and tannic acid, commonly found in many plants, having a common gallic acid moiety can block AHL-mediated intercellular interactions among bacteria by disrupting the signalling cascade (Sarabhai et al. 2013, Nazzaro et al. 2013). For example, berries and pomegranates are highly rich in ellagitanins (e.g., punicalagin, ellagic acid), showing concentrations higher than 300 mg/100 g (Nazzaro et al. 2013). These ellagitanins are hydrolyzed to ellagic acid by the human gut microbiota and ultimately release urolithin-A and urolithin-B metabolites that accumulate in the intestine, imparting important biological effects. Urolithin A and urolithin B can diminish QS-related processes by up to 40% and decrease the levels of AHLs secreted by the entheropathogen *Yersinia entherocolitica* by up to 45% (Giménez-Bastida et al. 2012). Besides, another phenolic compound, 4',5'-O-dicaffeoylquinic acid (4',5'-ODCQA), can act as a pump inhibitor with a potential of targeting efflux systems in a good number of Gram-positive human pathogenic bacteria (Fiamegos et al. 2011). Chlorogenic, rutin, and vanillic acids (found in berries), resveratrol (grapes), daidzein (soy), kinurenic acid (honey), dimethyl-esculetin (artemisia), and pomegranate extract have all been explored for QS inhibition and show promising effects (Nazzaro et al. 2013). Mohamed et al. (2014) reported that mangostanaxanthone I and α-mangostin from the pericarp of *Garcinia mangostana* established QS inhibition in *C. violaceum* ATCC 12472 with MIC values 2 and 3 μg/mL, respectively, when compared to catechin (MIC 2 μg/mL).

5.3.5 Essential Oils

Essential oils (EOs) are defined as valuable natural compounds derived from aromatic plants, and as they have shown immense biological, nutritional, and pharmaceutical importance, they have been utilized in traditional pharmacopoeia for millennia (Elshafie and Camele 2017). Several EOs have exhibited intriguing antimicrobial activities against a wide range of significantly pathogenic fungi and bacteria that cause plant and human diseases (Mancini et al. 2014). Among those medically important EOs, many have even demonstrated promising antibiofilm and anti-QS properties (Poli et al. 2018).

EOs consist chiefly of two types of constituents, terpenoids (monoterpene, di-terpene, and ses-quiterpene), and phenylpropanoids (Nieto 2017). The terpenoid group may in turn contain different compounds present in variable percentages in EOs, such as terpinene, p-cymene, limonene, pinene, sabinene, linalool, geraniol, menthol, thymol, citronellol, carvone, carvacrol, eugenyl acetate, geranyl acetate, geranial, 1,8-cineole, and neral (Camele et al. 2019). Phenylpropanoids may include different aromatic compounds, such as cinnamaldehyde, cinnamyl alcohol, methyl cinnamate, and eugenol (Hyldgaard et al. 2012). Amidst this great variety of compounds, only a fraction are actually responsible for potential anti-QS activity. Among the medicinal and aromatic plants exploited for extracting bioactive EOs, the plants from the family Lamiaceae (which includes *Origanum vulgare* L., *Majorana hortensis* L., *Thymus vulgaris* L., *Salvia officinalis* L., *Lavandula stoechas* L., *Hyptis suaveolens* L., and *Rosmarinus officinalis* L.), as well as many other species, such as *Verbena officinalis* L. (Verbenaceae), *Carum carvi* L. (Apiaceae), *Citrus clementina* Hort. ex Tan. (Rutaceae), *Murraya koenigii* L. Sprengel (Rutaceae) are considered important where at least one of the constituents has promising anti-QS properties to combat different pathogenic microorganisms (Camele et al. 2019).

EOs from *O. vulgare*, with its main bioactive constituent, carvacrol, efficiently hinder biofilm formation and the QS mechanism, demonstrating a potential antibacterial action against a variety of pathogens such as *S. enterica* subsp. *typhimurium* and *S. aureus* (Asfour 2018). Furthermore, carvacrol hindered civil gene expression, which codes for the AHL synthase as well as QS-regulated processes such as violacein biosynthesis and chitinase activity, eventually establishing anti-QS and antibiofilm effect in *C. violaceum* (Burt et al. 2014). Other individual constituents of oregano EO, such as limonene, eugenol, linalool, (E)-citral, 1,8-cineole, and γ-terpinene, have also displayed anti-QS effects (Al-Haidari et al. 2016). In addition, EOs from lemon, clary sage, marjoram, mandarin, mint, and juniper have also been examined to reveal anti-QS properties and thereby prevent biofilm production, against numerous bacterial strains like *B. cereus, E. coli, P. aeruginosa, C. violaceum, Listeria monocytogenes, Serratia* sp., and *Pichia anomala* (Camele et al. 2019). When evaluated for the effects on growth and QS using the sensor strain *C. violaceum* CV026, AHL-producing *E. coli* ATTC 31298 and the grapevine-colonizing Ezf 10–17 strains, lavender, geranium, rose, and rosemary oils, were found to be the most potent QS inhibitors; citrus and eucalyptus oils only moderately affected pigment production in CV026, whereas orange, juniper, and chamomile oils were ineffective (Szabó et al. 2010). Poli et al. (2018) discovered that *Mentha suaveolens* ssp. *insularis* inhibits violacein production and biofilm formation in *C. violaceum*. Also, the *Carum copticum* EO revealed anti-QS activity against *C. violaceum* (Snoussi et al. 2018). Cinnamaldehyde, derived from the cinnamon EO and its derivatives, exhibited an extensive modulation of QS mediated by AI-2 and hence also of the formation of biofilm in various pathogens such as *P. aeruginosa* and *Vibrio* species (Subramanian et al. 2022). Mint essential oil was tested to show inhibition of biofilm development in *Candida albicans* (Benzaid et al. 2019) whereas marjoram EO showed anti-QS effect against *C. violaceum*, the positive sensor strain for AHL-mediated QS (Kerekes et al., 2013). When Olivero-Verbel et al. (2014) evaluated the anti-QS and antibacterial potential of EOs isolated from *Lippia alba* on the Tn-5 mutant of *C. violaceum* CV026, and also on the growth of *S. aureus* ATCC 25923, two EOs, one with geranial:neral chemotype and the other with limonene:carvone chemotype, were the most effective as QS inhibitors, causing inhibition of violacein pigment production in the sensor bacteria, and the geranial:neral chemotype oil also produced growth inhibition against *S. aureus* strain. Other studies also demonstrated that EOs from several medicinal plants, such as *Citrus reticulate, Eucalyptus radiate, E. globules*, and *Thymus vulgar*, as well as organic extracts from many, such as *Rubus rosaefolius, Centella asiatica, Areca catechu*, and *Sclerocarya birrea*, have shown anti-QS effects (Bouyahya et al. 2017). However, the exact mechanism of anti-QS activity along with antimicrobial efficacy of EOs and also of medicinal plant extracts is yet to be explored in many cases, as possibilities are that the effects may come from a single major component in these products, from synergistic effects between major chemotypes, or even from minor components that may have an additive effect in some types of mechanisms (Bouyahya et al. 2017). Nonetheless, all these studies showed that EOs have the potential to be exploited in the pharmaceutical sector to develop novel therapies for major human infections.

5.4 CONCLUSIONS

With increasing concerns about the accelerated pace of emergence of more threatening pathogenic strains with antibiotic resistance, many studies have been pursued over the last few decades to understand more about the processes utilized by bacteria to communicate and regulate virulence features. Inevitably, this information influx is steadily leading to search and discovery of new compounds, along with their effects on microbial pathogenicity. It is obvious that the relationship between QS and bacterial virulence is a potential area for the development of novel, efficient antivirulence medicines and therapeutic protocols. Studies have already made it clear that inhibiting virulence via QS inhibition, especially using phytocompounds, is conceivable, affordable, and somewhat practical. Compared to conventional antibiotics, phytocompounds have the greater benefit of being natural agents possessing massive structural and functional diversity that can be consistently exploited to expand advanced therapeutic strategies with increased safety, reliability, activity, and stability, but without stirring up any selection pressure. Still, further study is needed to identify their mechanism of action as well as the optimum amounts of anti-QS agents that may be safely used. Moreover, future research should also aim at converting the potent anti-QS phytochemicals into drugs and making them commercially available for human administration while retaining their experimental efficacy. For achieving better outcome, combination therapy with phytochemicals and commercial antibiotics can be considered as

TABLE 5.1
Anti-QS Activity of Plant-Derived Compounds

Plants	Bioactive compound(s)	Anti-QS activity	Test organism(s)	References
Tannic acid and Quercetin	*Alnus japonica*	Inhibition of biofilm formation; repression of the intercellular adhesion genes icaA and icaD; reduction of haemolysis	*S. aureus* strains	Murata et al. 2010
Emodin	*Rheum palmatum, Polygonum cuspidatum*	Down-regulating *argA*, *cidA*, *dltB*, *icaA*, *sarA*, and *sortaseA* biofilm-forming genes	*S. aureus*	Yan et al. 2017
Ginkgolic acids	*Ginkgo biloba*	Inhibition of biofilm formation	*E. coli* O157:H7 (EHEC) *S. aureus*	Murata et al. 2010
Phytol	*P. betle*	Down-regulation of QS genes and swarming activity	*S. marcescens*	Srinivasan et al. 2016
Glycosyl flavonoids	*Cecropia pachystachya*	Inhibition of violacein production; inhibition of bioluminescence	C. *violaceum* ATCC 31532 and *E. coli* pSB403 and *Pseudomonas putida* (isoF WT)	Brango-Vanegas et al. 2014
Alkaloids and flavonoids	*Berberis artistata, Camellia sinensis*, and *Holarrhena antidysenterica*	Inhibition of acyl homoserine lactone production and biofilm formation	Carbapenem-resistant *E. coli*	Thakur et al. 2016
Eugenol and linalool	*Ocimum tenuiflorum*	Inhibits the production of pyocyanin, elastase	*P. aeruginosa*	Lahiri et al. 2021
Curcumin	*C. longa*	Inhibits the production of various QS molecules	*E. coli*	Arita-Morioka et al. 2018
Licoricone, glycyrin, gylzyrin (flavonoids)	*Glycyrrhiza glabra*	Significant reduction of QS-regulated virulence factors such as surface motility, twitching biofilm formation, and production of antioxidant enzymes catalase and SOD	*Acinetobacter baumannii*	Bhargava et al. 2015
Epigallocatechin-3-gallate	*C. sinesis*	Helps in disrupting biofilm and down-regulates the mechanism of QS	*S. mutans*	Murata et al. 2010

(*Continued*)

TABLE 5.1 CONTINUED

Plants	Bioactive compound(s)	Anti-QS activity	Test organism(s)	References
Sesquiterpene lactones	*Vernonia blumeoides*	Inhibition of violacein production	*C. violaceum*	Aliyu et al. 2016
Zingerone	*Z. officinale*	Helps in the mechanism of inhibiting the production of pyocyanin, elastase, and protease; also brings about inhibition of swarming and twitching motility	*P. aeruginosa*	Vijayakumar and Ramanathan 2020
Limonene	*Citrus reticulata*	Inhibition of biofilms formation at 0.1 mg/mL; inhibition of biofilm cell viability (41%) and AHL production (33%)	*P. aeruginosa*	Luciardi et al. 2016
Epigallocatechin gallate; epicatechin	*R. rugosa*	Reduction of violacein production in *C. violaceum*; inhibition of swarming motility and biofilm formation in *E. coli* and *P. aeruginosa*.	*C. violaceum*, *E. coli* K12 and *P. aeruginosa* PAO1	Zhang et al. 2002
5-Hydroxymethylfurfuryl	*Musa acuminata*	Helps in down-regulating the genes associated with the process of QS and reduces virulence	*P. aeruginosa*	Vijayakumar and Ramanathan 2020
Carvacrol and thymol	*T. vulgare*	Significant inhibition of AHL-related flgA gene expression and biofilm formation	*Pseudomonas fluorescens* KM121	Myszka et al. 2016
Ellagic acid (benzoic acid)	*T. chebula*	Reduction of biofilms formation	*B. cepacia*	Huber et al. 2003
Allicin	*A. sativum*	Interferes with the mechanism of QS, thus helps in down-regulating virulence factors	*P. aeruginosa*	Xu et al. 2020
Sesquiterpene lactones	*V. blumeoides*	Inhibition of violacein production	*C. violaceum*	Aliyu et al. 2016
Pulverulentone A	*Callistemon citrinus*	Prevents formation of biofilm by inhibiting production of staphyloxanthin	Methicillin-resistant *S. aureus*	Shehabeldine et al. 2020

another research direction to be explored for the sake of prompt advancement in the biomedical sectors.

REFERENCES

Abreu, A.C., Borges, A., Simões, L.C., Saavedra, M.J. and Simões, M., 2013. Antibacterial activity of phenyl isothiocyanate on Escherichia coli and Staphylococcus aureus. *Medicinal Chemistry*, *9*(5), 756–761.

Abreu, A.C., McBain, A.J. and Simoes, M., 2012. Plants as sources of new antimicrobials and resistance-modifying agents. *Natural Product Reports*, *29*(9), 1007–1021.

Adnan, M., Patel, M., Deshpande, S., Alreshidi, M., Siddiqui, A.J., Reddy, M.N., Emira, N. and De Feo, V., 2020. Effect of adiantum philippense extract on biofilm formation, adhesion with its antibacterial activities against foodborne pathogens, and characterization of bioactive metabolites: an in vitro-in silico approach. *Frontiers in Microbiology*, *11*, 823.

Adnan, M., Siddiqui, A.J., Hamadou, W.S., Ashraf, S.A., Hassan, M.I., Snoussi, M., Badraoui, R., Jamal, A., Bardakci, F., Awadelkareem, A.M. and Sachidanandan, M., 2021. Functional and structural characterization of pediococcus pentosaceus-derived biosurfactant and its biomedical potential against bacterial adhesion, quorum sensing, and biofilm formation. *Antibiotics*, *10*(11), 1371.

Ahmad, I., Khan, M.S.A., Husain, F.M., Zahin, M. and Singh, M., 2011. Bacterial quorum sensing and its interference: methods and significance. In *Microbes and Microbial Technology* 127–161. Springer, New York, NY.

Al-Haidari, R., Shaaban, M., Ibrahim, S. and Gamal, A.M., 2016. Anti-quorum sensing activity of some medicinal plants. *African Journal of Traditional, Complementary, and Alternative Medicines, 13*, 67–71.

Aliyu, A.B., Koorbanally, N.A., Moodley, B., Singh, P. and Chenia, H.Y., 2016. Quorum sensing inhibitory potential and molecular docking studies of sesquiterpene lactones from Vernonia blumeoides. *Phytochemistry*, *126*, 23–33.

Arita-Morioka, K.I., Yamanaka, K., Mizunoe, Y., Tanaka, Y., Ogura, T. and Sugimoto, S., 2018. Inhibitory effects of myricetin derivatives on curli-dependent biofilm formation in Escherichia coli. *Scientific Reports*, *8*(1), 1–11.

Asfour, H.Z., 2018. Anti-quorum sensing natural compounds. *Journal of Microscopy and Ultrastructure*, *6*(1), 1.

Aswathanarayan, J.B. and Vittal, R.R., 2018. Inhibition of biofilm formation and quorum sensing mediated phenotypes by berberine in Pseudomonas aeruginosa and Salmonella typhimurium. *RSC Advances*, *8*(63), 36133–36141.

Bacha, K., Tariku, Y., Gebreyesus, F., Zerihun, S., Mohammed, A., Weiland-Bräuer, N., Schmitz, R.A. and Mulat, M., 2016. Antimicrobial and anti-quorum sensing activities of selected medicinal plants of Ethiopia: implication for development of potent antimicrobial agents. *BMC Microbiology*, *16*(1), 1–9.

Baldé, M.A., Tuenter, E., Traoré, M.S., Matheeussen, A., Cos, P., Maes, L., Camara, A., Haba, N.L., Gomou, K., Diallo, M.S.T. and Baldé, E.S., 2020. Antimicrobial investigation of ethnobotanically selected guinean plant species. *Journal of Ethnopharmacology*, *263*, 113232.

Benzaid, C., Belmadani, A., Djeribi, R. and Rouabhia, M., 2019. The effects of Mentha× piperita essential oil on C. albicans growth, transition, biofilm formation, and the expression of secreted aspartyl proteinases genes. *Antibiotics*, *8*(1), 10.

Bettenworth, V., Steinfeld, B., Duin, H., Petersen, K., Streit, W.R., Bischofs, I. and Becker, A., 2019. Phenotypic heterogeneity in bacterial quorum sensing systems. *Journal of Molecular Biology*, *431*(23), 4530–4546.

Bhargava, N., Singh, S.P., Sharma, A., Sharma, P. and Capalash, N., 2015. Attenuation of quorum sensing-mediated virulence of Acinetobacter baumannii by Glycyrrhiza glabra flavonoids. *Future Microbiology*, *10*(12), 1953–1968.

Borges, A., Abreu, A.C., Dias, C., Saavedra, M.J., Borges, F. and Simões, M., 2016. New perspectives on the use of phytochemicals as an emergent strategy to control bacterial infections including biofilms. *Molecules*, *21*(7), 877.

Bouyahya, A., Dakka, N., Et-Touys, A., Abrini, J. and Bakri, Y., 2017. Medicinal plant products targeting quorum sensing for combating bacterial infections. *Asian Pacific Journal of Tropical Medicine*, *10*(8), 729–743.

Brackman, G., Breyne, K., De Rycke, R., Vermote, A., Van Nieuwerburgh, F., Meyer, E., Van Calenbergh, S. and Coenye, T., 2016. The quorum sensing inhibitor hamamelitannin increases antibiotic susceptibility of Staphylococcus aureus biofilms by affecting peptidoglycan biosynthesis and eDNA release. *Scientific Reports*, *6*(1), 1–14.

Brackman, G., Cos, P., Maes, L., Nelis, H.J. and Coenye, T., 2011. Quorum sensing inhibitors increase the susceptibility of bacterial biofilms to antibiotics in vitro and in vivo. *Antimicrobial Agents and Chemotherapy*, *55*(6), 2655–2661.

Brango-Vanegas, J., Costa, G.M., Ortmann, C.F., Schenkel, E.P., Reginatto, F.H., Ramos, F. A., Arévalo-Ferro, C. and Castellanos, L., 2014. Glycosylflavonoids from Cecropia pachystachya Trécul are quorum sensing inhibitors. *Phytomedicine*, *21*(5), 670–675.

Breslin, A., 2017. The chemical composition of green plants. *Sciencing, Leaf Group Ltd, 76.*

Burt, S.A., Ojo-Fakunle, V.T., Woertman, J. and Veldhuizen, E.J., 2014. The natural antimicrobial carvacrol inhibits quorum sensing in Chromobacterium violaceum and reduces bacterial biofilm formation at sub-lethal concentrations. *PLoS One*, *9*(4), e93414.

Camele, I., Elshafie, H.S., Caputo, L. and De Feo, V., 2019. Anti-quorum sensing and antimicrobial effect of mediterranean plant essential oils against phytopathogenic bacteria. *Frontiers in Microbiology*, 2619.

Cetin-Karaca, H. and Newman, M.C., 2015. Antimicrobial efficacy of plant phenolic compounds against Salmonella and Escherichia Coli. *Food Bioscience*, *11*, 8–16.

Ćirić, A.D., Petrović, J.D., Glamočlija, J.M., Smiljković, M.S., Nikolić, M.M., Stojković, D.S. and Soković, M.D., 2019. Natural products as biofilm formation antagonists and regulators of quorum sensing functions: a comprehensive review update and future trends. *South African Journal of Botany*, *120*, 65–80.

Cushnie, T.T., Cushnie, B. and Lamb, A.J., 2014. Alkaloids: an overview of their antibacterial, antibiotic-enhancing and antivirulence activities. *International Journal of Antimicrobial Agents*, *44*(5), 377–386.

Cushnie, T.T. and Lamb, A.J., 2011. Recent advances in understanding the antibacterial properties of flavonoids. *International Journal of Antimicrobial Agents*, *38*(2), 99–107.

Díaz-Nuñez, J.L., García-Contreras, R. and Castillo-Juárez, I., 2021. The new antibacterial properties of the plants: quo vadis studies of anti-virulence phytochemicals? *Frontiers in Microbiology*, *12*, 1054.

Doğaner, B.A., Yan, L.K. and Youk, H., 2016. Autocrine signaling and quorum sensing: extreme ends of a common spectrum. *Trends in Cell Biology*, *26*(4), 262–271.

Dwivedi, D. and Singh, V., 2016. Effects of the natural compounds embelin and piperine on the biofilm-producing property of *Streptococcus mutans*. *Journal of Traditional and Complementary Medicine*, *6*(1), 57–61.

Elshafie, H. S. and Camele, I., 2017. An overview of the biological effects of some mediterranean essential oils on human health (review article). *BioMed Research International*, 1–14. doi: 10.1155/2017/9268468

Fiamegos, Y.C., Kastritis, P.L., Exarchou, V., Han, H., Bonvin, A.M., Vervoort, J., Lewis, K., Hamblin, M.R. and Tegos, G.P., 2011. Antimicrobial and efflux pump inhibitory activity of caffeoylquinic acids from Artemisia absinthium against gram-positive pathogenic bacteria. *PLoS One*, *6*(4), e18127.

Ghosh, R., Tiwary, B.K., Kumar, A. and Chakraborty, R., 2014. Guava leaf extract inhibits quorum-sensing and Chromobacterium violaceum induced lysis of human hepatoma cells: whole transcriptome analysis reveals differential gene expression. *PLoS One*, *9*(9), e107703.

Ghosh, S., Lahiri, D., Nag, M., Dey, A., Pandit, S., Sarkar, T., Pati, S., Abdul Kari, Z., Ishak, A.R., Edinur, H.A. and Ray, R.R., 2022. Phytocompound mediated blockage of quorum sensing cascade in ESKAPE pathogens. *Antibiotics*, *11*(1), 61.

Gilabert, M., Marcinkevicius, K., Andujar, S., Schiavone, M., Arena, M.E. and Bardón, A., 2015. Sesqui-and triterpenoids from the liverwort Lepidozia chordulifera inhibitors of bacterial biofilm and elastase activity of human pathogenic bacteria. *Phytomedicine*, *22*(1), 77–85.

Giménez-Bastida, J.A., Truchado, P., Larrosa, M., Espín, J.C., Tomás-Barberán, F. A., Allende, A. and García-Conesa, M.T., 2012. Urolithins, ellagitannin metabolites produced by colon microbiota, inhibit quorum sensing in Yersinia enterocolitica: phenotypic response and associated molecular changes. *Food Chemistry*, *132*(3), 1465–1474.

Goo, E., Majerczyk, C.D., An, J.H., Chandler, J.R., Seo, Y.S., Ham, H., Lim, J.Y., Kim, H., Lee, B., Jang, M.S. and Greenberg, E.P., 2012. Bacterial quorum sensing, cooperativity, and anticipation of stationary-phase stress. *Proceedings of the National Academy of Sciences*, *109*(48), 19775–19780.

Gopu, V., Kothandapani, S. and Shetty, P.H., 2015. Quorum quenching activity of Syzygium cumini (L.) Skeels and its anthocyanin malvidin against Klebsiella pneumoniae. *Microbial Pathogenesis*, *79*, 61–69.

Hagen, S.J. and Son, M., 2017. Origins of heterogeneity in Streptococcus mutans competence: interpreting an environment-sensitive signaling pathway. *Physical Biology*, *14*(1), 015001.

Heilmann, S., Krishna, S. and Kerr, B., 2015. Why do bacteria regulate public goods by quorum sensing?—how the shapes of cost and benefit functions determine the form of optimal regulation. *Frontiers in Microbiology*, *6*, 767.

Huber, B., Eberl, L., Feucht, W. and Polster, J., 2003. Influence of polyphenols on bacterial biofilm formation and quorum-sensing. *Zeitschrift für Naturforschung C*, *58*(11–12), 879–884.

Hyldgaard, M., Mygind, T. and Meyer, R.L., 2012. Essential oils in food preservation: mode of action, synergies, and interactions with food matrix components. *Frontiers in Microbiology*, *3*, 12.

Kalia, V.C. ed., 2018. *Biotechnological applications of quorum sensing inhibitors*. Springer, Singapore.

Kerekes, E.B., Deák, É., Takó, M., Tserennadmid, R., Petkovits, T., Vágvölgyi, C. and Krisch, J., 2013. Anti-biofilm forming and anti-quorum sensing activity of selected essential oils and their main components on food-related micro-organisms. *Journal of Applied Microbiology*, *115*(4), 933–942.

Khan, M.S., Qais, F. A. and Ahmad, I., 2018. Quorum sensing interference by natural products from medicinal plants: significance in combating bacterial infection. *Biotechnological Applications of Quorum Sensing Inhibitors*, 417–445.

Khan, M.S.A., Zahin, M., Hasan, S., Husain, F.M. and Ahmad, I., 2009. Inhibition of quorum sensing regulated bacterial functions by plant essential oils with special reference to clove oil. *Letters in Applied Microbiology*, *49*(3), 354–360.

Lahiri, D., Nag, M., Dutta, B., Dey, S., Mukherjee, D., Joshi, S.J. and Ray, R.R., 2021. Antibiofilm and anti-quorum sensing activities of eugenol and linalool from Ocimum tenuiflorum against Pseudomonas aeruginosa biofilm. *Journal of Applied Microbiology, 131*(6), 2821–2837.

LaSarre, B. and Federle, M.J., 2013. Exploiting quorum sensing to confuse bacterial pathogens. *Microbiology and Molecular Biology Reviews, 77*(1), 73–111.

Li, S., Chan, K.K.W., Hua, M.Z., Gölz, G. and Lu, X., 2021. Inhibition of AI-2 quorum sensing and biofilm formation in campylobacter jejuni by decanoic and lauric acids. *Frontiers in Microbiology, 12.*

Luciardi, M.C., Blázquez, M.A., Cartagena, E., Bardón, A. and Arena, M.E., 2016. Mandarin essential oils inhibit quorum sensing and virulence factors of Pseudomonas aeruginosa. *LWT-Food Science and Technology, 68*, 373–380.

Ludwiczuk, A., Skalicka-Woźniak, K. and Georgiev, M.I., 2017. *Terpenoids: pharmacognosy.* Academic Press, pp. 233–266.

Mancini, E., Camele, I., Elshafie, H.S., De Martino, L., Pellegrino, C., Grulova, D. and De Feo, V., 2014. Chemical composition and biological activity of the essential oil of Origanum vulgare ssp. hirtum from different areas in the Southern Apennines (Italy). *Chemistry & Biodiversity, 11*(4), 639–651.

Mohamed, G.A., Ibrahim, S.R., Shaaban, M.I. and Ross, S.A., 2014. Mangostanaxanthones I and II, new xanthones from the pericarp of Garcinia mangostana. *Fitoterapia, 98*, 215–221.

Monte, J., Abreu, A.C., Borges, A., Simões, L.C. and Simões, M., 2014. Antimicrobial activity of selected phytochemicals against *Escherichia coli* and *Staphylococcus aureus* and their biofilms. *Pathogens, 3*(2), 473–498.

.Murata, R.M., Branco-de-Almeida, L.S., Franco, E.M., Yatsuda, R., dos Santos, M.H., de Alencar, S.M., Koo, H. and Rosalen, P.L., 2010. Inhibition of streptococcus mutans biofilm accumulation and development of dental caries in vivo by 7-epiclusianone and fluoride. *Biofouling, 26*(7),865–872.

Myszka, K., Schmidt, M.T., Majcher, M., Juzwa, W., Olkowicz, M. and Czaczyk, K., 2016. Inhibition of quorum sensing-related biofilm of Pseudomonas fluorescens KM121 by Thymus vulgare essential oil and its major bioactive compounds. *International Biodeterioration & Biodegradation, 114*, 252–259.

Nazzaro, F., Fratianni, F. and Coppola, R., 2013. Quorum sensing and phytochemicals. *International Journal of Molecular Sciences, 14*(6), 12607–12619.

Nieto, G., 2017. Biological activities of three essential oils of the Lamiaceae family. *Medicines (Basel). 4*(3), 63.

O'Connor, S.E., 2010. 1.25—alkaloids. *Comprehensive Natural Products II, 1*, 977–1007.

Olivero-Verbel, J., Barreto-Maya, A., Bertel-Sevilla, A. and Stashenko, E.E., 2014. Composition, anti-quorum sensing and antimicrobial activity of essential oils from Lippia alba. *Brazilian Journal of Microbiology, 45*(3), 759–767.

Ouyang, J., Sun, F., Feng, W., Sun, Y., Qiu, X., Xiong, L., Liu, Y. and Chen, Y., 2016. Quercetin is an effective inhibitor of quorum sensing, biofilm formation and virulence factors in Pseudomonas aeruginosa. *Journal of Applied Microbiology, 120*(4), 966–974.

Paluch, E., Rewak-Soroczyńska, J., Jędrusik, I., Mazurkiewicz, E. and Jermakow, K., 2020. Prevention of biofilm formation by quorum quenching. *Applied Microbiology and Biotechnology, 104*(5), 1871–1881.

Panche, A.N., Diwan, A.D. and Chandra, S.R., 2016. Flavonoids: An overview. *Journal of Nutritional Science, 5*, e47.

Papenfort, K. and Bassler, B.L., 2016. Quorum sensing signal–response systems in gram-negative bacteria. *Nature Reviews Microbiology, 14*(9), 576–588.

Pejin, B., Ciric, A., Glamoclija, J., Nikolic, M. and Sokovic, M., 2015. In vitro anti-quorum sensing activity of phytol. *Natural Product Research*, *29*(4), 374–377.

Perveen, S., 2018. Introductory Chapter: Terpenes and Terpenoids [Internet]. Terpenes and Terpenoids. IntechOpen. http://dx.doi.org/10.5772/intechopen.79683

Poli, J.P., Guinoiseau, E., de Rocca Serra, D., Sutour, S., Paoli, M., Tomi, F., Quilichini, Y., Berti, L. and Lorenzi, V., 2018. Anti-quorum sensing activity of 12 essential oils on Chromobacterium violaceum and specific action of cis-cis-p-menthenolide from corsican mentha suaveolens ssp. insularis. *Molecules*, *23*(9), 2125.

Quave, C.L., Lyles, J.T., Kavanaugh, J.S., Nelson, K., Parlet, C.P., Crosby, H.A., Heilmann, K.P. and Horswill, A.R., 2015. Castanea sativa (European Chestnut) leaf extracts rich in ursene and oleanene derivatives block Staphylococcus aureus virulence and pathogenesis without detectable resistance. *PLoS One*, *10*(8), e0136486.

Rajkumari, J., Meena, H., Gangatharan, M. and Busi, S., 2017. Green synthesis of anisotropic gold nanoparticles using hordenine and their antibiofilm efficacy against Pseudomonas aeruginosa. *IET Nanobiotechnology*, *11*(8), 987–994.

Rezzonico, F., Smits, T.H. and Duffy, B., 2012. Detection of AI-2 receptors in genomes of Enterobacteriaceae suggests a role of type-2 quorum sensing in closed ecosystems. *Sensors*, *12*(5), 6645–6665.

Samanta, A., Das, G. and Das, S.K., 2011. Roles of flavonoids in plants. *Carbon*, *100*(6), 12–35.

Sarabhai, S., Sharma, P. and Capalash, N., 2013. Ellagic acid derivatives from Terminalia chebula Retz. downregulate the expression of quorum sensing genes to attenuate Pseudomonas aeruginosa PAO1 virulence. *PLoS One*, *8*(1), e53441.

Shehabeldine, A.M., Ashour, R.M., Okba, M.M. and Saber, F.R., 2020. Callistemon citrinus bioactive metabolites as new inhibitors of methicillin-resistant Staphylococcus aureus biofilm formation. *Journal of Ethnopharmacology*, *254*, 112669.

Silva, L.N., Zimmer, K.R., Macedo, A.J. and Trentin, D.S., 2016. Plant natural products targeting bacterial virulence factors. *Chemical Reviews*, *116*(16), 9162–9236.

Snoussi, M., Noumi, E., Punchappady-Devasya, R., Trabelsi, N., Kanekar, S., Nazzaro, F., Fratianni, F., Flamini, G., De Feo, V. and Al-Sieni, A., 2018. Antioxidant properties and anti-quorum sensing potential of Carum copticum essential oil and phenolics against Chromobacterium violaceum. *Journal of Food Science and Technology*, *55*(8), 2824–2832.

Srinivasan, R., Devi, K.R., Kannappan, A., Pandian, S.K. and Ravi, A.V., 2016. Piper betle and its bioactive metabolite phytol mitigates quorum sensing mediated virulence factors and biofilm of nosocomial pathogen Serratia marcescens in vitro. *Journal of Ethnopharmacology*, *193*, 592–603.

Subramanian, K., Selvaraj, H., Sampath Renuga, P. and Aruni, W., 2022. Anti-quorum sensing in pathogenic microbes using plant-based bioactive phytochemicals. *Advances in Materials Science and Engineering*, *2022*.

Szabó, M.Á., Varga, G.Z., Hohmann, J., Schelz, Z., Szegedi, E., Amaral, L. and Molnár, J., 2010. Inhibition of quorum-sensing signals by essential oils. *Phytotherapy Research: An International Journal Devoted to Pharmacological and Toxicological Evaluation of Natural Product Derivatives*, *24*(5), 782–786.

Thakur, P., Chawla, R., Tanwar, A., Chakotiya, A.S., Narula, A., Goel, R., Arora, R. and Sharma, R.K., 2016. Attenuation of adhesion, quorum sensing and biofilm mediated virulence of carbapenem resistant Escherichia coli by selected natural plant products. *Microbial Pathogenesis*, *92*, 76–85.

Tiwary, B.K., Ghosh, R., Moktan, S., Ranjan, V.K., Dey, P., Choudhury, D., Dutta, S., Deb, D., Das, A.P. and Chakraborty, R., 2017. Prospective bacterial quorum sensing inhibitors from Indian medicinal plant extracts. *Letters in Applied Microbiology*, *65*(1), 2–10.

Vandeputte, O.M., Kiendrebeogo, M., Rajaonson, S., Diallo, B., Mol, A., El Jaziri, M. and Baucher, M., 2010. Identification of catechin as one of the flavonoids from *Combretum albiflorum* bark extract that reduces the production of quorum-sensing-controlled virulence factors in *Pseudomonas aeruginosa* PAO1. *Applied and Environmental Microbiology, 76*(1), 243–253.

Vandeputte, O.M., Kiendrebeogo, M., Rasamiravaka, T., Stevigny, C., Duez, P., Rajaonson, S., Diallo, B., Mol, A., Baucher, M. and El Jaziri, M., 2011. The flavanone naringenin reduces the production of quorum sensing-controlled virulence factors in Pseudomonas aeruginosa PAO1. *Microbiology, 157*(7), 2120–2132.

Vasavi, H.S., Arun, A.B. and Rekha, P.D., 2016. Anti-quorum sensing activity of flavonoid-rich fraction from Centella asiatica L. against Pseudomonas aeruginosa PAO1. *Journal of Microbiology, Immunology and Infection*, *49*(1), 8–15.

Vasquez, J.K. and Blackwell, H.E., 2019. Simplified autoinducing peptide mimetics with single-nanomolar activity against the Staphylococcus aureus AgrC quorum sensing receptor. *ACS Infectious Diseases*, *5*(4), 484–492.

Vijayakumar, K. and Ramanathan, T., 2020. Musa acuminata and its bioactive metabolite 5-Hydroxymethylfurfural mitigates quorum sensing (las and rhl) mediated biofilm and virulence production of nosocomial pathogen Pseudomonas aeruginosa in vitro. *Journal of Ethnopharmacology*, *246*, 112242.

Vikram, A., Jayaprakasha, G.K., Jesudhasan, P.R., Pillai, S.D. and Patil, B.S., 2010. Suppression of bacterial cell–cell signalling, biofilm formation and type III secretion system by citrus flavonoids. *Journal of Applied Microbiology*, *109*(2), 515–527.

Vikram, A., Jesudhasan, P.R., Jayaprakasha, G.K., Pillai S.D. and Patil B.S., 2011. Citrus limonoids interfere with *Vibrio harveyi* cell–cell signalling and biofilm formation by modulating the response regulator LuxO. *Microbiology, 157*, 99–110.

Vikram, A., Jesudhasan, P.R., Pillai, S.D. and Patil, B.S., 2012. Isolimonic acid interferes with Escherichia coli O157: H7 biofilm and TTSS in QseBC and QseA dependent fashion. *BMC Microbiology*, *12*(1), 1–13.

Wang, B., Zhao, A., Novick, R.P. and Muir, T.W., 2015. Key driving forces in the biosynthesis of autoinducing peptides required for staphylococcal virulence. *Proceedings of the National Academy of Sciences*, *112*(34), 10679–10684.

Wang, C., Liu, X., Wang, J., Zhou, J., Cui, Z. and Zhang, L.H., 2016. Design and characterization of a polyamine derivative inhibiting the expression of type III secretion system in Pseudomonas aeruginosa. *Scientific Reports*, *6*(1), 1–13.

World Health Organization (WHO), 2020. *WHO consolidated guidelines on tuberculosis module 4: Treatment drug-resistant tuberculosis treatment*. WHO, Geneva.

Xu, L., Shao, C., Li, G., Shan, A., Chou, S., Wang, J., Ma, Q. and Dong, N., 2020. Conversion of broad-spectrum antimicrobial peptides into species-specific antimicrobials capable of precisely targeting pathogenic bacteria. *Scientific Reports*, *10*(1), 1–9.

Yan, X., Gu, S., Shi, Y., Cui, X., Wen, S. and Ge, J., 2017. The effect of emodin on Staphylococcus aureus strains in planktonic form and biofilm formation in vitro. *Archives of Microbiology*, *199*(9), 1267–1275.

Yin, S., Davis, R.A., Shelper, T., Sykes, M.L., Avery, V.M., Elofsson, M., Sundin, C. and Quinn, R.J., 2011. Pseudoceramines A–D, new antibacterial bromotyrosine alkaloids from the marine sponge Pseudoceratina sp. *Organic & Biomolecular Chemistry*, *9*(19), 6755–6760.

Zhang, H.B., Wang, L.H. and Zhang, L.H., 2002. Genetic control of quorum-sensing signal turnover in Agrobacterium tumefaciens. *Proceedings of the National Academy of Sciences*, *99*(7), 4638–4643.

Zhang, J., Feng, T., Wang, J., Wang, Y. and Zhang, X.H., 2019. The mechanisms and applications of quorum sensing (QS) and quorum quenching (QQ). *Journal of Ocean University of China*, *18*(6), 1427–1442.

Zhang, J., Rui, X., Wang, L., Guan, Y., Sun, X. and Dong, M., 2014. Polyphenolic extract from Rosa rugosa tea inhibits bacterial quorum sensing and biofilm formation. *Food Control*, *42*, 125–131.

Zhang, Y., Wang, J.S., Wei, D.D., Gu, Y.C., Wang, X.B. and Kong, L.Y., 2013. Bioactive terpenoids from the fruits of Aphanamixis grandifolia. *Journal of Natural Products*, *76*(6), 1191–1195.

Zhou, J.W., Luo, H.Z., Jiang, H., Jian, T.K., Chen, Z.Q. and Jia, A.Q., 2018. Hordenine: a novel quorum sensing inhibitor and antibiofilm agent against Pseudomonas aeruginosa. *Journal of Agricultural and Food Chemistry*, *66*(7), 1620–1628.

6 Microbiologically Synthesized Nanoparticles in Inhibition of Quorum Sensing

*Drisha Roy, Moupriya Nag, Debasmita Bhattacharya, and Dibyajit Lahiri**

6.1 INTRODUCTION

Biofilm has gained a reputation as one of the infamous causes of chronic infections and successfully established itself as a global health concern (Costerton et al., 1999; de la Fuente-Núñez et al., 2013). The alarming rise in chronic infections is a serious threat worldwide due to the antimicrobial and antibiotic-resistant properties of multidrug-resistant bacteria. It is observed that bacterial cells are capable of causing both acute and chronic infections (Furukawa et al., 2006). Infectious diseases are spread in the medical sector due to biofilm formation in medical implants such as urinary catheters, contact lenses, prostheses, heart valves, and vascular grafts (Jackson et al., 2002; Metwalli et al., 2013; Minardi et al., 2012; Wiley et al., 2012). Chronic infections are mostly linked with native host tissues causing nosocomial infection, otitis media, cystic fibrosis, and bacterial prostatitis (\l \o "(ManLink):Burmølle, M., Thomsen, T. R., Fazli, M., Dige, I., Christensen, L., Homøe, P., Tvede, M., Nyvad, B., Tolker-Nielsen, T., Givskov, M., Moser, C., Kirketerp-Møller, K., Johansen, H. K., Høiby, N., Jensen, P. Ø., Sørensen, S. J., & Bjarnsholt, T. (2010). Biofilms in chronic infections – a matter of opportunity – monospecies biofilms in multispecies infections. FEMS Immunology & Medical Microbiology, 59(3), 324–336. https://doi.org/10.1111/j.1574-695X.2010.00714.x

UserName - DateTime: ain-23-06-2023 08:37:10Burmølle et al., 2010). The communication system among the bacterial cells present in the biofilm community is established through the process of quorum sensing (QS). The sessile cells interact with each other by the synthesis of signalling molecules which help them to set up a colony. Biofilm lifestyle provides protection to the inhabitant

DOI: 10.1201/9781003300557-6

cells from harsh and stressful external environmental conditions (e.g., change in pH, temperature, and nutrient density) which in some cases includes enforcing resistance to antimicrobials and antibiotics (de Kievit & Iglewski, 2000). The QS system involves manufacturing and recognition of small diffusible molecules in their residing environment. When the cellular density increases, the accumulation of the signal molecules in the matrix attains a threshold level which in turn triggers the cascading reaction mechanism of the QS system. This complex circuit of networking is responsible for the expression of virulent factors along with biofilm formation, motility, conjugation and so on (Blair et al., 2015). To obliterate the already existing biofilm, it is essential to target its ulterior structure with the assistance of compounds that possess the characteristic feature to infiltrate and eradicate the existing community. In this milieu, metallic nanoparticles gained a reputation as an essential part of recent studies because of their unique physicochemical and morphological features. The employment of metallic nanoparticles was brought into account as an alternative for antimicrobial and antibiotic multidrug-resistant pathogens. The incessant growth of multidrug resistance has prompted researchers to explore alternative pathways toward the discovery of novel antimicrobial and antibiofilm agents. Metal nanoparticles (NPs) have shown potential as effective antimicrobial agents. The branch of science surrounding nano-sized particles (1–100 nm) with large surface areas and distinct morphological characters is termed nanotechnology. In the past decade, nanosciences have emerged and caught the attention of scientific researchers worldwide. The properties of nanoparticles are dependent on their morphological features and are important in their successful implementation in fields, such as the treatment of diseases, pharmacy, agriculture, electronics, energy, and space (Koul et al., 2021). The widespread implementation of nanoparticles has provided an ecofriendly way to handle a broad spectrum of problematic situations in the medical field as well as the prevention of biofilm-related infectious diseases.

6.2 QUORUM SENSING AND BIOFILM FORMATION

Autoinducers (Ais) are signal molecules synthesized and released by bacterial cells in their living environment (Waters & Bassler, 2005). The signalling molecules accumulate until they reach a critical concentration, at which point they bind with the receptor protein, causing coordinated changes in the expression of certain genes (Abisado et al., 2018; Paluch et al., 2020). QS appears to bacterial cells as a 'switch' that they may deploy depending on the cell population to go from a planktonic to a biofilm-mediated existence. As the colony's cell density grows, the number of chemical signal molecules produced exceeds the threshold required to turn on the switch. QS was initially detected in a marine bacterium. *Vibrio fischeri*, at high cell density, according to research by Nealson and colleagues (Nealson et al., 1970). The intricate network of QS is in charge of virulence factor expression, motility, sporulation, bioluminescence, and biofilm development (Roux et al., 2009). Quorum sensing requires the establishment of three critical phases: the manufacturing of chemical signals, the detection of synthesized signalling molecules, and gene expression. First, the Ais are created under feverish circumstances as a result of the stimulation of a stressful environment. At low cell density, lesser concentrations of Ais are generated, considerably below the stimulatory concentration of an autoinducer required for recognition. The subsequent identification and responses are triggered by a greater concentration of AI during higher cell density. Second, Ais are identified only when they exceed a certain threshold by receptors found in the cytoplasm or membranes. Third, The receptor protein modulates the transcriptional activity of the signalling molecule it detects, which in turn influences receptor protein synthesis; hence, they are accurately referred to as autoinducers (J. Lee & Zhang, 2015). The detection of Ais activates gene expression profiles required for cooperative behaviour by bacterial cells, as well as additional AI synthesis via a feedback mechanism, enhancing population synchronization (Rutherford & Bassler, 2012).

There are three primary QS systems in use: (i) the acyl-homoserine lactone (AHL) QS system in Gram-negative bacteria; (ii) the auto-inducing peptide (AIP) QS system in Gram-positive bacteria; and (iii) the autoinducer-2 (AI-2) QS system present in both Gram-negative and Gram-positive bacteria (Brackman & Coenye, 2014). The first investigation to relate QS to biofilm development found that *P. aeruginosa* lasI mutants were reported by (Davies et al., 1998). Interfering with the communication channel between QS and biofilm formation might be a key target for the development of anti-QS drugs. QS inhibition is also being evaluated as a possible target for reducing bacterial pathogenicity and the likelihood of antibiotic resistance. Nanoparticles have several targets, culminating in broad-spectrum activity against pathogenic microbes.

6.3 QUORUM QUENCHING

Quorum quenching describes the various methods for inhibiting the synthesis of signalling molecules in order to reduce the spread of biofilm synthesis. The enzymatic degradation of QS system signalling molecules halts their accumulation in the extracellular environment and causes drastic changes in gene expression. The quenching of autoinducer molecule production prevents the expression of virulence, including biofilm production by drug-resistant bacteria (Paluch et al., 2020). Quorum sensing inhibitors (QSIs) are active substances that have the ability to quench the quorum (Carradori et al., 2020). There are several mechanisms for inhibiting the quorum sensing pathway: (i) inhibiting the production of the signal molecule by interfering with Lux-I synthase proteins (Lade et al., 2014); (ii) signal molecule enzymatic degradation by AHL acylase and AHL lactonase (Lade et al., 2014); (iii) experimenting with signal receptor analogues (Koch et al., 2005); and (iv) interfering with the AHL-LuxR complex (G. Chen et al., 2011; Rampioni et al., 2014). These strategies can be used to achieve the required inhibition algorithm of the AHL-regulated QS network in both Gram-positive and Gram-negative bacteria (Paluch et al., 2020). The best-known mechanism of quorum quenching is the enzymatic degradation of AHL molecules, which can be catalyzed by four distinct groups of enzymes: lactonases and acylases hydrolyze the HSL ring and amide bond of AHL, respectively, while reductases and oxidases modify but do not degrade AHL (Rehman & Leiknes, 2018). NPs can inhibit bacterial growth by effectively penetrating the exopolysaccharides of a biofilm matrix. The NPs penetrate the biofilm and disrupt the quorum-sensing gene cascades, impeding cell-to-cell communication and thus inhibiting biofilm synthesis. This review focuses on the microbial nanotechniques used to create various metallic and nonmetallic nanoparticles, as well as their 'signal jamming' effects in inhibiting biofilm formation (Lahiri et al., 2021).

6.4 OVERVIEW OF NANOPARTICLES

The recent progress in nanobiotechnology has caught the attention of many researchers due to its broad-spectrum application in various fields, such as bioremediation, medicine, diagnostics, and agriculture. The interesting qualities of nanoparticles superior to those of bulk materials are the underlying factor for their collective action compared to their bulk form (Wagh et al., 2013). Because of their larger surface area in comparison to their size, nanoparticles are more reactive than other bioactive substances, and their large surface area enhances their function as drug carriers. Nanoparticles typically range in size from 1 to 100 nm and have a wide range of morphological features as well as distinct chemical and physical properties. These distinct chemical and physical features allow them to interact with biological systems in novel ways and contribute to antibacterial activity (R. Singh & Nalwa, 2011). There is an inverse link between nanoparticle size and antimicrobial activity, with most nanoparticles in the size range of 1–10 nm eliciting the greatest antimicrobial activity (Elechiguerra et al., 2005). The antibacterial activity of nanoparticles is primarily determined by their water solubility, particle size, and metal ion release (Raghunath & Perumal, 2017). Noble metal nanoparticles, such as silver, gold, and copper nanoparticles, have demonstrated enhanced antibacterial activity against specific bacterial species (Bankier et al., 2019; Slavin et al., 2017). Nanomaterials are classified into four categories based on their dimensions:

- 0-D nanosized nanocrystals such as metallic and semiconductor nanoparticles
- 1-D nanomaterials such as nanowires, nanobots, and nanotubes
- 2-D nanomaterials such as nanocomposites and nanoplates, and
- 3-D nanomaterials such as bulkers (Nadaroğlu et al., 2017).

Owing to the potential of nanomolecules and nanocomposites such as Ag- or ZnO-based compounds to prevent microcolony development, they have been reportedly used to reduce biofilm generation and structural changes, thereby establishing their reputation as excellent quorum-quenching molecules. The action of nanoparticles might be connected to the degradation of receptor proteins or the suppression of AI molecule synthesis, resulting in lower production of virulence factors such as elastase, pyocyanin, etc. (Radzig et al., 2013). The concept of nanobiotechnology emerged at the beginning of the 20th century, with the goal of exploring numerous new pathways in the development of nanomedicine and the construction of a sustainable environment (Patil & Chandrasekaran, 2020). Nanoparticles are excellent for many applications in diverse sectors due to their unique size, shape, structure, and specialized biological, physical, and chemical capabilities. These sectors include:

- Medicine and diagnostics: cancer therapy, drug delivery, disease diagnosis, antibacterial activity, etc.
- Food and agriculture: food packaging, food processing, nanopesticides, etc., and
- Environment: bioremediation, water treatment, nanocatalyst, biodegradable polymers, pollutant scavenger, etc.

(Li et al., 2017; Mandeep & Shukla, 2020)

6.4.1 Biosynthesis of Nanoparticles

Microbial synthesis provides a 'greener' and more sustainable modern alternative which establishes a link between two different branches of microbial biotechnology and nanotechnology (Li et al., 2011). Bacterial species such as *Pseudomonas aeruginosa, P. deceptionensis, Bacillus licheniformis, Pyrococcus furiosus, Pyrobaculum islandicum,* among others, have been exploited by researchers for the biofabrication of different metal and metal oxide nanoparticles, like silver (Ag),

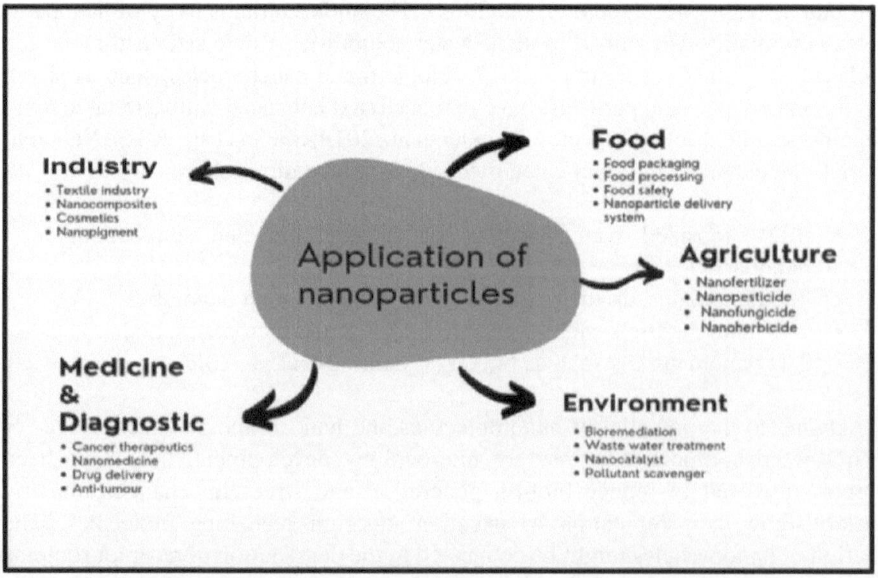

FIGURE 6.1 Application of NPs.

gold (Au), iron (Fe), nickel (Ni), zinc (Zn), and other important NPs (Castro et al., 2014; Jang et al., 2015; Salem & Fouda, 2021; Sanaeimehr et al., 2018; Saravanan et al., 2018; Shivaji et al., 2011; P. Singh et al., 2016). When compared to synthetic approaches, one benefit of biogenic synthesis over conventional chemical synthesis is the safer and easier handling of microbial cultures, as well as the simpler downstream processing of biomass (Rai et al., 2011). Green NP synthesis is a bottom-up strategy in which NPs are generated as a result of the oxidation/reduction process of metallic ions by secreted biomolecules such as enzymes, proteins, sugars, carbohydrates, and so on (Prabhu & Poulose, 2012). To state some examples, the marine yeast *Yarrowia lipolytica* catalyzes the synthesis of Au NP (Agnihotri et al., 2009; Apte, Girme, et al., 2013; Apte, Sambre, et al., 2013) and fungi like *Phanerochaete chrysosporium, Penicillium* species, *and Neurospora crassa* reportedly synthesized silver, gold, and bimetallic nanoparticles (Castro-Longoria et al., 2011; Du et al., 2011; Vigneshwaran et al., 2006). Hasan (2015) reported the utilization of both prokaryotes and eukaryotes for the synthesis of metallic nanoparticles (Au, Ag, palladium (Pa), platinum (Pt), zirconium (Zr), Fe, and cadmium (Cd)) and metal oxide nanoparticles (titanium oxide, zinc oxide). These microorganisms include bacteria, actinomycetes, fungi, yeast, algae, microalgae, and viruses. Microbes may adapt to greater metal concentrations and have the ability to degrade inorganic elements into NPs via extracellular or intracellular channels (Fariq et al., 2017; Salem & Fouda, 2021)). They collect metal ions from their surroundings and use enzymatic reduction to convert them to elemental form (Li et al., 2011). The synthesis of nanoparticles can be

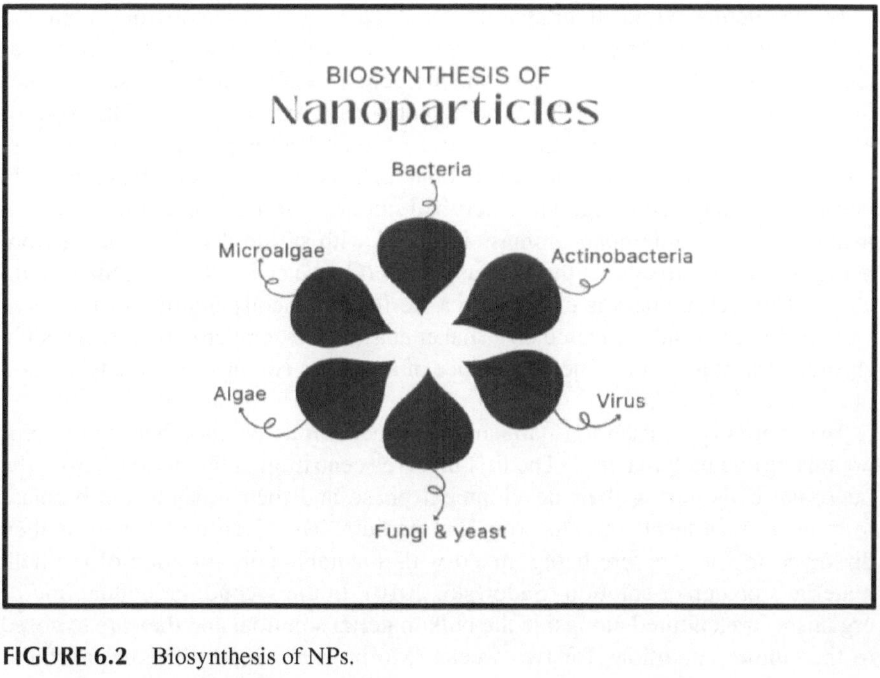

FIGURE 6.2 Biosynthesis of NPs.

both intracellular and extracellular, depending on the location of the nanoparticles (Hulkoti & Taranath, 2014; Prasad, 2016). To optimize the biosynthesis methods, a thorough knowledge of the process of microbial NP production has to be obtained, since different types of microorganisms interact with metallic ions in different ways. The size, shape, and morphology of the produced NPs are eventually affected by the biochemical processing and interaction activities of a given microorganism, as well as the influence of ambient factors such as temperature and pH (Makarov et al., 2014).

6.4.2 Extracellular and Intracellular Synthesis of NPs

The intracellular and extracellular synthesis of nanoparticles (NP) by microbes from metals, metal oxides, and metalloids have been widely reviewed in the literature (Patil & Chandrasekaran, 2020). Depending on the kind of microbe, nanoparticles are generated either intracellularly or extracellularly (Mohamed et al., 2019). Drug carriers for targeted delivery, cancer treatment, gene therapy, DNA analysis, antibacterial agents, biosensors, separation science, and magnetic resonance imaging have all made use of biosynthesized nanoparticles (Li et al., 2011). In the extracellular fabrication of NPs, microbial cultures are grown in suitable media. The broth is incubated and then it undergoes centrifugation. The collected supernatant is then utilized for the synthesis of NPs. In a separate container, the reductase enzyme–containing supernatant is permitted to interact with the metal ions. Metal ions are bioreduced in a cell-free supernatant, which results in the formation of NPs.

We can define extracellular synthesis in another way, as nanoparticles that are synthesized outside the cell with effluxed biomolecules, including enzymes, proteins, antioxidants, hormones, ions, and pigments, all of which play essential roles in the reduction process of NP. These biological compounds serve the role of capping and reducing agents throughout the biofabrication process (Ghosh et al., 2022). The cellular mechanism of microbial cells is exploited for NP synthesis in intracellular manufacturing. The microbial cultures are kept in a suitable liquid medium, and the microbial biomass is rinsed with sterile distilled water before being centrifuged to extract the biomass pellet (Castro et al., 2014; M. Shah et al., 2015). The wet biomass is exposed to a sterilized aqueous solution of metals at various dilutions and incubated on a shaker at a suitable temperature until a visible chromatic shift is occurs. The appearance of a specific colour shows the formation of NPs (Koul et al., 2021).

Two methods are used to establish an intracellular association between microorganisms and bulk material. The first involves centrifuging the biomass from the microbial cells during their development phase and then washing the biomass to remove trace levels of salts from the medium. The cleansed biomass is then dissolved in water before being mixed with a suitable concentration of the bulk material's precursor solution (Sadowski, 2010). In the second, individual microorganisms are cultured alongside the bulk material solution, and they are exposed to the culture conditions for two weeks (Merin et al., 2010). The extracellular

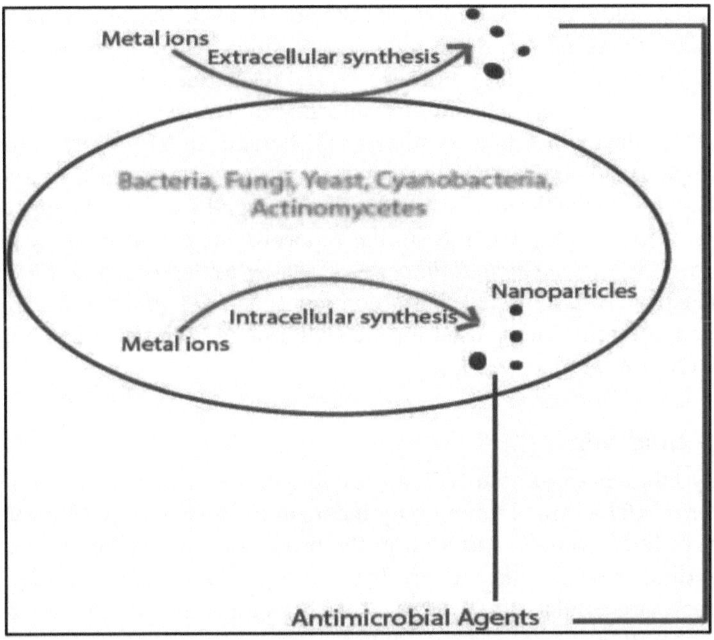

FIGURE 6.3A Extracellular and intracellular synthesis of nanoparticles.

and intracellular biocatalytic synthesis mechanisms are similar because in both, oxidoreductase enzymes such as NADPH-dependent sulphite reductase flavoprotein subunit α, NADH-dependent nitrate reductase, cysteine desulphhydrase, and cellular transporters all play key roles (Grasso et al., 2019). In essence, extracellular production of nanoparticles entails accumulating metal ions on the cell surface and reducing them in the presence of enzymes, whereas intracellular synthesis typically involves ions being carried into the microbial cell and manufacturing nanoparticles in the vicinity of enzymes.

6.5 BACTERIA-MEDIATED SYNTHESIS OF NANOPARTICLES

6.5.1 Metal NPs

6.5.1.1 Silver NPs

A wide range of bacterial species have been reported for the biosynthesis of silver nanoparticles. The synthesis of AgNPs has gained momentum due to their applications in a wide range of fields. It has been reported that facultative bacteria such as *P. stutzeri* and *P. aeruginosa* can thrive under elevated concentrations of heavy metal (Bridges et al., 1979; Haefeli et al., 1984). Studies conducted by the team of Klaus et al. (1999) reveal that the bacterial strain *P. stutzeri* AG259 was used in the early fabrication of silver nanoparticles of dimensions less than

200 nm. The study also shed a light on the ability of the bacteria to accumulate various sizes of Ag NPs in the cellular periplasmic space (up to 200 nm) when sited in a concentrated solution of $AgNO_3$ (50 μM). Bacteria such as *Klebsiella pneumonia, E. coli, Bacillus subtilis,* and *Brevibacterium frigoritolerans* have been used for the extracellular synthesis of spherical Ag NPs (Gurunathan et al., 2009; Kalpana & Lee, 2013; Velmurugan et al., 2014). Another bacterial strain, *B. cereus*, was used in the synthesis of spherical Ag NPs of 20–40 nm as reported by Sunkar & Nachiyar (2012). A similar report on the extracellular synthesis of Ag NP from *Serratia nematodiphila* was given by Malarkodi et al. (2013). Nair and Pradeep in 2002 reported that a common bacterial strain of *Lactobacillus* present in buttermilk can be used for the synthesis of both silver and gold nanoparticles (Nair & Pradeep, 2002).

6.5.1.2 Gold NPs

The early fabrication of gold NPs was done with the use of the bacterial strain of *B. subtilis* 168 and was reported by the team of Beveridge & Murray (1980). The intracellular synthesis resulted in the production of octahedral Au NPs of 5–25 nm dimension. Another report, from Correa-Llantén et al. (2013), demonstrated the extracellular synthesis of quasi-hexagonal Au NPs of size 5–50 nm from *Geobacillus* species. Many studies were conducted on the effect of pH on the morphological features of Au NPs. A study was conducted by He et al. (2007) in which the effect of pH was also noticed with the use of *Rhodopseudomonas capsulate*. They reported the synthesis of spherical Au NPs of 10–20 nm in a solution of pH 7.0 and Au nanoplates in a solution of pH 4.0. Erasmus et al. (2014) reported both extracellular and intracellular synthesis of gold NPs using *Thermos scotoductus* SA-01 which resulted in the production of triangular and hexagonal NPs.

6.5.2 Metal Oxides

Metal oxide NPs such as Ag_2O, CuO, ZnO, TiO_2, MnO_2, MgO, and Fe_2O_3 have also gained popularity due to their wide range of applications in biomedicine, the food industry, optoelectronics, biosensors, and as antimicrobial agents (Koul et al., 2021). Spherical TiO_2 NPs of 8–35 nm were extracellularly synthesized using *Lactobacillus* species, as reported by Jha et al. (2009). Perez-Gonzalez et al. (2010) demonstrated the extracellular fabrication of Fe_3O_4 NPs using *Shewanella oneidensis* of 40–50 nm in the shape of rectangular, rhombic, and hexagonal. Another study, done by Ogunyemi et al. (2020), reports the biofabrication of three metal oxide NPs – ZnO, MnO_2 and MgO – via a rhizophytic bacterial strain, *Paenibacillus polymyxa* Sx3. TEM and SEM analysis revealed the size and shape of the synthesized ZnO NPs, MgO NPs, and MnO_2 NPs. The three metal oxide NPs showed significant antimicrobial activity against the bacterium *Xanthomonas oryzae* pv. *Oryzae* (Xoo) causes bacterial leaf blight in rice (Ogunyemi et al., 2020).

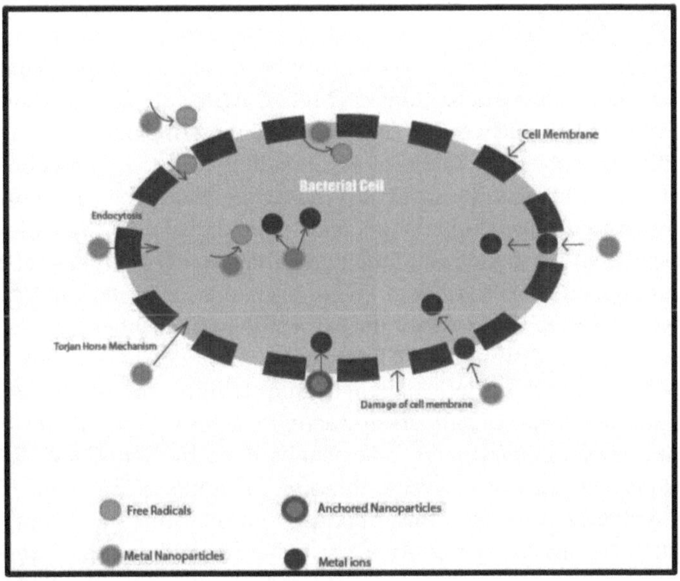

FIGURE 6.3B Mechanism of cell damage by NPs.

6.5.3 Cyanobacteria

Cyanobacteria possess immense potential for the biofabrication of NPs. They are capable of synthesizing metal NPs, metal oxide NPs, and semiconductor NPs. Lengke et al. (2007) demonstrated the biosynthesis of Ag NPs using the filamentous cyanobacterium *Plectonema boryanum* UTEX 485 via enzymatic reduction. It was found that the Ag NPs were spherical or octahedral, with sizes ranging between 1 and 200 nm. Husain et al. (2015) demonstrated synthesis of triangular Ag NPs using *Arthrospira platensis* and *A. maxima* at a temperature of 30°C of sizes 46 and 61 nm, respectively. Husain et al. (2015) conducted a study on 30 species of cyanobacteria for Ag NP synthesis. The results indicated that *Cylindrospermum stagnale* was the best in synthesizing nanosized NPs ranging from 38–40 nm in size. Lenartowicz et al. (2017) reported that *Anabaena laxa* catalyzed the reduction of hydrogen tetrachloroaurate (III) to form Au NPs. It was also observed that *A. laxa* helped in the intracellular fabrication of different sizes of Au NPs, with varied morphologies in the presence of three different auric chloride concentrations. Parial et al. (2012) found that *Phormidium valderianum* is capable of biosynthesizing Au NPs of varied morphology such as hexagonal (25 nm), triangular (24 nm), and nanosphere (15 nm) under the effect of different pH values. Metal oxide NPs are also synthesized using *Cyanobacteria*. For example, Rahman et al. (2010) demonstrated the synthesis of ZnO NPs using *Anabaena* species and *Nostoc* species, while *Spirulina platensis* and *P. cyanobacterium* could synthesize CuO NPs.

6.5.4 Actinobacteria

The filamentous bacteria *Actinomycetes* have been considered for green synthesis due to their metabolic resourcefulness (Md. A. Ali et al., 2020). They have the capability of extracellularly or intracellularly synthesizing nanoparticles and various bioactive compounds (Gahlawat & Choudhury, 2019). Extracellular is preferred over intracellular synthesis in the commercial field. The enzymes produced by the actinomycetes have high commercial value due to their higher level of stability and specificity than the rest of the other green sources (Prakash et al., 2013). In a report obtained from Karthik et al. (2014), a marine bacterium strain, *Streptomyces* species LK-3, was used for the extracellular biosynthesis of Ag NPs. The researchers observed that the NADH-dependent nitrate reductase took part in the reduction of Ag ions which helped them to gain stability. The synthesized Ag NPs possessed elevated levels of antiparasitic activity against *Rhipicephalus microplus* and *Haemaphysalis bispinosa*. Another study, done by Ranjitha & Rai (2017), used *Streptomyces griseoruber* for synthesis of Au NPs of 5–50 nm. The extracellularly synthesized Au NPs thus obtained showed different shapes, like hexagonal, spherical, and triangular. Another study revealed the use of *Rhodococcus* species for the reduction of aqueous $AgNO_3$, which ultimately resulted in the biofabrication of Ag NPs (Otari et al., 2012). Buszewski et al. (2018) demonstrated green synthesis of silver nanoparticles with the use of the acidophilic actinobacteria S*treptacidiphilus durhamensis*. The NPs are spherical and 8–48 nm in size. The biosynthesized Ag NPs showed antimicrobial activity in *S. aureus*, *Proteus mirabilis,* and *P. aeruginosa*. Actinobacteria are also used in the synthesis of other types of metallic NPs. For example, a study conducted by Hassan et al. (2018) observed that copper NPs were biosynthesized with the deployment of *Streptomyces capillispiralis* Ca-1, isolated from a medicinal plant, *Convolvulus arvensis* L. FTIR analysis of the NPs confirmed the presence of various bioactive functional groups which helped in the formation of CuNPs. The biosynthesized NPs showed various biomedical applications along with antimicrobial and antifungal activity.

6.5.5 Algae and Microalgae

The extracellular or intracellular biosynthesis of nanoparticles from algae is dependent on the conditions of the culture. The morphological features of the biofabricated nanoparticles can be modified with the adjustment and controlling of different parameters such as light, temperature, agitation, and synthesis time of the fungal culture (Birla et al., 2013). We can know from the study done by Siddiqi & Husen (2016) that biosynthesis of metal and metal oxide nanoparticles happens with the deployment of algal species such as *Fucus vesiculosus, Chlamydomonas reinhardtii, Sargassum muticum, Chlorella vulgaris, Ecklonia cava, Spirulina platensis, Oscillatoria willei*, and *Stoechospermum marginatum*, among others. In a study by da Silva Ferreira et al. (2017), silver chloride (AgCl) NPs were extracellularly biofabricated using *C. vulgaris*. They used dried biomass of the green unicellular alga for the fabrication of the AgCl NPs. It was determined from the FTIR results that the algal proteins acted as the capping/stabilizing agents of the biosynthesized AgCl NPs. From the TEM results, the size and shape of the nanoparticles were determined. The NPs were spherical in shape and 9.8 nm in size. The biofabricated AgCl NPs showed strong antimicrobial activity against the pathogenic bacteria *S. aureus* and *K. pneumoniae*. In another study, conducted by Mahdavi et al. (2013), biosynthesis of iron oxide nanoparticles was done with the help of an aqueous extract of the brown alga *S. muticum* at a temperature of 25°C. The study revealed that the presence of polysaccharides in the algal extract helped in the reduction of $FeCl_3$ to Fe_3O_4 nanoparticles. Sharma et al. (2018) demonstrated that the biofabrication of TiO_2 NPs and a TiO_2-graphene oxide nanocomposite is possible with the aid of a green alga, *Chlorella pyrenoidosa*. The XRD and SEM analysis revealed the morphological characteristics of the synthesized NPs. The TiO_2 NPs were 50 nm in size and spherical in shape while the TiO_2-graphene oxide nanocomposite formed a sheet-like structure. *C. vulgaris* was used in the synthesis of spherical Pd NPs of size 5–20 nm as demonstrated by Garole et al. (2019).

Microalgae can be correctly termed 'nano-biofactories' for the synthesis of metallic nanoparticles. The synthesis of NPs can be done with both live and dead biomass of the microalgae. Biosynthesis of NPs has been achieved with microalgae such as *S. platensis*, *C. vulgaris*, and *Lyngbya majuscule* (El-Sheekh & El-Kassas, 2016; Mubarak et al., 2011). The extracellular biosynthesis of ZnO NPs was carried out using an extract of the *Chlorella* green microalgae demonstrated by Khalafi et al. (2019). The SEM and TEM results revealed the characteristics of the NPs. The 20 nm NPs were monodispersed and hexagonal, and exhibited remediation properties against organosulphur pollutants. Several marine microalgae are also deployed for the biosynthesis of Ag NPs, such as *Tetraselmis gracilis*, *Chaetoceros calcitrans*, *Isochrysis galbana*, and *Chlorella salina* (Merin et al., 2010).

6.5.6 Viruses

There exist several reports and studies that verify the usage and exploitation of viruses for the biosynthesis of NPs. To give an example, M13 bacteriophage has been successfully used as a template for the fabrication of semiconductor NPs of cadmium sulphide (CdS) and zinc sulphide (ZnS) (Mao et al., 2003). In a study carried out by Cunha et al. (2018) it was shown that the virus *Rhodotorula glutinis* was deployed for the biosynthesis of spherical Ag NPs of size 15 nm. The tobacco mosaic virus has been used for the biofabrication of Pd NP (3–4 nm) and Au NP (5 nm). The synthesized Pd NPs were extracted in the form of carbon nanotubes, while the Au NPs were spherical (Fan et al., 2013; Kobayashi et al., 2012). To cite a recent example, Thangavelu et al. (2020) were successful in the synthesis of a hybrid Au-Ag nanocomposite with the aid of using the plant pathogenic squash leaf curl China virus (SLCCNV). The virus acted as a biotemplate for the biofabrication of the hybrid metallic nanomaterials. The SLCCNV was brought in presence of $HAuCl_4$ and $AgNO_3$ precursors and exposed to sunlight for approximately

5 min which resulted in the formation of Au-Ag nanomaterials in a 'green and clean,' eco-friendly way. The hybrid nanomaterial found its application in the biomedical sector due to its admirable biocompatibility function. Another study demonstrated the synthesis of cadmium sulphide (CdS) nanocrystal quantum dots on the procapsid of bacteriophage P22 (Kale et al., 2013). Helical nanocarriers were successfully biofabricated using the potato virus X as reported by Le et al. (2017). They were 13 nm in size.

6.5.7 Fungi and Yeast

NPs can also be biofabricated with the help of a wide variety of fungi. Fungi have the ability to withstand high concentrations of metal ions due to the elevated potential of the cell wall to attach with metal ions. This provides them with an upper hand in yielding comparatively more NPs than bacterial cells (P. Singh et al., 2016). This method is cost effective and also efficient for the green synthesis of NPs, so these organisms have been used for reduction and stabilization during NPsynthesis. Fungi can be easily synthesized on a large scale while maintaining their uniformity in shape and size (Ahluwalia et al., 2014; Azmath et al., 2016). Fungi possess the ability to synthesize enzymes and proteins needed for the fabrication of nanoparticles which makes them a convenient source for NP manufacturing (Alghuthaymi et al., 2015). For example, a study carried out by Husseiny et al. (2015) employed *Fusarium oxysporum* for the biosynthesis of Ag NPs and demonstrated the effect of temperature and substrate concentration on the NPs. The size of the Ag NPs was smaller at 50°C and increased with increase in temperature. The biosynthesis of NPs using fungi can be both extracellular and intracellular (Guilger-Casagrande & Lima, 2019). To give another example, Ag NPs were extracellularly synthesized using *Arthroderma fulvum*, as reported by Wang et al. (2016). The NPs were 13–18 nm in size and spherical in shape. To state another example, a study conducted by Mishra et al. (2014) revealed that extracellular biosynthesis of Au NPs can be carried out with the help of *Trichoderma viride* and *Hypocrea lixii*. The study also revealed how the temperature and pH exert an effect on the biofabricated Au NPs. The duration and temperature of synthesis from the cell-free extract were also shown. *T. viride* fabricated the Au NPs within 10 min at 30°C, while *H. lixii* was able to synthesize the Au NPs at 100°C. Although *T. viride* failed to synthesize NPs at pH 5, they were able to synthesize Au NPs at a higher pH range (7 and 9). The biosynthesized Au NPs were in the size range of 20 and 30 nm. The Au NPs possessed strong biocatalytic and antimicrobial activities. Silver nanoparticles (Ag NPs) of size 2.86, 25.89, and 48.43 nm were synthesized using the filtrate of *Rhizopus stolonifer* at respective temperatures of 40°, 20°, and 60°C (AbdelRahim et al., 2017). Similarly, Metuku et al. (2014) were successful in fabricating Ag NPs using the fungi *Schizophyllum radiatum*. The study revealed that the fungi-mediated Ag NPs were 10 to 40 nm in size and produced through the process of extracellular biomineralization. The NPs exhibited excellent antimicrobial properties against both Gram-negative and Gram-positive bacteria. As reported by Kitching et al. (2015), NADH-dependent oxidoreductase from fungi can be of use in the biosynthesis of gold NPs (Au NPs). Antibacterial and antidermatophytic activity was exhibited in the synthesized NPs against *B. subtilis, E. coli, K. pneumonia, Pseudomonas* species, *Trichophyton mentagrophytes*, and *Trichophyton* species (Busi & Rajkumari, 2019). In a recent study by Ramos et al. (2020) it was reported that the Ag NPs were extracellularly biosynthesized from the fungi *Trichoderma* species, extracted from seeds of Brazil nut (*Bertholletia excelsa).* The biofabricated Ag NPs exhibited advanced antibacterial activity against Gram-negative bacteria.

Yeast can tolerate high concentrations of heavy metals and can be deployed for the bioaccumulation of metal ions by living microorganisms. This is a very cost effective and convenient method for the green synthesis of nanoparticles. Elahian et al. (2017) effectively demonstrated the use of the genetically altered yeast strain *Pichia pastoris* for the biofabrication of Ag NPs. The synthesized NPs were uniform in size and stable in nature. Korbekandi et al. (2016) demonstrated that Ag NPs were biosynthesized using *Saccharomyces cerevisiae* through the method of biotransformation. Compared to dried yeast culture, freshly prepared yeast culture resulted in greater yield of NPs. The biotransformed Ag NPs were found in the interior part of the cell, within the cell membrane, and in the exterior part of the yeast cells. They were spherical in shape and 2–20 nm in size with the majority having particles 5.4 nm in diameter. The early synthesis of Au NPs from the yeast *Magnusiomyces ingens* LH-F1 was demonstrated in a study done by Zhang et al. (2016). They reported Au NPs were produced in various sizes of hexagons, spheres, and triangles. The biofabricated Au NPs exhibited strong reducing properties against nitrophenols.

6.6 NANOPARTICLES AS ANTI-QS AGENTS

To discover neoteric compounds that could aid in disrupting quorum sensing, researchers have trotted down the avenue of unconventional modus operandi. In the recent past, nanoparticles gained a reputation as a convenient and efficient alternative to deal with the quorum-quenching technique. Studies have shown QS inhibition attributes of nanoparticles and established their role as quorum-sensing inhibitors (QSI) (Miller et al., 2015; Radzig et al., 2013). Nanoparticles interfere with the cell circuit, disrupting the quorum-sensing machinery and thus eventually helping to hinder the developmental process of biofilm formation (Lahiri et al., 2021). Nanoparticles, owing to their small size, can take over the bacterial cell by infiltrating the 3D matrix of the sessile communities. The nanoparticles affix themselves to the cell wall and cell membrane of the bacterial cell. This way the nanoparticles accumulate inside the cell membranes and ensure an incessant ejection of ions inside the cell. Nanoparticles have taken a front seat in exploring different means to combat bacterial infections (Al-Shabib et al., 2016; Gómez-Gómez et al., 2019; Samanta et al., 2017; Wagh et al., 2013). NPs act as potent inhibitors of QS by affecting the mechanism of cell-cell communication or inhibiting the signals associated with the QS molecular circuit. This hinders the synthesis of various types of signalling molecules and prevents the formation of the molecule receptor complex. The ultimate result is the cessation of the signal transduction cascade (Sadekuzzaman et al., 2015). There exist many methods to modify and alter the quorum-sensing circuit and bring about quorum quenching (QQ) – for example, enzymatic degradation of signal molecules, suppressing the production of the signal molecule, jamming the molecule-receptor complex formation, and meddling with signal receptor analogues (Sadekuzzaman et al., 2015). The NPs can utilize one of the aforementioned strategies to exhibit anti-QS properties. *Chromobacterium violaceum,* a Gram-negative bacterium, has been chosen as a model organism. *C. violaceum* produces a purple pigment mediated through the quorum-sensing machinery with the aid of two necessary genes: cviI, an autoinducer synthase, and cviR, a cytoplasmic receptor. A study done by Gómez-Gómez et al. (2019), showed the use of metal and metal oxide NPs (ZnO NPs, TiO_2 NPs, and Ag NPs) on two different strains of the model organism *C. violaceum*. Results revealed the effect of NPs on the violacein production and thus its effect on the QS circuit.

6.6.1 Silver NPs

The potential of silver nanoparticles (Ag NPs) as QQ agents is mainly due to their excellent antimicrobial properties (X. Chen & Schluesener, 2008). One of the most extensively researched metal nanoparticles, silver, has been shown to block virulence factors that are controlled by QS in both Gram-positive and Gram-negative bacteria (Wagh et al., 2013). Ag NPs' broad range of antibacterial properties, which are caused by their large surface-area-to-volume ratio, have been a major contributor to their success (Brandt et al., 2012; J. S. Kim et al., 2007; Lara et al., 2011). Not much is yet known about the quorum quenching mechanism of Ag NPs, but to summarize the existing knowledge, it revolves around three common pathways. The first pathway initiates through the accumulation of free silver ions, which influences the synthesis of ATP and DNA replication. The second one involves the production of reactive oxygen species (ROS), and the third one is the disintegration of the cytoplasmic membrane (Qais et al., 2018). According to studies done by S.G. Ali, Ansari, Sajid Jamal, et al. (2017), Ag NPs may stop the manufacture of signalling molecules by jamming the LasI/Rhl I synthase. *P. aeruginosa* has been used to illustrate the anti-QS impact of Ag NPs. The results of molecular docking tests showed that Ag NPs are tightly bound to surrounding residues in the active sites of several proteins, including the transcriptional regulator proteins LasI/RhlI synthase and LasR/RhlI. Ag NPs' inhibition of LasI/RhlI synthase prevented the synthesis of signalling molecules, preventing QS. Additionally, the LasR/RhlR system was inactivated by interfering with transcriptional regulatory proteins, which eventually prevented the expression of virulence genes that were controlled by QS. By targeting the genes responsible for producing signalling molecules (LasI/RhlI) or by inactivating the genes required for the synthesis of transcriptional regulatory proteins (such as LasR/RhlR), the Ag NPs have the ability to exhibit anti-QS property (Lahiri et al., 2021). Evidence from various studies has demonstrated the antibiofilm and anti-QS activities of Ag NPs. Research by Jagtap & Priolkar (2013) using silver nanowires showed promising anti-QS action by lowering the development of biofilms in *P. aeruginosa* and the generation of violacein. At a concentration of 4 μg ml^{-1} of silver nanowires, it was found that 80% of pigment formation was blocked. Another analytical study, conducted by S.G. Ali, Ansari, Sajid Jamal, et al. (2017) revealed that silver nanoparticles exhibited anti-QS activity at a concentration of 15 μg/ml against *C. violaceum*. There was a noticeable and effective reduction of the virulence factors of *P. aeruginosa*. In all examined drug-resistant clinical isolates of *P. aeruginosa*, the formation of pyocyanin was suppressed up to 74.64%, the production of proteases was lowered up to 47.3%, and the activity of haemolysin was decreased to 47.7%. Confocal laser scanning microscopy also showed that the production of biofilm was significantly decreased by 70.9–79.7%. Ag NPs derived from *Vetiveria zizanioides* were found to suppress the expression of QS-dependent genes. Without hindering bacterial growth, *V. zizanioides*-based Ag NPs reduced QS-dependent virulence factors in *S. marcescens*, including biofilm formation, protease, prodigiosin, lipase, and exopolysaccharide synthesis (Ravindran et al., 2018). B.R. Singh et al. (2015) myofabricated Ag NPs from *Rhizopus arrhizus* metabolites and studied their impact on the virulence of *P. aeruginosa* and QS-regulated biofilm formation. After cells were exposed to various concentrations of mfAg NPs, the synthesis of signalling molecules was reduced by 2–76%. According to the study, the expression of the lasA and lasB genes was down-regulated by mfAg NPs in up to 79–84% of the cases. Additionally, gene quantification analysis showed that lasI, lasR, rhlI, rhlR, and fabH2 genes were down-regulated by mfAg NPs at a concentration of 25 g/ml by 71%, 63%, 51%, and 36%, respectively. Analysis of results obtained from SEM (scanning electron microscopy) and CLSM (confocal laser scanning microscopy) data revealed that Ag NPs exhibited a reduction in biofilm formation at 5–25 mg/l (B.R. Singh et al., 2015; R. Singh et al., 2015).

6.6.2 Gold NPs

Due to its uncomplicated manufacturing process, ease of use, and comparably low toxicity to other commonly used nanomaterials, gold NPs (Au NPs) have a wide range of applications (Lahiri et al., 2021). The utilization of AuNPs has drawn the attention of several researchers recently because of their catalytic capabilities, which are widely employed in the fields of diagnostics and biologics (Giasuddin et al., 2013; Jain et al., 2012; Mesbahi, 2010). Several papers affirm the antibiofilm activity of gold nanoparticles. For example, Sathyanarayanan et al. (2013), demonstrated that Au NPs at higher concentrations (>50 μg mL−1) were successful in reducing biofilm formation by *S. aureus* and *P. aeruginosa*. On the other hand, morphological changes in the development of biofilm formation of *Legionella pneumophila* were reported by Raftery et al. (2013, 2014) when exposed to very low concentration of Au NPs. In a research study, functional acyl homoserine lactone lactonase protein-coated Au NPs (AiiA Au NPs) were evaluated against *Proteus* species (Vinoj et al., 2015). The results revealed that N-hexanoyl-L-homoserine lactone (C6-HSL) has undergone degradation with the help of 2 μM AiiA Au NPs which ultimately resulted in the exhibition of quorum-quenching activity. Owing to the property of N-acyl homoserine lactonase to hydrolyze the lactone moiety and open the HSL ring, signalling molecules were unable to form a bond with the transcriptional regulator LuxR. This lack of bond formation hampers the QS-mediated gene expression in pathogens (Bai et al., 2008; Kaufmann et al., 2005). AiiA Au NPs also inhibited the metabolic activity, synthesis of extrapolymeric substance (EPS), biofilm formation, and hydrophobicity of the targeted bacterial strains. Au NPs showed efficient antimicrobial activity against various types of microbes, like methicillin-resistant *S. aureus* (MRSA), *Salmonella typhi*, and *Bacillus Calmette-Guérin* (Bindhu & Umadevi, 2014; Lima et al., 2013; Zhao et al., 2010). They were also able to inhibit the metabolic activities and production of EPS, preventing the formation of biofilm and changing the hydrophobicity of the bacterial cells (Samanta et al., 2017). In the recent past, a study done by Ramasamy et al. (2016) demonstrated the successful inhibition of biofilm formation as well as invasion of dental pulp stem cells by *P. aeruginosa* and *Candida albicans* with the deployment of Au NPs. Gentamicin-doped phosphatidylcholine-coated Au nanoparticles effectively inhibited the growth of biofilms in both Gram-positive and Gram-negative bacteria (Mu et al., 2016). In another study, it was shown that nanogold, when applied in a concentration of 100 μg mL−1, was efficient in the inhibition of biofilm formed by the multiple-antibiotic-resistant *Enterococcus faecalis* (Vijayakumar et al., 2017).

6.6.3 Silicon Dioxide NPs

The mechanism of QS was also inhibited significantly using silicon dioxide NPs (SiO_2 NPs) which are found naturally in the form of quartz that is present as a major element within minerals, rocks, and sands. These NPs are widely used in biomedicines due to their small particle size and biocompatibility (Malvindi et al., 2012). The working mechanism of SiO_2 NPs proceeds with the generation of ROS, which brings about damage to DNA and eventually leads to cell death (Chu et al., 2011, 2012). As a result, they are viewed as beneficial compounds with antibacterial properties (Tang & Cheng, 2013). The elimination of signalling molecules from the biofilm neighbourhood with the help of NPs can be used to cease cell-to-cell communication. In a study, functionalized SiO_2 NPs were coated with β-cyclodextrin, which acts as a quorum-sensing inhibitor molecule. The prepared NPs were able to inhibit the QS-mediated synthesis of AHL molecules in *V. fischeri* (Miller et al., 2015). The study revealed that the presence of β-cyclodextrin associated with SiO_2 NPs helped in the uptake of AHL molecules from the surrounding population and ultimately reduced the bioluminescence. When administered in combination with functionalized SiO_2 NPs (50 nm), quantitative PCR revealed down-regulation of the luxA and luxR genes in the presence of 133 nM β-cyclodextrin.

6.6.4 Titanium Dioxide NPs

Titanium dioxide nanoparticles (TiO_2 NPs) have a wide range of applications, such as in photocatalysis of organic dyes, dye-sensitized solar cells, gas sensors, and photochromic appliances, and also exhibit antimicrobial activities (Banerjee, 2011). TiO_2-based nanoparticles provide a number of benefits, including ease of manufacture, nontoxicity, and cost efficiency (Xie et al., 2011). Md. S. A. S. Shah et al. (2008) revealed that when exposed to nonlethal UV radiation, TiO_2 nanoparticles break down organic molecules by producing and continuously releasing superoxide ions and hydroxyl radicals. This effectively prevents the formation of MRSA. In a study done by Naik & Kowshik (2014), it was demonstrated that silver-titanium nanocomposite (AgCl-TiO_2 NPs) exhibited anti-QS activity against *C. violaceum*. At 100 μg/ml, violacein production was 82% suppressed, while in nutritional broth, complete eradication (100%) of biofilm formation was observed at 300 μg/ml. Homoserine lactone (HSL) degradation by AgCl-TiO_2 NPs was discovered to be one of the fundamental mechanisms of QS inhibition. Silver present in ATNPs significantly suppressed the formation of violacein in a dose-dependent manner, while TiO_2 in AgCl-TiO_2 nanoparticle complex enhanced the regulated release of silver by providing a better-supporting matrix. When bacterial culture was prepared in the presence of ATNPs, there existed no AIs. The removal of signalling molecules confirmed the mechanism of the quorum-quenching process. In a study by Blecher et al. (2011), it was shown that the cell membrane of *E. coli* is damaged by thin film composite (TFC) membranes coated with TiO_2-NP, which prevents bacteria from adhering to the surface of the TFC membrane and prevents the development of biofilms.

6.6.5 Zinc Oxide NPs

Numerous investigations have shown that the ZnO NPs demonstrated potent anti-QS properties. The precise working mechanism behind their QQ activity is not yet brought to light. ZnO NPs have the ability to reduce the expression of genes that control QS in a variety of Gram-negative bacteria. In a study done by Al-Shabib et al. (2016), it was demonstrated that ZnO NPs biofabricated with the aid of *Nigella sativa* serves the purpose of quorum-quenching activity. Inhibition of QS-regulated virulence factors, such as violacein production in *C. violaceum* and protease, elastase, and pyocyanin synthesis in *P. aeruginosa*, was revealed in the research. The broad-spectrum inhibition of QS-regulated functions was recently demonstrated in *Listeria monocytogenes, P. aeruginosa, E. coli*, and *C. violaceum*, and it was shown that subinhibitory quantities of ZnO NPs caused the lasB gene to be down-regulated up to 91%, 93%, 82%, and 83%, respectively, which eventually reduced the amount of AHLs in the surrounding environment. Numerous studies have shown how effective ZnO NPs are as an anti-QS agent. The QS genes within Gram-negative bacterial cells can be down-regulated by these NPs. Researchers showed that ZnO NPs were able to inhibit QS in *P. aeruginosa* by down-regulating lasR, lasI, rhl I, and rhl R (Saleh et al., 2019). Lee et al. (2014) described that ZnO NPs were responsible for the efflux of the czc operon mediated zinc cation efflux pump and various other regulators of transcription such as type III repressor ptrA and porin gene opdT. They were also able to bring about repression in the production of pyocyanin-associated phz operon. Their study demonstrated that ZnO NPs were efficient and successful in inhibiting *P. aeruginosa* biofilm formation without hampering the planktonic growth. It was also revealed in the study that various virulence factors such as pyocyanin and pyochelin were successfully inhibited by ZnO NPs (J.-H. Lee et al., 2014). A recent study demonstrates the effectiveness of ZnO NPs against the pqs and las system of QS as well as their ability to decrease the swimming and swarming motility of *P. aeruginosa* (Khan et al., 2020).

6.6.6 Chitosan NPs

Chitosan and its derivatives have a variety of properties, including biodegradability, the ability to form gels, a high capacity for adsorption, antifungal and antibacterial actions, and anticancer effects (George et al., 2020). The biofilm network created by nosocomial bacterial species like *S. aureus* and *P. aeruginosa* are typically the target of studies on antibiofilm action. SEM was used to study biofilm architecture on silicone surfaces (Zhou et al., 2008). Chitosan has been noted to have stronger selective toxicity and broad-spectrum antibacterial activity. Although the exact mechanism of antimicrobial activity is not fully understood, it has been suggested that chitosan's molecular architecture is crucial for its antimicrobial properties (Kaya et al., 2015). Chitosan nanocapsules are utilized to enhance a range of antimicrobial drugs' QS-inhibiting properties. Biofilms formed by different varieties of normal and clinical strains of *S. aureus* were reported to be resistant to ChNP-DNase-Oxa (Tan et al., 2018). Flavonoid-loaded nanocapsules worked in one of two ways: preventing diffusion of AHL molecules or hindering AI synthase. A nontoxic variety of chitosan exhibited quorum-quenching properties by inhibiting the diffusion of AHL molecules (Qin et al., 2017).

6.6.7 Other NPs

There aren't many studies that discuss the antibiofilm activity of other metal nanoparticles. To state an example, Raftery et al. (2014) successfully demonstrated that with the help of the citrate-coated PtNPs, biofilm formed by *L. pneumophila* can be eradicated. In another study, Shakibaie et al. (2015) used *Bacillus* sp. MSh-1 for preparation of selenium nanoparticles. When used against clinically isolated nosocomial infections that produce biofilm, the biosynthesized nanoparticles showed outstanding antibiofilm activity. They also reduced the biofilm formation in *S. aureus, P. mirabilis,* and *P. aeruginosa*, by 42%, 53.4%, and 34.3%, respectively. Efficient inhibition of oral biofilm formation was proved for Cu NPs (Ren et al., 2009). Selenium and tellurium nanoparticles biofabricated with the help of *Stenotrophomonas maltophilia* SeITE02 and *Ochrobactrum* species MPV1 showed excellent antibiofilm and antibacterial activity against *P. aeruginosa* PAO1, *E. coli* JM109, and *S. aureus* ATCC-25923 (Zonaro et al., 2015)

TABLE 6.1
NPs as Quorum Quenchers

Nanoparticles	Mechanism	Test organism	References
AgCl-TiO_2 nanocomposite	Inhibition of violacein of *C. violaceum*, inhibition of biofilm formation and degradation of homoserine lactone	*C. violaceum*	(Naik & Kowshik, 2014)
Ag nanowires	Inhibition of violacein of *C. violaceum*, inhibition of biofilm of *P. aeruginosa*	*C. violaceum* and *P. aeruginosa*	(Wagh et al., 2013)
Ag NP	Inhibition of violacein of *C. violaceum*	*C. violaceum*	(Anju & Sarada, n.d.)
Ag NP	Jamming LasI/Rhl I synthase, Inhibition of violacein of *C. violaceum*, inhibition of production of pyocyanin, protease, haemolysin, and biofilm of *P. aeruginosa*	*C. violaceum* and *P. aeruginosa*	(S. G. Ali, Ansari, Khan, et al., 2017)
AiiA-Au NP	Degradation of N-hexanoyl-L-homoserine lactone with N-acyl homoserine lactonase protein	*Proteus* sp.	(Vinoj et al., 2015)
Au NP	Inhibition of metabolic activities, EPS formation, inhibition of biofilm formation, altering hydrophobicity of the bacterial cell	*P. aeruginosa*	(Samanta et al., 2017)
Au-NP	Inhibition of biofilm formation, invasion of dental pulp stem cells	*P. aeruginosa and C. albicans*	(Ramasamy et al., 2016)

(Continued)

TABLE 6.1 CONTINUED

Nanoparticles	Mechanism	Test organism	References
B-cyclodextrin functionalized silicon SiO_2 NP	Inhibition of bioluminescence of *V. fischeri*, down-regulation of *luxA* and *luxR* gene of *V. fischeri*	*V. fischeri*	(Miller et al., 2015)
Chitosan	Retarding diffusion of AHLs	*E. coli*	(Qin et al., 2017)
Chitosan-encapsulated kaempherol tripolyphosphate nanocapsules	Inhibition of SI synthase, inhibition of AHL diffusion	*C. violaceum*	(Ilk et al., 2017)
MfAg	Inhibition of violacein of *C. violaceum*, inhibition of virulence factors such as LasA protease, LasB elastase, pyocyanin, pyochelin, and rhamnolipid of *P. aeruginosa*	*C. violaceum* and *P. aeruginosa*	(B. R. Singh et al., 2015)
Selenium NP and telerium NP	Inhibition of biofilm formation	*P. aeruginosa* PA01, *E. coli* JM109, *S. aureus* ATCC-25923	(Zonaro et al., 2015)
Silver-coated carbon nanotubes	Down-regulation of *sdiA* (a QS gene) and many virulence genes (*safC*, *ychP*, *sseA* and *sseG*) of *S. aureus*	*S. aureus*	(Chaudhari et al., 2015)
TiO_2-coated TFC membrane	Damages the cell membrane of *E. coli*, inhibition of biofilm formation	*E. coli*	(Blecher et al., 2011)

ZnO NP	Inhibition of violacein of *C. violaceum*, inhibition of elastase, total protease, pyocyanin production, exopolysaccharide production and swarming motility of *L. monocytogenes*, *P. aeruginosa*, *E. coli* and *C. violaceum*, disruption of mature biofilm, down-regulation in *pqsA*	*L. monocytogenes*, *P. aeruginosa*, *E. coli*, and *C. violaceum*	(HYPERLINK "15064-4537-FullBook.docx" \l "Ref_225_FILE150644537006" \o "(ManLink):Al-Shabib, N. A., Husain, F. M., Ahmed, F., Khan, R. A., Ahmad, I., Alsharaeh, E., Khan, M. S., Hussain, A., Rehman, M. T., Yusuf, M., Hassan, I., Khan, J. M., Ashraf, G. M., Alsalme, A., Al-Ajmi, M. F., Tarasov, V. V., & Aliev, G. (2016). Biogenic synthesis of Zinc oxide nanostructures from Nigella sativa seed: Prospective role as food packaging material inhibiting broad-spectrum quorum sensing and biofilm. Scientific Reports, 6(1), 36761. https://doi.org/10.1038/srep36761 UserName - DateTime: ain-23-06-2023 08:54:36"Al-Shabib et al., 2016)
ZnO NP	Down-regulation of *lasR, lasI, rhl I, rhl R*	*P. aeruginosa*	(Saleh et al., 2019)
ZnO NP	Reduction of swimming and swarming motility, efficacy against *pqs* and *las* system of QS	*P. aeruginosa*	(Khan et al., 2020)
ZnO NP-coated glass surface	Generation of ROS	*E. coli and S. aureus*	(Hajipour et al., 2012)

6.7 DEMERITS AND SAFETY CONCERNS OF NANOPARTICLES

Every coin has two sides. Nanoparticles are potentially beneficial in a wide spectrum of fields. However, deployment of nanoparticles also comes with its own set of negative consequences on many environmental components. This cytotoxicity of nanoparticles limits their widespread usage. Several studies have revealed that NPs can be absorbed by microorganisms in soils, sediments, and plant roots. Later, these NPs travel from the roots to other regions of the plant, where they accumulate (Gardea-Torresdey et al., 2014). 'Green' nanoparticles can be synthesized by using a number of different synthetic media and environmental conditions to mimic the growth of biological organisms like bacteria, algae, and fungi. The reproducibility of size, shape, and charge properties of such synthesized nanoparticles is still a big question mark. As bacteria, plant products, or their waste items are ingested by diverse creatures such as protozoa, arthropods, annelids, molluscs, fish, insects, birds, and mammals, the level of NPs shifts from one trophic level to another. This proposes a negative consequence that may be passed down to the next generation (De la Torre Roche et al., 2015; J. I. Kim et al., 2016; Shimizu et al., 2009). Several recent investigations reveal the multi-organ toxicity of therapeutically administered antimicrobial nanoparticles by inducing ROS formed due to the interaction of nanoparticles with cells. ROS induces many metabolic changes, such as mitochondrial failure, enhanced ketogenesis, fatty acid b-oxidation, and glycolysis, which results in hepatotoxicity, nephrotoxicity, and pulmonary toxicity (de Jong, 2008; Lei et al., 2008). This chapter provides new insights into the mechanisms by which these commonly employed metal-based nanoparticles disturb bacterial QS-based communication and clearly evidences the potential risk of releasing nanoparticles to the environment, especially for microbial communities, which play a key role in many environmental and technological processes. A high concentration of these metal-based nanoparticles alone is able to strongly inhibit growth and biofilm formation. The application of nanoparticles should be accompanied by the knowledge of possible risks for direct or indirect effects on the various sectors of the environment.

6.8 CONCLUSION AND PERSPECTIVE FOR THE FUTURE

There has been recent development of functionalized nanomaterials with potential applications in various healthcare domains. For covering the nanoparticle (NP) core surface with a capping agent, the surface can be functionalized using either a covalent modification strategy via a standard organic synthesis procedure or using an adsorption process (Chandra Kanth et al., 2020; Darwish & Mohammadi, 2018). The biofabrication of NPs using genetically altered organisms offers several advantages, including environmental safety, economic effectiveness, and NP manufacturing practicality. To enhance NP synthesis, issues like reproducibility, NP purity and separation, and recombinant strain survival must still be addressed. The capacity of nanoparticles to suppress QS suggests that they have a high potential for treating illnesses caused by bacterial biofilms. The molecular mechanism of QQ action has just recently begun to be unravelled, and the specifics of the multistep QS signal transduction disruption by nanoparticles are currently being researched. Metal nanoparticles' antibacterial activity, as well as their antibiofilm potential, has been widely reported. Owing to their unique physicochemical features and high surface-area-to-volume ratio delivery, nano-factories have emerged as a novel class of antimicrobial agents and antimicrobial delivery vehicles in recent years. As a result, the potential to disrupt QS networks using nanofactories might have real-world significance. The limited use of NPs is certainly related to their lack of specificity in comparison to traditional antibiotics. A complete understanding of the mechanism of QQ action and consequently NPs' interference with bacterial virulence will be extremely useful for therapeutic applications. Before extensive clinical usage of nanoparticles may be approved, more comprehensive research is needed to understand their favourable and their harmful impacts. Predictions suggest that there will be a significant advancement in the field of nanobiotechnology, with many noteworthy, fascinating, and trailblazing discoveries. To achieve safer and more sustainable fabrication of nanoparticles with the aid of existing green alternatives, we must consider some aspects in the near future (Md. A. Ali et al., 2020). These include:

- (i) opting for metals with low toxicity such as zinc, iron, and magnesium to carry out tests on plants and microbes
- (ii) determining MIC in vitro and in vivo
- (iii) paying attention to direct and indirect effect of utilizing NPs on the overall growth of plants, and
- (iv) paying attention to direct and indirect effect of utilizing NP on different aspects of the environment.

REFERENCES

AbdelRahim, K., Mahmoud, S. Y., Ali, A. M., Almaary, K. S., Mustafa, A. E.-Z. M. A., & Husseiny, S. M. (2017). Extracellular biosynthesis of silver nanoparticles using Rhizopus stolonifer. *Saudi Journal of Biological Sciences*, *24*(1), 208–216. https://doi.org/10.1016/j.sjbs.2016.02.025

Abisado, R. G., Benomar, S., Klaus, J. R., Dandekar, A. A., & Chandler, J. R. (2018). Bacterial quorum sensing and microbial community interactions. *MBio*, *9*(3), e02331–17. https://doi.org/10.1128/mBio.02331-17

Agnihotri, M., Joshi, S., Kumar, A. R., Zinjarde, S., & Kulkarni, S. (2009). Biosynthesis of gold nanoparticles by the tropical marine yeast Yarrowia lipolytica NCIM 3589. *Materials Letters*, *63*(15), 1231–1234. https://doi.org/10.1016/j.matlet.2009.02.042

Ahluwalia, V., Kumar, J., Sisodia, R., Shakil, N. A., & Walia, S. (2014). Green synthesis of silver nanoparticles by Trichoderma harzianum and their bio-efficacy evaluation against Staphylococcus aureus and Klebsiella pneumonia. *Industrial Crops and Products*, *55*, 202–206. https://doi.org/10.1016/j.indcrop.2014.01.026

Alghuthaymi, M. A., Almoammar, H., Rai, M., Said-Galiev, E., & Abd-Elsalam, K. A. (2015). Myconanoparticles: Synthesis and their role in phytopathogens management. *Biotechnology & Biotechnological Equipment*, *29*(2), 221–236. https://doi.org/10.1080/13102818.2015.1008194

Ali, M. A., Ahmed, T., Wu, W., Hossain, A., Hafeez, R., Islam Masum, Md. M., Wang, Y., An, Q., Sun, G., & Li, B. (2020). Advancements in plant and microbe-based synthesis of metallic nanoparticles and their antimicrobial activity against plant pathogens. *Nanomaterials*, *10*(6), 1146. https://doi.org/10.3390/nano10061146

Ali, S. G., Ansari, M. A., Khan, H. M., Jalal, M., Mahdi, A. A., & Cameotra, S. S. (2017). *Crataeva nurvala* nanoparticles inhibit virulence factors and biofilm formation in clinical isolates of *Pseudomonas aeruginosa*: Nanoparticles as an antivirulent. *Journal of Basic Microbiology*, *57*(3), 193–203. https://doi.org/10.1002/jobm.201600175

Ali, S. G., Ansari, M. A., Sajid Jamal, Q. Mohd., Khan, H. M., Jalal, M., Ahmad, H., & Mahdi, A. A. (2017). Antiquorum sensing activity of silver nanoparticles in P. aeruginosa: An in silico study. *In Silico Pharmacology*, *5*(1), 12. https://doi.org/10.1007/s40203-017-0031-3

Al-Shabib, N. A., Husain, F. M., Ahmed, F., Khan, R. A., Ahmad, I., Alsharaeh, E., Khan, M. S., Hussain, A., Rehman, M. T., Yusuf, M., Hassan, I., Khan, J. M., Ashraf, G. M., Alsalme, A., Al-Ajmi, M. F., Tarasov, V. V., & Aliev, G. (2016). Biogenic synthesis of Zinc oxide nanostructures from Nigella sativa seed: Prospective role as food packaging material inhibiting broad-spectrum quorum sensing and biofilm. *Scientific Reports*, *6*(1), 36761. https://doi.org/10.1038/srep36761

Anju, S., & Sarada, J. (n.d.). Quorum sensing inhibiting activity of silver nanoparticles synthesized by Bacillus isolate. *International Journal of Pharmacy and Biological Sciences*, *8*.

Apte, M., Girme, G., Bankar, A., RaviKumar, A., & Zinjarde, S. (2013). 3, 4-dihydroxy-L-phenylalanine-derived melanin from Yarrowia lipolytica mediates the synthesis of silver and gold nanostructures. *Journal of Nanobiotechnology*, *11*(1), 2. https://doi.org/10.1186/1477-3155-11-2

Apte, M., Sambre, D., Gaikawad, S., Joshi, S., Bankar, A., Kumar, A. R., & Zinjarde, S. (2013). Psychrotrophic yeast Yarrowia lipolytica NCYC 789 mediates the synthesis of antimicrobial silver nanoparticles via cell-associated melanin. *AMB Express*, *3*(1), 32. https://doi.org/10.1186/2191-0855-3-32

Azmath, P., Baker, S., Rakshith, D., & Satish, S. (2016). Mycosynthesis of silver nanoparticles bearing antibacterial activity. *Saudi Pharmaceutical Journal*, *24*(2), 140–146. https://doi.org/10.1016/j.jsps.2015.01.008

Bai, F., Han, Y., Chen, J., & Zhang, X.-H. (2008). Disruption of quorum sensing in Vibrio harveyi by the AiiA protein of Bacillus thuringiensis. *Aquaculture*, *274*(1), 36–40. https://doi.org/10.1016/j.aquaculture.2007.11.024

Banerjee, A. (2011). The design, fabrication, and photocatalytic utility of nanostructured semiconductors: Focus on TiO2-based nanostructures. *Nanotechnology, Science and Applications*, *35*. https://doi.org/10.2147/NSA.S9040

Bankier, C., Matharu, R. K., Cheong, Y. K., Ren, G. G., Cloutman-Green, E., & Ciric, L. (2019). Synergistic antibacterial effects of metallic nanoparticle combinations. *Scientific Reports*, *9*(1), 16074. https://doi.org/10.1038/s41598-019-52473-2

Beveridge, T. J., & Murray, R. G. (1980). Sites of metal deposition in the cell wall of Bacillus subtilis. *Journal of Bacteriology*, *141*(2), 876–887. https://doi.org/10.1128/jb.141.2.876-887.1980

Bindhu, M. R., & Umadevi, M. (2014). Antibacterial activities of green synthesized gold nanoparticles. *Materials Letters*, *120*, 122–125. https://doi.org/10.1016/j.matlet.2014.01.108

Birla, S. S., Gaikwad, S. C., Gade, A. K., & Rai, M. K. (2013). Rapid synthesis of silver nanoparticles from *Fusarium oxysporum* by optimizing physicocultural conditions. *The Scientific World Journal*, *2013*, 1–12. https://doi.org/10.1155/2013/796018

Blair, J. M. A., Webber, M. A., Baylay, A. J., Ogbolu, D. O., & Piddock, L. J. V. (2015). Molecular mechanisms of antibiotic resistance. *Nature Reviews Microbiology*, *13*(1), 42–51. https://doi.org/10.1038/nrmicro3380

Blecher, K., Nasir, A., & Friedman, A. (2011). The growing role of nanotechnology in combating infectious disease. *Virulence*, *2*(5), 395–401. https://doi.org/10.4161/viru.2.5.17035

Brackman, G., & Coenye, T. (2014). Quorum sensing inhibitors as anti-biofilm agents. *Current Pharmaceutical Design*, *21*(1), 5–11. https://doi.org/10.2174/13816128206 66140905114627

Brandt, O., Mildner, M., Egger, A. E., Groessl, M., Rix, U., Posch, M., Keppler, B. K., Strupp, C., Mueller, B., & Stingl, G. (2012). Nanoscalic silver possesses broad-spectrum antimicrobial activities and exhibits fewer toxicological side effects than silver sulfadiazine. *Nanomedicine: Nanotechnology, Biology and Medicine*, *8*(4), 478–488. https://doi.org/10.1016/j.nano.2011.07.005

Bridges, K., Kidson, A., Lowbury, E. J., & Wilkins, M. D. (1979). Gentamicin- and silver-resistant pseudomonas in a burns unit. *BMJ*, *1*(6161), 446–449. https://doi.org/10.1136/bmj.1.6161.446

Burmølle, M., Thomsen, T. R., Fazli, M., Dige, I., Christensen, L., Homøe, P., Tvede, M., Nyvad, B., Tolker-Nielsen, T., Givskov, M., Moser, C., Kirketerp-Møller, K., Johansen, H. K., Høiby, N., Jensen, P. Ø., Sørensen, S. J., & Bjarnsholt, T. (2010). Biofilms in chronic infections – a matter of opportunity – monospecies biofilms in multispecies infections. *FEMS Immunology & Medical Microbiology*, *59*(3), 324–336. https://doi.org/10.1111/j.1574-695X.2010.00714.x

Busi, S., & Rajkumari, J. (2019). Microbially synthesized nanoparticles as next generation antimicrobials: Scope and applications. In *Nanoparticles in pharmacotherapy* (pp. 485–524). Elsevier. https://doi.org/10.1016/B978-0-12-816504-1.00008-9

Buszewski, B., Railean-Plugaru, V., Pomastowski, P., Rafińska, K., Szultka-Mlynska, M., Golinska, P., Wypij, M., Laskowski, D., & Dahm, H. (2018). Antimicrobial activity of biosilver nanoparticles produced by a novel Streptacidiphilus durhamensis strain. *Journal of Microbiology, Immunology and Infection*, *51*(1), 45–54. https://doi.org/10.1016/j.jmii.2016.03.002

Carradori, S., Di Giacomo, N., Lobefalo, M., Luisi, G., Campestre, C., & Sisto, F. (2020). Biofilm and quorum sensing inhibitors: The road so far. *Expert Opinion on Therapeutic Patents*, *30*(12), 917–930. https://doi.org/10.1080/13543776.2020.1830059

Castro, L., Blázquez, M. L., Muñoz, J. ángel, González, F. G., & Ballester, A. (2014). Mechanism and applications of metal nanoparticles prepared by bio-mediated process. *Reviews in Advanced Sciences and Engineering*, *3*(3), 199–216. https://doi.org/10.1166/rase.2014.1064

Castro-Longoria, E., Vilchis-Nestor, A. R., & Avalos-Borja, M. (2011). Biosynthesis of silver, gold and bimetallic nanoparticles using the filamentous fungus Neurospora crassa. *Colloids and Surfaces B: Biointerfaces*, *83*(1), 42–48. https://doi.org/10.1016/j.colsurfb.2010.10.035

Chandra Kanth, P., Verma, S. K., & Gour, N. (2020). Functionalized nanomaterials for biomedical and agriculture industries. In *Handbook of functionalized nanomaterials for industrial applications* (pp. 231–265). Elsevier. https://doi.org/10.1016/B978-0-12-816787-8.00010-7

Chaudhari, A. A., Jasper, S. L., Dosunmu, E., Miller, M. E., Arnold, R. D., Singh, S. R., & Pillai, S. (2015). Novel pegylated silver coated carbon nanotubes kill Salmonella but they are non-toxic to eukaryotic cells. *Journal of Nanobiotechnology*, *13*(1), 23. https://doi.org/10.1186/s12951-015-0085-5

Chen, G., Swem, L. R., Swem, D. L., Stauff, D. L., O'Loughlin, C. T., Jeffrey, P. D., Bassler, B. L., & Hughson, F. M. (2011). A strategy for antagonizing quorum sensing. *Molecular Cell*, *42*(2), 199–209. https://doi.org/10.1016/j.molcel.2011.04.003

Chen, H., Fujita, M., Feng, Q., Clardy, J., & Fink, G. R. (2004). Tyrosol is a quorum-sensing molecule in *Candida albicans. Proceedings of the National Academy of Sciences*, *101*(14), 5048–5052. https://doi.org/10.1073/pnas.0401416101

Chen, X., & Schluesener, H. J. (2008). Nanosilver: A nanoproduct in medical application. *Toxicology Letters*, *176*(1), 1–12. https://doi.org/10.1016/j.toxlet.2007.10.004

Chu, Z., Huang, Y., Li, L., Tao, Q., & Li, Q. (2012). Physiological pathway of human cell damage induced by genotoxic crystalline silica nanoparticles. *Biomaterials*, *33*(30), 7540–7546. https://doi.org/10.1016/j.biomaterials.2012.06.073

Chu, Z., Huang, Y., Tao, Q., & Li, Q. (2011). Cellular uptake, evolution, and excretion of silica nanoparticles in human cells. *Nanoscale*, *3*(8), 3291. https://doi.org/10.1039/c1nr10499c

Correa-Llantén, D. N., Muñoz-Ibacache, S. A., Castro, M. E., Muñoz, P. A., & Blamey, J. M. (2013). Gold nanoparticles synthesized by Geobacillus sp. Strain ID17 a thermophilic bacterium isolated from Deception Island, Antarctica. *Microbial Cell Factories*, *12*(1), 75. https://doi.org/10.1186/1475-2859-12-75

Costerton, J. W., Stewart, P. S., & Greenberg, E. P. (1999). Bacterial biofilms: A common Cause of persistent infections. *Science*, *284*(5418), 1318–1322. https://doi.org/10.1126/science.284.5418.1318

Cunha, F. A., Cunha, M. D. C. S. O., da Frota, S. M., Mallmann, E. J. J., Freire, T. M., Costa, L. S., Paula, A. J., Menezes, E. A., & Fechine, P. B. A. (2018). Biogenic synthesis of multifunctional silver nanoparticles from Rhodotorula glutinis and Rhodotorula mucilaginosa: Antifungal, catalytic and cytotoxicity activities. *World Journal of Microbiology and Biotechnology*, *34*(9), 127. https://doi.org/10.1007/s11274-018-2514-8

Darwish, M., & Mohammadi, A. (2018). Functionalized nanomaterial for environmental techniques. In C. M. Hussain, & A. Kumar Mishra (eds.), *Nanotechnology in environmental science* (p. 36). Wiley.

da Silva Ferreira, V., ConzFerreira, M. E., Lima, L. M. T. R., Frasés, S., de Souza, W., & Sant'Anna, C. (2017). Green production of microalgae-based silver chloride nanoparticles with antimicrobial activity against pathogenic bacteria. *Enzyme and Microbial Technology*, *97*, 114–121. https://doi.org/10.1016/j.enzmictec.2016.10.018

Davies, D. G., Parsek, M. R., Pearson, J. P., Iglewski, B. H., Costerton, J. W., & Greenberg, E. P. (1998). The involvement of cell-to-cell signals in the development of a bacterial biofilm. *Science*, *280*(5361), 295–298. https://doi.org/10.1126/science.280.5361.295

de Jong. (2008). Drug delivery and nanoparticles: Applications and hazards. *International Journal of Nanomedicine*, *133*. https://doi.org/10.2147/IJN.S596

de Kievit, T. R., & Iglewski, B. H. (2000). Bacterial quorum sensing in pathogenic relationships. *Infection and Immunity*, *68*(9), 4839–4849. https://doi.org/10.1128/IAI.68.9.4839-4849.2000

de la Fuente-Núñez, C., Reffuveille, F., Fernández, L., & Hancock, R. E. (2013). Bacterial biofilm development as a multicellular adaptation: Antibiotic resistance and new therapeutic strategies. *Current Opinion in Microbiology*, *16*(5), 580–589. https://doi.org/10.1016/j.mib.2013.06.013

De la Torre Roche, R., Servin, A., Hawthorne, J., Xing, B., Newman, L. A., Ma, X., Chen, G., & White, J. C. (2015). Terrestrial trophic transfer of bulk and nanoparticle La2O3 does not depend on particle size. *Environmental Science & Technology*, *49*(19), 11866–11874. https://doi.org/10.1021/acs.est.5b02583

Du, L., Xian, L., & Feng, J.-X. (2011). Rapid extra-/intracellular biosynthesis of gold nanoparticles by the fungus Penicillium sp. *Journal of Nanoparticle Research*, *13*(3), 921–930. https://doi.org/10.1007/s11051-010-0165-2

Elahian, F., Reiisi, S., Shahidi, A., & Mirzaei, S. A. (2017). High-throughput bioaccumulation, biotransformation, and production of silver and selenium nanoparticles using genetically engineered Pichia pastoris. *Nanomedicine: Nanotechnology, Biology and Medicine*, *13*(3), 853–861. https://doi.org/10.1016/j.nano.2016.10.009

Elechiguerra, J. L., Burt, J. L., Morones, J. R., Camacho-Bragado, A., Gao, X., Lara, H. H., & Yacaman, M. J. (2005). Interaction of silver nanoparticles with HIV-1. *Journal of Nanobiotechnology*, *3*(1), 6. https://doi.org/10.1186/1477-3155-3-6

El-Sheekh, M. M., & El-Kassas, H. Y. (2016). Algal production of nano-silver and gold: Their antimicrobial and cytotoxic activities: A review. *Journal of Genetic Engineering and Biotechnology*, *14*(2), 299–310. https://doi.org/10.1016/j.jgeb.2016.09.008

Erasmus, M., Cason, E. D., van Marwijk, J., Botes, E., Gericke, M., & van Heerden, E. (2014). Gold nanoparticle synthesis using the thermophilic bacterium Thermus scotoductus SA-01 and the purification and characterization of its unusual gold reducing protein. *Gold Bulletin*, *47*(4), 245–253. https://doi.org/10.1007/s13404-014-0147-8

Fan, X. Z., Pomerantseva, E., Gnerlich, M., Brown, A., Gerasopoulos, K., McCarthy, M., Culver, J., & Ghodssi, R. (2013). *Tobacco mosaic virus*: A biological building block for micro/nano/bio systems. *Journal of Vacuum Science & Technology A: Vacuum, Surfaces, and Films*, *31*(5), 050815. https://doi.org/10.1116/1.4816584

Fariq, A., Khan, T., & Yasmin, A. (2017). Microbial synthesis of nanoparticles and their potential applications in biomedicine. *Journal of Applied Biomedicine*, *15*(4), 241–248. https://doi.org/10.1016/j.jab.2017.03.004

Furukawa, S., Kuchma, S. L., & O'Toole, G. A. (2006). Keeping their options open: Acute versus persistent infections. *Journal of Bacteriology*, *188*(4), 1211–1217. https://doi.org/10.1128/JB.188.4.1211-1217.2006

Gahlawat, G., & Choudhury, A. R. (2019). A review on the biosynthesis of metal and metal salt nanoparticles by microbes. *RSC Advances*, *9*(23), 12944–12967. https://doi.org/10.1039/C8RA10483B

Gardea-Torresdey, J. L., Rico, C. M., & White, J. C. (2014). Trophic transfer, transformation, and impact of engineered nanomaterials in terrestrial environments. *Environmental Science & Technology*, *48*(5), 2526–2540. https://doi.org/10.1021/es4050665

Garole, V. J., Choudhary, B. C., Tetgure, S. R., Garole, D. J., & Borse, A. U. (2019). Palladium nanocatalyst: Green synthesis, characterization, and catalytic application. *International Journal of Environmental Science and Technology*, *16*(12), 7885–7892. https://doi.org/10.1007/s13762-018-2173-1

George, D., Maheswari, P. U., & Begum, K. M. M. S. (2020). Chitosan-cellulose hydrogel conjugated with L-histidine and zinc oxide nanoparticles for sustained drug delivery: Kinetics and in-vitro biological studies. *Carbohydrate Polymers*, *236*, 116101. https://doi.org/10.1016/j.carbpol.2020.116101

Ghosh, S., Nag, M., Lahiri, D., Ghosh, S., Dey, A., & Ray, R. R. (2022). Biogenic nanoparticles from cyanobacteria and their applications. In *Agricultural nanobiotechnology* (pp. 79–103). Elsevier. https://doi.org/10.1016/B978-0-323-91908-1.00003-1

Giasuddin, A., Jhuma, K., & Haq, A. M. (2013). Use of gold nanoparticles in diagnostics, surgery and medicine: A review. *Bangladesh Journal of Medical Biochemistry*, *5*(2), 56–60. https://doi.org/10.3329/bjmb.v5i2.13346

Gómez-Gómez, B., Arregui, L., Serrano, S., Santos, A., Pérez-Corona, T., & Madrid, Y. (2019). Unravelling mechanisms of bacterial quorum sensing disruption by metal-based nanoparticles. *Science of The Total Environment*, *696*, 133869. https://doi.org/10.1016/j.scitotenv.2019.133869

Grasso, G., Zane, D., & Dragone, R. (2019). Microbial nanotechnology: Challenges and prospects for green biocatalytic synthesis of nanoscale materials for sensoristic and biomedical applications. *Nanomaterials*, *10*(1), 11. https://doi.org/10.3390/nano10010011

Guilger-Casagrande, M., & Lima, R. de. (2019). Synthesis of silver nanoparticles mediated by fungi: A review. *Frontiers in Bioengineering and Biotechnology, 7*, 287. https://doi.org/10.3389/fbioe.2019.00287

Gurunathan, S., Kalishwaralal, K., Vaidyanathan, R., Venkataraman, D., Pandian, S. R. K., Muniyandi, J., Hariharan, N., & Eom, S. H. (2009). Biosynthesis, purification and characterization of silver nanoparticles using Escherichia coli. *Colloids and Surfaces B: Biointerfaces, 74*(1), 328–335. https://doi.org/10.1016/j.colsurfb.2009.07.048

Haefeli, C., Franklin, C., & Hardy, K. (1984). Plasmid-determined silver resistance in Pseudomonas stutzeri isolated from a silver mine. *Journal of Bacteriology, 158*(1), 389–392. https://doi.org/10.1128/jb.158.1.389-392.1984

Hajipour, M. J., Fromm, K. M., Akbar Ashkarran, A., Jimenez de Aberasturi, D., Larramendi, I. R. de, Rojo, T., Serpooshan, V., Parak, W. J., & Mahmoudi, M. (2012). Antibacterial properties of nanoparticles. *Trends in Biotechnology, 30*(10), 499–511. https://doi.org/10.1016/j.tibtech.2012.06.004

Hasan, S. (2015). A review on nanoparticles: Their synthesis and types. *Research Journal of Recent Sciences, 4*, 1–3.

Hassan, S. E.-D., Salem, S. S., Fouda, A., Awad, M. A., El-Gamal, M. S., & Abdo, A. M. (2018). New approach for antimicrobial activity and bio-control of various pathogens by biosynthesized copper nanoparticles using endophytic actinomycetes. *Journal of Radiation Research and Applied Sciences, 11*(3), 262–270. https://doi.org/10.1016/j.jrras.2018.05.003

He, S., Guo, Z., Zhang, Y., Zhang, S., Wang, J., & Gu, N. (2007). Biosynthesis of gold nanoparticles using the bacteria Rhodopseudomonas capsulata. *Materials Letters, 61*(18), 3984–3987. https://doi.org/10.1016/j.matlet.2007.01.018

Hornby, J. M., & Nickerson, K. W. (2004). Enhanced production of farnesol by *Candida albicans* treated with four azoles. *Antimicrobial Agents and Chemotherapy, 48*(6), 2305–2307. https://doi.org/10.1128/AAC.48.6.2305-2307.2004

Hulkoti, N. I., & Taranath, T. C. (2014). Biosynthesis of nanoparticles using microbes—a review. *Colloids and Surfaces B: Biointerfaces, 121*, 474–483. https://doi.org/10.1016/j.colsurfb.2014.05.027

Husain, S., Sardar, M., & Fatma, T. (2015). Screening of cyanobacterial extracts for synthesis of silver nanoparticles. *World Journal of Microbiology and Biotechnology, 31*(8), 1279–1283. https://doi.org/10.1007/s11274-015-1869-3

Husseiny, S. M., Salah, T. A., & Anter, H. A. (2015). Biosynthesis of size controlled silver nanoparticles by Fusarium oxysporum, their antibacterial and antitumor activities. *Beni-Suef University Journal of Basic and Applied Sciences, 4*(3), 225–231. https://doi.org/10.1016/j.bjbas.2015.07.004

Ilk, S., Sağlam, N., Özgen, M., & Korkusuz, F. (2017). Chitosan nanoparticles enhances the anti-quorum sensing activity of kaempferol. *International Journal of Biological Macromolecules, 94*, 653–662. https://doi.org/10.1016/j.ijbiomac.2016.10.068

Jackson, D. W., Suzuki, K., Oakford, L., Simecka, J. W., Hart, M. E., & Romeo, T. (2002). Biofilm formation and dispersal under the influence of the global regulator CsrA of *Escherichia coli. Journal of Bacteriology, 184*(1), 290–301. https://doi.org/10.1128/JB.184.1.290-301.2002

Jagtap, S., & Priolkar, K. R. (2013). Evaluation of ZnO nanoparticles and study of ZnO–TiO2 composites for lead free humidity sensors. *Sensors and Actuators B: Chemical, 183*, 411–418. https://doi.org/10.1016/j.snb.2013.04.010

Jain, S., Hirst, D. G., & O'Sullivan, J. M. (2012). Gold nanoparticles as novel agents for cancer therapy. *The British Journal of Radiology, 85*(1010), 101–113. https://doi.org/10.1259/bjr/59448833

Jang, H. D., Kim, S. K., Chang, H., Choi, J.-H., Cho, B.-G., Jo, E. H., Choi, J.-W., & Huang, J. (2015). Three-dimensional crumpled graphene-based platinum–gold alloy nanoparticle composites as superior electrocatalysts for direct methanol fuel cells. *Carbon*, *93*, 869–877. https://doi.org/10.1016/j.carbon.2015.06.009

Jha, A. K., Prasad, K., & Kulkarni, A. R. (2009). Synthesis of TiO2 nanoparticles using microorganisms. *Colloids and Surfaces B: Biointerfaces*, *71*(2), 226–229. https://doi.org/10.1016/j.colsurfb.2009.02.007

Kale, A., Bao, Y., Zhou, Z., Prevelige, P. E., & Gupta, A. (2013). Directed self-assembly of CdS quantum dots on bacteriophage P22 coat protein templates. *Nanotechnology*, *24*(4), 045603. https://doi.org/10.1088/0957-4484/24/4/045603

Kalpana, D., & Lee, Y. S. (2013). Synthesis and characterization of bactericidal silver nanoparticles using cultural filtrate of simulated microgravity grown Klebsiella pneumoniae. *Enzyme and Microbial Technology*, *52*(3), 151–156. https://doi.org/10.1016/j.enzmictec.2012.12.006

Karthik, L., Kumar, G., Kirthi, A. V., Rahuman, A. A., & Bhaskara Rao, K. V. (2014). Streptomyces sp. LK3 mediated synthesis of silver nanoparticles and its biomedical application. *Bioprocess and Biosystems Engineering*, *37*(2), 261–267. https://doi.org/10.1007/s00449-013-0994-3

Kaufmann, G. F., Sartorio, R., Lee, S.-H., Rogers, C. J., Meijler, M. M., Moss, J. A., Clapham, B., Brogan, A. P., Dickerson, T. J., & Janda, K. D. (2005). Revisiting quorum sensing: Discovery of additional chemical and biological functions for 3-oxo-*N*-acylhomoserine lactones. *Proceedings of the National Academy of Sciences*, *102*(2), 309–314. https://doi.org/10.1073/pnas.0408639102

Kaya, M., Baran, T., Asan-Ozusaglam, M., Cakmak, Y. S., Tozak, K. O., Mol, A., Mentes, A., & Sezen, G. (2015). Extraction and characterization of chitin and chitosan with antimicrobial and antioxidant activities from cosmopolitan Orthoptera species (Insecta). *Biotechnology and Bioprocess Engineering*, *20*(1), 168–179. https://doi.org/10.1007/s12257-014-0391-z

Khalafi, T., Buazar, F., & Ghanemi, K. (2019). Phycosynthesis and enhanced photocatalytic activity of zinc oxide nanoparticles toward organosulfur pollutants. *Scientific Reports*, *9*(1), 6866. https://doi.org/10.1038/s41598-019-43368-3

Khan, M. F., Husain, F. M., Zia, Q., Ahmad, E., Jamal, A., Alaidarous, M., Banawas, S., Alam, Md. M., Alshehri, B. A., Jameel, M., Alam, P., Ahamed, M. I., Ansari, A. H., & Ahmad, I. (2020). Anti-quorum sensing and anti-biofilm activity of zinc oxide nanospikes. *ACS Omega*, *5*(50), 32203–32215. https://doi.org/10.1021/acsomega.0c03634

Kim, J. I., Park, H.-G., Chang, K.-H., Nam, D. H., & Yeo, M.-K. (2016). Trophic transfer of nano-TiO2 in a paddy microcosm: A comparison of single-dose versus sequential multi-dose exposures. *Environmental Pollution*, *212*, 316–324. https://doi.org/10.1016/j.envpol.2016.01.076

Kim, J. S., Kuk, E., Yu, K. N., Kim, J.-H., Park, S. J., Lee, H. J., Kim, S. H., Park, Y. K., Park, Y. H., Hwang, C.-Y., Kim, Y.-K., Lee, Y.-S., Jeong, D. H., & Cho, M.-H. (2007). Antimicrobial effects of silver nanoparticles. *Nanomedicine: Nanotechnology, Biology and Medicine*, *3*(1), 95–101. https://doi.org/10.1016/j.nano.2006.12.001

Kitching, M., Ramani, M., & Marsili, E. (2015). Fungal biosynthesis of gold nanoparticles: Mechanism and scale up. *Microbial Biotechnology*, *8*(6), 904–917. https://doi.org/10.1111/1751-7915.12151

Klaus, T., Joerger, R., Olsson, E., & Granqvist, C.-G. (1999). Silver-based crystalline nanoparticles, microbially fabricated. *Proceedings of the National Academy of Sciences*, *96*(24), 13611–13614. https://doi.org/10.1073/pnas.96.24.13611

Kobayashi, M., Tomita, S., Sawada, K., Shiba, K., Yanagi, H., Yamashita, I., & Uraoka, Y. (2012). Chiral meta-molecules consisting of gold nanoparticles and genetically engineered tobacco mosaic virus. *Optics Express*, *20*(22), 24856. https://doi.org/10.1364/OE.20.024856

Koch, B., Liljefors, T., Persson, T., Nielsen, J., Kjelleberg, S., & Givskov, M. (2005). The LuxR receptor: The sites of interaction with quorum-sensing signals and inhibitors. *Microbiology*, *151*(11), 3589–3602. https://doi.org/10.1099/mic.0.27954-0

Korbekandi, H., Mohseni, S., Mardani Jouneghani, R., Pourhossein, M., & Iravani, S. (2016). Biosynthesis of silver nanoparticles using *Saccharomyces cerevisiae*. *Artificial Cells, Nanomedicine, and Biotechnology*, *44*(1), 235–239. https://doi.org/10.3109/21691401.2014.937870

Koul, B., Poonia, A. K., Yadav, D., & Jin, J.-O. (2021). Microbe-mediated biosynthesis of nanoparticles: Applications and future prospects. *Biomolecules*, *11*(6), 886. https://doi.org/10.3390/biom11060886

Lade, H., Paul, D., & Kweon, J. H. (2014). Quorum quenching mediated approaches for control of membrane biofouling. *International Journal of Biological Sciences*, *10*(5), 550–565. https://doi.org/10.7150/ijbs.9028

Lahiri, D., Nag, M., Sheikh, H. I., Sarkar, T., Edinur, H. A., Pati, S., & Ray, R. R. (2021). Microbiologically-synthesized nanoparticles and their role in silencing the biofilm signaling cascade. *Frontiers in Microbiology*, *12*, 636588. https://doi.org/10.3389/fmicb.2021.636588

Lara, H. H., Garza-Treviño, E. N., Ixtepan-Turrent, L., & Singh, D. K. (2011). Silver nanoparticles are broad-spectrum bactericidal and virucidal compounds. *Journal of Nanobiotechnology*, *9*(1), 30. https://doi.org/10.1186/1477-3155-9-30

Le, D. H. T., Lee, K. L., Shukla, S., Commandeur, U., & Steinmetz, N. F. (2017). Potato virus X, a filamentous plant viral nanoparticle for doxorubicin delivery in cancer therapy. *Nanoscale*, *9*(6), 2348–2357. https://doi.org/10.1039/C6NR09099K

Lee, J., & Zhang, L. (2015). The hierarchy quorum sensing network in Pseudomonas aeruginosa. *Protein & Cell*, *6*(1), 26–41. https://doi.org/10.1007/s13238-014-0100-x

Lee, J.-H., Kim, Y.-G., Cho, M. H., & Lee, J. (2014). ZnO nanoparticles inhibit Pseudomonas aeruginosa biofilm formation and virulence factor production. *Microbiological Research*, *169*(12), 888–896. https://doi.org/10.1016/j.micres.2014.05.005

Lei, R., Wu, C., Yang, B., Ma, H., Shi, C., Wang, Q., Wang, Q., Yuan, Y., & Liao, M. (2008). Integrated metabolomic analysis of the nano-sized copper particle-induced hepatotoxicity and nephrotoxicity in rats: A rapid in vivo screening method for nanotoxicity. *Toxicology and Applied Pharmacology*, *232*(2), 292–301. https://doi.org/10.1016/j.taap.2008.06.026

Lenartowicz, M., Marek, P. H., Madura, I. D., & Lipok, J. (2017). Formation of variously shaped gold nanoparticles by anabaena laxa. *Journal of Cluster Science*, *28*(5), 3035–3055. https://doi.org/10.1007/s10876-017-1275-0

Lengke, M. F., Fleet, M. E., & Southam, G. (2007). Biosynthesis of silver nanoparticles by filamentous cyanobacteria from a silver(I) nitrate complex. *Langmuir*, *23*(5), 2694–2699. https://doi.org/10.1021/la0613124

Li, C., Zhou, L., Yang, H., Lv, R., Li, X., Zhang, Y., Chen, Z., & Lin, F. (2017.). Self-assembled exopolysaccharide nanoparticles for bioremediation and green synthesis of noble metal nanoparticles. *ACS Applied Materials & Interfaces*, *9*(27), 22808–22818. https://doi.org/10.1021/acsami.7b02908

Li, X., Xu, H., Chen, Z.-S., & Chen, G. (2011). Biosynthesis of nanoparticles by microorganisms and their applications. *Journal of Nanomaterials*, *2011*, 1–16. https://doi.org/10.1155/2011/270974

Lima, E., Guerra, R., Lara, V., & Guzmán, A. (2013). Gold nanoparticles as efficient antimicrobial agents for Escherichia coli and Salmonella typhi. *Chemistry Central Journal*, *7*(1), 11. https://doi.org/10.1186/1752-153X-7-11

Mahdavi, M., Namvar, F., Ahmad, M., & Mohamad, R. (2013). Green biosynthesis and characterization of magnetic iron oxide (Fe3O4) nanoparticles using seaweed (Sargassum muticum) aqueous extract. *Molecules*, *18*(5), 5954–5964. https://doi.org/10.3390/molecules18055954

Makarov, V. V., Love, A. J., Sinitsyna, O. V., Makarova, S. S., Yaminsky, I. V., Taliansky, M. E., & Kalinina, N. O. (2014). "Green" nanotechnologies: Synthesis of metal nanoparticles using plants. *Acta Naturae*, *6*(1), 35–44. https://doi.org/10.32607/20758251-2014-6-1-35-44

Malarkodi, C., Rajeshkumar, S., Paulkumar, K., Vanaja, M., Jobitha, G. D. G., & Annadurai, G. (2013). Bactericidal activity of bio mediated silver nanoparticles synthesized by Serratia nematodiphila. *Drug Invention Today*, *5*(2), 119–125. https://doi.org/10.1016/j.dit.2013.05.005

Malvindi, M. A., Brunetti, V., Vecchio, G., Galeone, A., Cingolani, R., & Pompa, P. P. (2012). SiO2 nanoparticles biocompatibility and their potential for gene delivery and silencing. *Nanoscale*, *4*(2), 486–495. https://doi.org/10.1039/C1NR11269D

Mandeep, & Shukla, P. (2020). Microbial nanotechnology for bioremediation of industrial wastewater. *Frontiers in Microbiology*, *11*, 590631. https://doi.org/10.3389/fmicb.2020.590631

Mao, C., Flynn, C. E., Hayhurst, A., Sweeney, R., Qi, J., Georgiou, G., Iverson, B., & Belcher, A. M. (2003). Viral assembly of oriented quantum dot nanowires. *Proceedings of the National Academy of Sciences*, *100*(12), 6946–6951. https://doi.org/10.1073/pnas.0832310100

Merin, D. D., Prakash, S., & Bhimba, B. V. (2010). Antibacterial screening of silver nanoparticles synthesized by marine micro algae. *Asian Pacific Journal of Tropical Medicine*, *3*(10), 797–799. https://doi.org/10.1016/S1995-7645(10)60191-5

Mesbahi, A. (2010). A review on gold nanoparticles radiosensitization effect in radiation therapy of cancer. *Reports of Practical Oncology & Radiotherapy*, *15*(6), 176–180. https://doi.org/10.1016/j.rpor.2010.09.001

Metuku, R. P., Pabba, S., Burra, S., Hima Bindu, N. S. V. S. S. S. L., Gudikandula, K., & Singara Charya, M. A. (2014). Biosynthesis of silver nanoparticles from Schizophyllum radiatum HE 863742.1: Their characterization and antimicrobial activity. *3 Biotech*, *4*(3), 227–234. https://doi.org/10.1007/s13205-013-0138-0

Metwalli, K. H., Khan, S. A., Krom, B. P., & Jabra-Rizk, M. A. (2013). Streptococcus mutans, Candida albicans, and the human mouth: A sticky situation. *PLoS Pathogens*, *9*(10), e1003616. https://doi.org/10.1371/journal.ppat.1003616

Miller, K. P., Wang, L., Chen, Y.-P., Pellechia, P. J., Benicewicz, B. C., & Decho, A. W. (2015). Engineering nanoparticles to silence bacterial communication. *Frontiers in Microbiology*, *6*. https://doi.org/10.3389/fmicb.2015.00189

Minardi, D., Cirioni, O., Ghiselli, R., Silvestri, C., Mocchegiani, F., Gabrielli, E., d'Anzeo, G., Conti, A., Orlando, F., Rimini, M., Brescini, L., Guerrieri, M., Giacometti, A., & Muzzonigro, G. (2012). Efficacy of tigecycline and rifampin alone and in combination against enterococcus faecalis biofilm infection in a rat model of ureteral stent. *Journal of Surgical Research*, *176*(1), 1–6. https://doi.org/10.1016/j.jss.2011.05.002

Mishra, A., Kumari, M., Pandey, S., Chaudhry, V., Gupta, K. C., & Nautiyal, C. S. (2014). Biocatalytic and antimicrobial activities of gold nanoparticles synthesized by Trichoderma sp. *Bioresource Technology*, *166*, 235–242. https://doi.org/10.1016/j.biortech.2014.04.085

Mohamed, A. A., Fouda, A., Abdel-Rahman, M. A., Hassan, S. E.-D., El-Gamal, M. S., Salem, S. S., & Shaheen, T. I. (2019). Fungal strain impacts the shape, bioactivity and multifunctional properties of green synthesized zinc oxide nanoparticles. *Biocatalysis and Agricultural Biotechnology*, *19*, 101103. https://doi.org/10.1016/j.bcab.2019.101103

Mu, H., Tang, J., Liu, Q., Sun, C., Wang, T., & Duan, J. (2016). Potent antibacterial nanoparticles against biofilm and intracellular bacteria. *Scientific Reports*, *6*(1), 18877. https://doi.org/10.1038/srep18877

Mubarak Ali, D., Sasikala, M., Gunasekaran, M., & Thajuddin, N. (2011). Biosynthesis and characterization of silver nanoparticles using marine cyanobacterium, Oscillatoria willei NTDM01. *Digest Journal of Nanomaterials and Biostructures, 6*(2), 385–390.

Nadaroğlu, H., Güngör, A. A., & Ince, S. (2017). Synthesis of nanoparticles by green synthesis method. *International Journal of Innovative Research and Reviews, 1*(1), 6–9.

Naik, K., & Kowshik, M. (2014). Anti-biofilm efficacy of low temperature processed AgCl–TiO2 nanocomposite coating. *Materials Science and Engineering: C, 34*, 62–68. https://doi.org/10.1016/j.msec.2013.10.008

Nair, B., & Pradeep, T. (2002). Coalescence of nanoclusters and formation of submicron crystallites assisted by *Lactobacillus* strains. *Crystal Growth & Design*, *2*(4), 293–298. https://doi.org/10.1021/cg0255164

Nealson, K. H., Platt, T., & Hastings, J. W. (1970). Cellular control of the synthesis and activity of the bacterial luminescent system. *Journal of Bacteriology*, *104*(1), 313–322. https://doi.org/10.1128/jb.104.1.313-322.1970

Ogunyemi, S. O., Zhang, M., Abdallah, Y., Ahmed, T., Qiu, W., Ali, Md. A., Yan, C., Yang, Y., Chen, J., & Li, B. (2020). The bio-synthesis of three metal oxide nanoparticles (ZnO, MnO2, and MgO) and Their antibacterial activity against the bacterial leaf blight pathogen. *Frontiers in Microbiology*, *11*, 588326. https://doi.org/10.3389/fmicb.2020.588326

Otari, S. V., Patil, R. M., Nadaf, N. H., Ghosh, S. J., & Pawar, S. H. (2012). Green biosynthesis of silver nanoparticles from an actinobacteria Rhodococcus sp. *Materials Letters*, *72*, 92–94. https://doi.org/10.1016/j.matlet.2011.12.109

Paluch, E., Rewak-Soroczyńska, J., Jędrusik, I., Mazurkiewicz, E., & Jermakow, K. (2020). Prevention of biofilm formation by quorum quenching. *Applied Microbiology and Biotechnology*, *104*(5), 1871–1881. https://doi.org/10.1007/s00253-020-10349-w

Parial, D., Patra, H. K., Dasgupta, A. K. R., & Pal, R. (2012). Screening of different algae for green synthesis of gold nanoparticles. *European Journal of Phycology*, *47*(1), 22–29. https://doi.org/10.1080/09670262.2011.653406

Patil, S., & Chandrasekaran, R. (2020). Biogenic nanoparticles: A comprehensive perspective in synthesis, characterization, application and its challenges. *Journal of Genetic Engineering and Biotechnology*, *18*(1), 67. https://doi.org/10.1186/s43141-020-00081-3

Perez-Gonzalez, T., Jimenez-Lopez, C., Neal, A. L., Rull-Perez, F., Rodriguez-Navarro, A., Fernandez-Vivas, A., & Iañez-Pareja, E. (2010). Magnetite biomineralization induced by Shewanella oneidensis. *Geochimica et Cosmochimica Acta*, *74*(3), 967–979. https://doi.org/10.1016/j.gca.2009.10.035

Prabhu, S., & Poulose, E. K. (2012). Silver nanoparticles: Mechanism of antimicrobial action, synthesis, medical applications, and toxicity effects. *International Nano Letters*, *2*(1), 32. https://doi.org/10.1186/2228-5326-2-32

Prakash, D., Nawani, N., Prakash, M., Bodas, M., Mandal, A., Khetmalas, M., & Kapadnis, B. (2013). Actinomycetes: A repertory of green catalysts with a potential revenue resource. *BioMed Research International*, *2013*, 1–8. https://doi.org/10.1155/2013/264020

Prasad, R. (Ed.). (2016). *Advances and applications through fungal nanobiotechnology*. Springer International Publishing. https://doi.org/10.1007/978-3-319-42990-8

Qais, F. A., Khan, M. S., & Ahmad, I. (2018). Nanoparticles as quorum sensing inhibitor: Prospects and limitations. In V. C. Kalia (Ed.), *Biotechnological applications of quorum sensing inhibitors* (pp. 227–244). Springer. https://doi.org/10.1007/978-981-10-9026-4_11

Qin, X., Engwer, C., Desai, S., Vila-Sanjurjo, C., & Goycoolea, F. M. (2017). An investigation of the interactions between an E. coli bacterial quorum sensing biosensor and chitosan-based nanocapsules. *Colloids and Surfaces B: Biointerfaces*, *149*, 358–368. https://doi.org/10.1016/j.colsurfb.2016.10.031

Radzig, M. A., Nadtochenko, V. A., Koksharova, O. A., Kiwi, J., Lipasova, V. A., & Khmel, I. A. (2013). Antibacterial effects of silver nanoparticles on gram-negative bacteria: Influence on the growth and biofilms formation, mechanisms of action. *Colloids and Surfaces B: Biointerfaces*, *102*, 300–306. https://doi.org/10.1016/j.colsurfb.2012.07.039

Raftery, T. D., Kerscher, P., Hart, A. E., Saville, S. L., Qi, B., Kitchens, C. L., Mefford, O. T., & McNealy, T. L. (2014). Discrete nanoparticles induce loss of *Legionella pneumophila* biofilms from surfaces. *Nanotoxicology*, *8*(5), 477–484. https://doi.org/10.3109/17435390.2013.796537

Raghunath, A., & Perumal, E. (2017). Metal oxide nanoparticles as antimicrobial agents: A promise for the future. *International Journal of Antimicrobial Agents*, *49*(2), 137–152. https://doi.org/10.1016/j.ijantimicag.2016.11.011

Rahman, A., Ismail, A., Jumbianti, D., Magdalena, S., & Sudrajat, H. (2010). Synthesis of copper oxide nano particles by using Phormidium cyanobacterium. *Indonesian Journal of Chemistry*, *9*(3), 355–360. https://doi.org/10.22146/ijc.21498

Rai, M., Gade, A., & Yadav, A. (2011). Biogenic nanoparticles: An introduction to what they are, how they are synthesized and their applications. In M. Rai & N. Duran (Eds.), *Metal nanoparticles in microbiology* (pp. 1–14). Springer. https://doi.org/10.1007/978-3-642-18312-6_1

Ramasamy, M., Lee, J.-H., & Lee, J. (2016). Potent antimicrobial and antibiofilm activities of bacteriogenically synthesized gold–silver nanoparticles against pathogenic bacteria and their physiochemical characterizations. *Journal of Biomaterials Applications*, *31*(3), 366–378. https://doi.org/10.1177/0885328216646910

Ramos, M. M., dos S. Morais, E., da S. Sena, I., Lima, A. L., de Oliveira, F. R., de Freitas, C. M., Fernandes, C. P., de Carvalho, J. C. T., & Ferreira, I. M. (2020). Silver nanoparticle from whole cells of the fungi Trichoderma spp. isolated from Brazilian Amazon. *Biotechnology Letters*, *42*(5), 833–843. https://doi.org/10.1007/s10529-020-02819-y

Rampioni, G., Leoni, L., & Williams, P. (2014). The art of antibacterial warfare: Deception through interference with quorum sensing–mediated communication. *Bioorganic Chemistry*, *55*, 60–68. https://doi.org/10.1016/j.bioorg.2014.04.005

Ranjitha, V. R., & Rai, V. R. (2017). Actinomycetes mediated synthesis of gold nanoparticles from the culture supernatant of Streptomyces griseoruber with special reference to catalytic activity. *3 Biotech*, *7*(5), 299. https://doi.org/10.1007/s13205-017-0930-3

Ravindran, D., Ramanathan, S., Arunachalam, K., Jeyaraj, G. P., Shunmugiah, K. P., & Arumugam, V. R. (2018). Phytosynthesized silver nanoparticles as antiquorum sensing and antibiofilm agent against the nosocomial pathogen *Serratia marcescens*: An *in vitro* study. *Journal of Applied Microbiology*, *124*(6), 1425–1440. https://doi.org/10.1111/jam.13728

Rehman, Z. U., & Leiknes, T. (2018). Quorum-quenching bacteria isolated from red sea sediments reduce biofilm formation by Pseudomonas aeruginosa. *Frontiers in Microbiology*, *9*, 1354. https://doi.org/10.3389/fmicb.2018.01354

Ren, G., Hu, D., Cheng, E. W. C., Vargas-Reus, M. A., Reip, P., & Allaker, R. P. (2009). Characterisation of copper oxide nanoparticles for antimicrobial applications. *International Journal of Antimicrobial Agents*, *33*(6), 587–590. https://doi.org/10.1016/j.ijantimicag.2008.12.004

Roux, A., Payne, S. M., & Gilmore, M. S. (2009). Microbial telesensing: Probing the environment for friends, foes, and food. *Cell Host & Microbe*, *6*(2), 115–124. https://doi.org/10.1016/j.chom.2009.07.004

Rutherford, S. T., & Bassler, B. L. (2012). Bacterial quorum sensing: Its role in virulence and possibilities for its control. *Cold Spring Harbor Perspectives in Medicine*, *2*(11), a012427. https://doi.org/10.1101/cshperspect.a012427

Sadekuzzaman, M., Yang, S., Mizan, M. F. R., & Ha, S. D. (2015). Current and recent advanced strategies for combating biofilms: Strategies for combating biofilm. *Comprehensive Reviews in Food Science and Food Safety*, *14*(4), 491–509. https://doi.org/10.1111/1541-4337.12144

Sadowski, Z. (2010). Biosynthesis and application of silver and gold nanoparticles. In D. Pozo (Ed.), *Silver nanoparticles*. InTech. https://doi.org/10.5772/8508

Saleh, M. M., Sadeq, R. A., Abdel Latif, H. K., Abbas, H. A., & Askoura, M. (2019). Zinc oxide nanoparticles inhibits quorum sensing and virulence in Pseudomonas aeruginosa. *African Health Sciences*, *19*(2), 2043. https://doi.org/10.4314/ahs.v19i2.28

Salem, S. S., & Fouda, A. (2021). Green synthesis of metallic nanoparticles and their prospective biotechnological applications: An overview. *Biological Trace Element Research*, *199*(1), 344–370. https://doi.org/10.1007/s12011-020-02138-3

Samanta, S., Singh, B. R., & Adholeya, A. (2017). Intracellular synthesis of gold nanoparticles using an ectomycorrhizal strain EM-1083 of laccaria fraterna and its nanoanti-quorum sensing potential against Pseudomonas aeruginosa. *Indian Journal of Microbiology*, *57*(4), 448–460. https://doi.org/10.1007/s12088-017-0662-4

Sanaeimehr, Z., Javadi, I., & Namvar, F. (2018). Antiangiogenic and antiapoptotic effects of green-synthesized zinc oxide nanoparticles using Sargassum muticum algae extraction. *Cancer Nanotechnology*, *9*(1), 3. https://doi.org/10.1186/s12645-018-0037-5

Saravanan, M., Barik, S. K., MubarakAli, D., Prakash, P., & Pugazhendhi, A. (2018). Synthesis of silver nanoparticles from Bacillus brevis (NCIM 2533) and their antibacterial activity against pathogenic bacteria. *Microbial Pathogenesis*, *116*, 221–226. https://doi.org/10.1016/j.micpath.2018.01.038

Sathyanarayanan, M. B., Balachandranath, R., Genji Srinivasulu, Y., Kannaiyan, S. K., & Subbiahdoss, G. (2013). The effect of gold and iron-oxide nanoparticles on biofilm-forming pathogens. *ISRN Microbiology*, *25*, 272086. https://doi.org/10.1155/2013/272086

Shah, M., Fawcett, D., Sharma, S., Tripathy, S., & Poinern, G. (2015). Green synthesis of metallic nanoparticles via biological entities. *Materials*, *8*(11), 7278–7308. https://doi.org/10.3390/ma8115377

Shah, M. S. A. S., Nag, M., Kalagara, T., Singh, S., & Manorama, S. V. (2008). Silver on PEG-PU-TiO2 polymer nanocomposite films: An excellent system for antibacterial applications. *Chemistry of Materials*, *20*(7), 2455–2460. https://doi.org/10.1021/cm7033867

Shakibaie, M., Forootanfar, H., Golkari, Y., Mohammadi-Khorsand, T., & Shakibaie, M. R. (2015). Anti-biofilm activity of biogenic selenium nanoparticles and selenium dioxide against clinical isolates of Staphylococcus aureus, Pseudomonas aeruginosa, and Proteus mirabilis. *Journal of Trace Elements in Medicine and Biology*, *29*, 235–241. https://doi.org/10.1016/j.jtemb.2014.07.020

Sharma, M., Behl, K., Nigam, S., & Joshi, M. (2018). TiO2-GO nanocomposite for photocatalysis and environmental applications: A green synthesis approach. *Vacuum*, *156*, 434–439. https://doi.org/10.1016/j.vacuum.2018.08.009

Shimizu, M., Tainaka, H., Oba, T., Mizuo, K., Umezawa, M., & Takeda, K. (2009). Maternal exposure to nanoparticulate titanium dioxide during the prenatal period alters gene expression related to brain development in the mouse. *Particle and Fibre Toxicology*, *6*(1), 20. https://doi.org/10.1186/1743-8977-6-20

Shivaji, S., Madhu, S., & Singh, S. (2011). Extracellular synthesis of antibacterial silver nanoparticles using psychrophilic bacteria. *Process Biochemistry*, *46*(9), 1800–1807. https://doi.org/10.1016/j.procbio.2011.06.008

Siddiqi, K. S., & Husen, A. (2016). Engineered gold nanoparticles and plant adaptation potential. *Nanoscale Research Letters*, *11*(1), 400. https://doi.org/10.1186/s11671-016-1607-2

Singh, B. R., Singh, B. N., Singh, A., Khan, W., Naqvi, A. H., & Singh, H. B. (2015). Mycofabricated biosilver nanoparticles interrupt Pseudomonas aeruginosa quorum sensing systems. *Scientific Reports*, *5*(1), 13719. https://doi.org/10.1038/srep13719

Singh, P., Kim, Y.-J., Zhang, D., & Yang, D.-C. (2016). Biological synthesis of nanoparticles from plants and microorganisms. *Trends in Biotechnology*, *34*(7), 588–599. https://doi.org/10.1016/j.tibtech.2016.02.006

Singh, R., & Nalwa, H. S. (2011). Medical applications of nanoparticles in biological imaging, cell labeling, antimicrobial agents, and anticancer nanodrugs. *Journal of Biomedical Nanotechnology*, *7*(4), 489–503. https://doi.org/10.1166/jbn.2011.1324

Singh, R., Shedbalkar, U. U., Wadhwani, S. A., & Chopade, B. A. (2015). Bacteriagenic silver nanoparticles: Synthesis, mechanism, and applications. *Applied Microbiology and Biotechnology*, *99*(11), 4579–4593. https://doi.org/10.1007/s00253-015-6622-1

Slavin, Y. N., Asnis, J., Häfeli, U. O., & Bach, H. (2017). Metal nanoparticles: Understanding the mechanisms behind antibacterial activity. *Journal of Nanobiotechnology*, *15*(1), 65. https://doi.org/10.1186/s12951-017-0308-z

Sunkar, S., & Nachiyar, C. V. (2012). Biogenesis of antibacterial silver nanoparticles using the endophytic bacterium Bacillus cereus isolated from Garcinia xanthochymus. *Asian Pacific Journal of Tropical Biomedicine*, *2*(12), 953–959. https://doi.org/10.1016/S2221-1691(13)60006-4

Tan, Y., Ma, S., Leonhard, M., Moser, D., Haselmann, G. M., Wang, J., Eder, D., & Schneider-Stickler, B. (2018). Enhancing antibiofilm activity with functional chitosan nanoparticles targeting biofilm cells and biofilm matrix. *Carbohydrate Polymers*, *200*, 35–42. https://doi.org/10.1016/j.carbpol.2018.07.072

Tang, L., & Cheng, J. (2013). Nonporous silica nanoparticles for nanomedicine application. *Nano Today*, *8*(3), 290–312. https://doi.org/10.1016/j.nantod.2013.04.007

Thangavelu, R. M., Ganapathy, R., Ramasamy, P., & Krishnan, K. (2020). Fabrication of virus metal hybrid nanomaterials: An ideal reference for bio semiconductor. *Arabian Journal of Chemistry*, *13*(1), 2750–2765. https://doi.org/10.1016/j.arabjc.2018.07.006

Velmurugan, P., Iydroose, M., Mohideen, M. H. A. K., Mohan, T. S., Cho, M., & Oh, B.-T. (2014). Biosynthesis of silver nanoparticles using Bacillus subtilis EWP-46 cell-free extract and evaluation of its antibacterial activity. *Bioprocess and Biosystems Engineering*, *37*(8), 1527–1534. https://doi.org/10.1007/s00449-014-1124-6

Vigneshwaran, N., Kathe, A. A., Varadarajan, P. V., Nachane, R. P., & Balasubramanya, R. H. (2006). Biomimetics of silver nanoparticles by white rot fungus, phaenerochaete chrysosporium. *Colloids and Surfaces B: Biointerfaces*, *53*(1), 55–59. https://doi.org/10.1016/j.colsurfb.2006.07.014

Vijayakumar, S., Vaseeharan, B., Malaikozhundan, B., Gopi, N., Ekambaram, P., Pachaiappan, R., Velusamy, P., Murugan, K., Benelli, G., Suresh Kumar, R., & Suriyanarayanamoorthy, M. (2017). Therapeutic effects of gold nanoparticles synthesized using Musa paradisiaca peel extract against multiple antibiotic resistant Enterococcus faecalis biofilms and human lung cancer cells (A549). *Microbial Pathogenesis*, *102*, 173–183. https://doi.org/10.1016/j.micpath.2016.11.029

Vinoj, G., Pati, R., Sonawane, A., & Vaseeharan, B. (2015). *In Vitro* cytotoxic effects of gold nanoparticles coated with functional acyl homoserine lactone lactonase protein from bacillus licheniformis and their antibiofilm activity against proteus species. *Antimicrobial Agents and Chemotherapy*, *59*(2), 763–771. https://doi.org/10.1128/AAC.03047-14

Wagh, M. S., Patil, R. H., Thombre, D. K., Kulkarni, M. V., Gade, W. N., & Kale, B. B. (2013). Evaluation of anti-quorum sensing activity of silver nanowires. *Applied Microbiology and Biotechnology*, *97*(8), 3593–3601. https://doi.org/10.1007/s00253-012-4603-1

Wang, L., He, D., Gao, S., Wang, D., Xue, B., & Yokoyama, K. (2016). Biosynthesis of silver nanoparticles by the fungus Arthroderma fulvum and its antifungal activity against genera of Candida, Aspergillus and Fusarium. *International Journal of Nanomedicine*, 1899. https://doi.org/10.2147/IJN.S98339

Waters, C. M., & Bassler, B. L. (2005). Quorum sensing: Cell-to-cell communication in bacteria. *Annual Review Cell Developmental Biology*, *21*, 319–346. https://doi.org/10.1146/annurev.cellbio.21.012704.131001

Wiley, L., Bridge, D. R., Wiley, L. A., Odom, J. V., Elliott, T., & Olson, J. C. (2012). Bacterial biofilm diversity in contact lens-related disease: Emerging role of *Achromobacter*, *Stenotrophomonas*, and *Delftia*. *Investigative Opthalmology & Visual Science*, *53*(7), 3896. https://doi.org/10.1167/iovs.11-8762

Xie, H., Mason, M. M., & Wise, J. P. (2011). Genotoxicity of metal nanoparticles. *Reviews on Environmental Health*, *26*(4). https://doi.org/10.1515/REVEH.2011.033

Zhang, X., Qu, Y., Shen, W., Wang, J., Li, H., Zhang, Z., Li, S., & Zhou, J. (2016). Biogenic synthesis of gold nanoparticles by yeast Magnusiomyces ingens LH-F1 for catalytic reduction of nitrophenols. *Colloids and Surfaces A: Physicochemical and Engineering Aspects*, *497*, 280–285. https://doi.org/10.1016/j.colsurfa.2016.02.033

Zhao, Y., Tian, Y., Cui, Y., Liu, W., Ma, W., & Jiang, X. (2010). Small molecule-capped gold nanoparticles as potent antibacterial agents that target gram-negative bacteria. *Journal of the American Chemical Society*, *132*(35), 12349–12356. https://doi.org/10.1021/ja1028843

Zhou, X., Meng, X., & Sun, B. (2008). An EAL domain protein and cyclic AMP contribute to the interaction between the two quorum sensing systems in Escherichia coli. *Cell Research*, *18*(9), 937–948. https://doi.org/10.1038/cr.2008.67

Zonaro, E., Lampis, S., Turner, R. J., Qazi, S. J. S., & Vallini, G. (2015). Biogenic selenium and tellurium nanoparticles synthesized by environmental microbial isolates efficaciously inhibit bacterial planktonic cultures and biofilms. *Frontiers in Microbiology*, *6*. https://doi.org/10.3389/fmicb.2015.

7 Plant-Mediated Synthesized Nanoparticles in Down-Regulating Quorum Sensing

Kavita Singh, Avnish Kumar, and Vinita Gupta

7.1 QUORUM SENSING

7.1.1 Introduction of Quorum Sensing

Quorum sensing (QS) is a communication mechanism between bacteria that allows specific processes to be controlled, such as biofilm formation, virulence factor expression, production of secondary metabolites, and stress adaptation mechanisms such as bacterial competition systems including secretion systems (SS). Molecular biologist Bonnie Bassler and her graduate student Justin Silpe (Silpe & Bassler, 2019) found that viruses can use QS chemicals (autoinducers) released by bacteria to determine when best to start multiplying and killing the viruses. Quorum sensing involves cells determining their population density due to the cells engaging in neighbour communication with/without self-communication. There is an ability of the cell to achieve self-communication (autocrine signaling), neighbour communication (including quorum sensing), and a mixture of the two (Youk & Lim, 2014).

Mid-1960s studies by Hungarian microbiologist Alexander Tomasz reported bacterial activity of quorum sensing in *Pneumococcus* (later known as *Streptococcus pneumoniae*). The pneumococcus was a model organism for the major discoveries regarding natural transformation, culminating with the discovery of DNA as hereditary material. Later research progressed to describe the competence factors to take up free DNA from culture supernatants. (This is considered a first report of quorum sensing in Gram-positive bacteria.) Molecular studies also suggested that various gene and/or protein systems exist for quorum-sensing signalling in the streptococci group. Rgg/SHP systems in *S. pyogenes* and *S. agalactiae* have revealed SHP production and regulation. The research paper of Nealson et al. (1970) was considered a first report on quorum sensing. Nealson's research describes a conditioning of the medium for growth of the bioluminescent marine bacterium *Vibrio fischeri*.

DOI: 10.1201/9781003300557-7

Similarly to Gram-positive bacteria, Gram-negative bacteria also use quorum sensing communication pathways to achieve genetic control of varied sequenced or nonsequenced physiological activities, such as symbiosis, virulence, competence, conjugation, antibiotic production, motility, sporulation, and biofilm formation. This regulation of gene expression is dependent on the concentration of autoinducers which ultimately fluctuates in response to cell-population density.

The signalling chemicals released by quorum sensing bacteria are called autoinducers. These chemicals increase with the increasing number of cells, and at an optimum concentration it becomes a regulatory molecule of bacterial genetic pathways. In Gram-negative bacteria, the autoinducer chemical is acyl homoserine lactone (AHL). The research into AHL-based QS appeared in the late 1960s. The bioluminescent bacterium *V. fischeri* generate light only with a large number (Greenberg, 1997) of bacteria in culture. Initially it was explained as a fault of culture media inhibiting luminescence, and researchers considered that only a large number of bacteria could remove the hindrance created by media (Kempner & Hanson, 1968). Later it was suggested that the bacteria could generate luminescence by accumulation of an activator molecule or autoinducer instead of removing the inhibitor (Nealson et al., 1970; Eberhard, 1972). It was also thought that the autoinducer could only be produced by the bacteria and a large number of bacteria was necessary to produce a sufficient concentration of autoinducers. Further, it is explained that via QS, bacterial cells could sense their density at an optimum concentration of autoinducers.

In the case of *V. fischeri*, the autoinducer molecule was identified chemically in 1981 by Eberhard et al. as N-(3-oxohexanoyl)-homoserine lactone(3-oxo-C6-HSL). The homoserine lactone molecules are produced by bacteria using the LuxI synthase enzyme and are secreted into the extracellular environment. Engebrecht et al. (1983) studied the genes associated with QS in *V. fischeri*. This *V. fischeri* research is now a model for similar quorum sensing systems.

7.1.1.1 Antibiotic Production in Erwinia Carotovora

For many years AHL-based QS was associated only with marine bacteria such as *V. fischeri* and *V. harveyi*. In the early 1990s, English researchers Barrie Bycroft and Paul Williams and George Salmond were studying responsible genes for carbapenem antibiotics production and biosynthetic pathway in two mutant groups of *E. carotovora*. One group of mutants could not make antibiotics on their own but could do so when cross-fed by a second group of mutants. Analysis suggested that the second type provided a signalling molecule to trigger antibiotic synthesis in the first group, and this signalling molecule was same one found in *V. fisheri* for generating bioluminescence (Bainton et al., 1992). Carbapenem production by bacteria is autoregulated by N-(3-Oxohexanoyl)-L-homoserine lactone.

7.1.1.1.1 Quorum Sensing in Pseudomonas Aeruginosa

At University of Rochester School of Medicine and Dentistry, New York, Gambello and Iglewski (1991) discovered a *V. fischeri*-like QS system in *P. aeruginosa*. In this bacteria AHL regulates and induces the production of elastase. *P.*

aeruginosa AHL is chemically 3-oxo-C12-HSL (Pearson et al., 1994). Latifi et al. (1995) demonstrated the presence of a QS system for regulating rhamnolipid, haemolysin, and other important virulence factors, and this was later confirmed by Winson et al. (1995), who showed the role of C4-HSL. However, Ochsner et al. (1994) had also discovered it independently.

After development of biosensors for AHL detection (Swift et al., 1993; Winson et al., 1998), AHL-based QS systems was also identified in *Enterobacter*, *Hafnia*, *Rahnella, Serratia*, and many other species now known to have quorum sensing ability.

7.1.1.1.2 Other QS Systems

Including AHLs, bacteria may use many other signalling molecules to sense cell density. The most studied are the peptides used by many Gram-positive bacteria and the AI-2 molecules first discovered in *V. harveyi*. The following are a few microorganisms and their QS signalling molecules (in parentheses): *Bacillus subtilis* (ComX, CSF [PhrC], PhrA,-E, -F, -K, -H); *Myxococcus xanthus* (A-signal, C-signal); *P. aeruginosa* (3O-C12-HSL, C4-HSL); *Staphylococcus aureus* (AIP-I, AIP-II, AIP-III, AIP-IV); *Streptococcus mutans* (CSP [ComC], XIP [ComS]); *S. pneumonia* (CSPs); *Vibrio harveyi* (HAI-1, CAI-1 A1–2) (Lu et al., 2022).

7.1.2 Types of Autoinducers

The signal molecules, autoinducers, are peptides or modified peptides which typically regulate signal import or transduction across the membrane and later steps of interaction between intracellular effectors. Normally, a nanomolar or lesser concentration of signal is sufficient for biological activities. The bacteria needs a threshold concentration of molecules to detect the signalling and activation of genes responsible for regulation of virulence, biofilm, horizontal gene transfer, and their ability to take up DNA. Three factors counted for standard quorum-sensing pathways are bacterial population, signal molecule, and corresponding genes. Based on their chemical nature, bacteria consist of three types of autoinducers, N-acyl homoserine lactones (AHLs), which have been discovered in about 70 species of Gram-negative bacteria; oligopeptides, which have been found in Gram-positive bacteria; and ribose-like S-4,5-dihydroxy-2,3-pentanedione (DPD)/autoinducer-2 (AI-2), which have been found in both Gram-negative and Gram-positive bacteria.

7.1.3 Mechanism of Quorum Sensing

The different types of bacteria apply different mechanisms of quorum sensing depending on their requirement. For example, the bacterium *P. aeruginosa* uses quorum sensing to overcome host defences by activating the regulatory genes of biofilm and virulence. This process supports pneumonia and blood infections in the host. On the other hand, nonpathogenic or symbiotic microbes use QS for regulating population sizes and symbiotic processes. In *Rhizobium leguminosarum* the nitrogen-fixing steps like nodulation efficiency, symbiosome development, exopolysaccharide production, and nitrogen fixation (González & Marketon, 2003) are regulated by QS to establish a perfect symbiosis.

7.1.4 Methods of Inhibition of Quorum Sensing

There is an urgent need for novel strategies for controlling infectious diseases by preventing antibiotic resistance. Newer approaches include inhibition of quorum sensing (QS) circuits by using small molecules.

Quorum sensing is facilitating the survival of bacteria in the host by developing pathogenicity. Therefore, blocking any step of the QS circuit with small molecules has become the novel strategy of preventing bacterial pathogenicity. The mechanism of QS inhibition may be executed by inhibition of AHL autoinducer synthesis, destruction of AHL molecules by QS quenching enzymes, and/or interference with signal receptors or blocing formation of the receptor complex. These strategies, similarly, can be applied to the inhibition of AIP-mediated quorum sensing in Gram-positive bacteria (Fernando et al., 2020).

7.2 METHODS OF INHIBITION OF QUORUM SENSING USING NANOPARTICLES

7.2.1 Inhibitors of Quorum Sensing

Quorum sensing can be inhibited in the following ways.

1 *Preventing the AHL molecule from binding to its receptor.* AHL and its receptor binding can be inhibited by using molecules that bind competitively to the receptor.

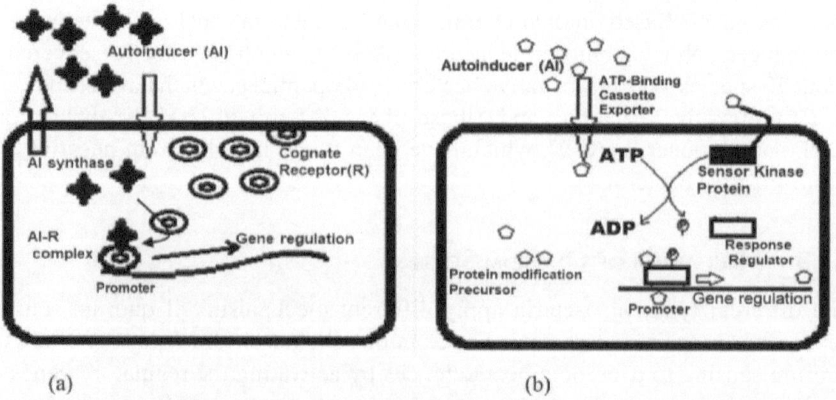

FIGURE 7.1 A graphical representation of QS in Gram-negative bacteria (a) and Gram-positive bacteria (b).

2 *Inhibition of AIs synthesis.* Enzymes or molecules that reduce the expression of AI-synthesizing genes could be potential quorum-sensing inhibitors.
3 *Inhibition of targets downstream of receptor binding.* There are potential inhibitors of QS signal transduction that inhibit AHL by binding competitively or noncompetitively to the corresponding receptor protein molecule. Natural or synthetic analogues may modify the acyl side chain, the lactone ring, or both these moieties of the AHL molecule. The modifications of the natural halogenated furanones are also used and prove to be potential antagonists.
4 *Sequestration of AIs using anti-AI-antibodies.* A monoclonal antibody against AI and/or against a rationally-designed hapten could be synthesized and used to block the generation of unwanted virulence factors – for instance, anti-AIP IV antibodies could restrict the QS signalling in *S. aureus* (Park et al., 2007; Zhou et al., 2020).
5 *Degradation of AIs using either catalytic antibodies (abzymes) or enzymes (such as lactonases).* The enzyme AHL-lactonase (AiiA; encoded by *aiiA* gene) is capable of paralyzing QS signals and blocking production of virulence factors in Gram-positive *Bacillus* species by inactivating AHL signals. The enzyme hydrolyzes the ester bond of the homoserine lactone ring (Fong et al., 2018). Phosphotriesterase-like lactonases, lactonases, acylases, and oxidoreductases target AHL-mediated and AI2-mediated quorum quenching. QS quenching molecules identified from natural compounds are polyphenols, ajoene, eugenol from plant sources, or synthetic molecules such as 5-fluorouracil (5-FU) or azithromycin.
6 *Inhibition of AI secretion/transport.* Bacterial phospholipid membranes contain eight different types of secretion systems for bacterial communication: T1SS, T2SS, T3SS, T4SS, T5SS, T6SS, T7SS, and T9SS. These coded secretion systems have now been better defined for structure, composition, and activity. The effector protein, alkaline protease AprA, of T1SS is regulated by QS. Additionally, QS also regulates the genes of the AprA inhibitor aprI and the structural genes aprDEF (Pena et al., 2019). Halogenated furanones, cinnamaldehyde, and cinnamaldehyde analogues decrease the DNA binding activity of the master regulator LuxRvh and therefore block QS channels in this organism.

Ethnobotanical plants are potential resources of defence molecules against bacterial antibiotic resistance and characterize newer claims and strategies as alternative medicine. Similarly, effective drug delivery with an advent of nanotechnology has attracted significant research interest. Nanotechnology is improving the treatment of diseases due to characteristics of nanoparticles such as their extremely small size and large relative surface area, plus their nontoxic and ecofriendly qualities. The plants and nanoparticles are well known for their antimicrobial qualities in traditional and conventional medicines. Plant-based metal nanoparticles have been reviewed for their mechanism of QS inhibition.

7.3 STUDY OF QUORUM-SENSING INHIBITION

7.3.1 Mechanism of Inhibition by Silver Nanoparticle (Ag NPs)

Biofilms are the microbial communities and extracellular polymeric substances (EPS) that act as a protective barrier. Plant-mediated synthesized Ag NPs are capable of reducing multiple QS-regulated pathways of Gram-negative bacterial pathogens. Ag NPs work against the development of biofilms and the QS-mediated virulence in *Pseudomonas aeruginosa, Streptococcus marcescens*, and *Chromobacterium violaceum*. Bacterial treatment with Ag NPs shows an inhibition of violacein pigment production (>80%), QS-controlled virulence (90%), and the formation of biofilms (>70%). Ag NPs reduce the formation of biofilms on the surfaces of glass cover slips by inhibiting the aggregation and formation of EPS (Qais et al., 2021).

7.3.2 Mechanism of Inhibition by Gold Nanoparticles

The antimicrobial activity of plant extract may derive from its active constituents alquinone, 3,5,7,4-tetrahydroxy flavone, and flavonoid glycosides, among others. Also reported was an antibacterial effect of crude plant extracts plus gold nanoparticles (Au NPs). This synergistic activity is justified on the grounds that the nanoparticle size enables efficient delivery. This efficient delivery across the membrane is due to the hydrophobic nature of bacterial cell membrane and hydrophilic nature of plant extracts conjugated with Au NPs. The nanoparticles entered into the cell interfere with the growth-signalling pathway, e.g., tyrosine phosphorylation. The modulation of signal transduction cascades may give rise to the initiation of programmed cell death, ultimately inhibiting bacterial replication (Fernando & Judan Cruz, 2020).

7.3.3 Mechanism of Inhibition by Zinc and Copper Nanoparticles

There are some reports (e.g., Sawai & Yoshikawa, 2004) on the antibacterial potential of CaO, MgO, and ZnO. A conductometric method has shown the production of reactive oxygen species on the surface of these oxides. The advantage of using these inorganic oxides as antimicrobial agents is that they contain mineral elements essential to humans and exhibit strong activity even when administered in small amounts (Sawai & Yoshikawa, 2004).

7.4 PLANT EXTRACTS PREPARATION METHOD

7.4.1 Maceration

In maceration extraction, coarsely powdered leaves, stem, or root bark is placed inside a container, then a solvent is poured on top until it completely covers the material. Then the container is closed and kept for at least three days, and shaken from time to time to ensure complete extraction. At the end of extraction, the micelle is separated by filtration or decantation. This method is very suitable for thermolabile plant material.

7.4.2 Infusion

The plant material is ground into fine powder, dipped in an extraction solvent (hot or cold), soaked, and kept for a short period of time. This method is suitable for extracting bioactive constituents that are readily soluble. Infusion is an appropriate method for the preparation of fresh extract before use.

7.4.3 Digestion

This extraction method also requires finely powdered plant material kept under moderate heat (50°C) in solvent. The applied heat decreases the viscosity of extraction solvent during the process and removes secondary metabolites. This method is suitable for plant materials that are readily soluble (Abubakar & Haque, 2020).

7.4.4 Decoction

This process uses aqueous solvent for the extraction of water-soluble and heat-stable plant material. Dried finely ground or powdered plant material is put in a clean container with water (usually at a ratio of 1:4 or 1:16). Then the plant material is stirred and heated for a short duration, usually about 15 min. The method is useful for extraction of water-soluble and heat-stable plant material (Abubakar & Haque, 2020).

7.4.5 Percolation

A percolator –narrow cone-shaped glass vessel with an opening at each end – is used for the percolation method. A dried finely ground or powdered plant material and a solvent mixture are kept for 4 h before being put into the percolator and incubated for 24 h. The next day, the lower end of the percolator is opened, allowing the extraction to drip out slowly. Some solvent may be added, drop by drop, to help push the drug material downward (Abubakar & Haque, 2020).

7.4.6 Soxhlet Extraction

The Soxhlet extraction requires a Soxhlet apparatus, made of glass supported by a heating mantel. For extraction, plant material should be washed and dried in shade and then ground into fine powder. The powder is put into a filter paper pocket (called a thimble) inside the extraction chamber and the bottom flask is filled 2/3 full of solvent. Then the solvent is heated to evaporate it. Thereafter, the vapour reaches the condenser through the solvent vapour tube and is condensed to liquid before coming into contact with the plant powder pocket. The liquid reaches the extraction chamber drop by drop and the thimble moves to the top level of the siphon. The solvent-extracted plant material flows back to the flask by siphon tube. The entire process is repeated until the solvent gets colour from the thimble (about 20–25 cycles). This method is suitable for plant material that is partially soluble in the chosen solvent and for plant materials with insoluble impurities (Abubakar & Haque, 2020).

7.4.7 Microwave-Assisted Extraction

The technique uses the mechanism of dipole rotation and ionic transfer by displacement of charged ions present in the solvent and drug material. This method is suitable for extraction of flavonoids. The technique uses microwave radiation to bombard an object, which can absorb electromagnetic energy and convert it into heat. Subsequently, the heat produced facilitates movement of the solvent into the drug matrix. Microwave-assisted extraction is useful for phenolic compounds and flavonoids. It minimizes the solvent amount and time to extraction. However, this method is not suitable for other compounds, such as tannins and anthocyanins, that may be degraded because of high temperature involved (Abubakar & Haque, 2020).

7.4.8 Ultrasound-Assisted Extraction

Ultrasonic sound energy uses a very high frequency (greater than 20 KHz) to disrupt plant cells and provide larger plant-cell surface area for solvent penetration. This method is useful for secondary metabolites. On combining with microwave-assisted extraction (MAE), ultrasonic heating as a pre-treatment technique for plant cells can increase the extraction yield. Intermittent sonication is more favourable than continuous sonication for more yield. However, it does not give reproducible results each time. It may generate free radicals into the samples.

7.5 PREPARATION OF PLANT EXTRACT-BASED NANOPARTICLES AND THEIR CHARACTERIZATION

This method is called green synthesis of nanoparticles when plant molecules are used in nanoparticle preparation. The green synthesis is considered to have fewer environmental risks and is cheaper than other methods of nanoparticle synthesis. The green synthesis of nanoparticles has emerged as a simple, safe, sustainable, reliable and ecofriendly protocol.

The various techniques are as follows.

1 UV-visible spectroscopy shows the kinetic behaviour of nanoparticles.
2 X-ray diffraction analysis gives the phase evolution and particle size of samples using Cu Kα radiation.
3 Transmission electron microscopy (TEM) analysis determines the exact size and shape of the synthesized nanoparticles.

4 Dynamic light scattering (DLS) is used to determine particle size distribution, particle zeta potential (related to the magnitude of the electrical charge at the particle surface), and molecular weight of large polymeric substances dispersed in water. DLS is used to analyze the zeta potential of the synthesized Ag NPs. For DLS measurements, powder Ag NPs are resuspended in distilled water and sonicated for a few minutes to properly disperse the particles in water. Zeta potential values are obtained from the triplicate analysis of the nanoparticles in the aqueous milieu.
5 Scanning electron microscopy (SEM) is used to obtain the morphological information of all the samples.
6 X-ray photoelectron spectroscopy (XPS) would be applied to understand the nature of nanoparticle surfaces as well as the structure and thickness of coatings on nanoparticles.

7.6 FUTURE OF PLANT-BASED NANOPARTICLES

Phytochemicals and other plant products can be obtained from leaves, stems, shoots, flowers, bark, seeds, and roots. A required molecule may enhanced by external stimulation of secondary metabolism. The phytochemicals and other metabolites of plants may produce nanoparticles of varied shape and size compared to those produced by other organisms (Iravani, 2011). Considering the size- and shape-based applications of plant-based nanoparticles, future studies must evaluate the safety risks and focus on eliminating toxic effects of nanoparticles, if any, on the environment or living beings.

Additionally, detailed methodology and better synthesis mechanisms should be developed for nanoparticle synthesis. It is also necessary to study the mechanisms responsible for the biological benefits shown by nanoparticles. It is becoming obvious that nanoparticles have a potential role in clinical practice, thus nanoparticle-borne toxicity must be explored in detail, including synergistic approaches. The resistance mechanism of microbes to nanoparticles has not been studied, but it should be investigated with adequate care. If there is any adaptation in microbes to develop resistance to nanoparticles, then it should be exploited in future studies.

7.7 CONCLUSION

Nanoparticle-based applications could be developed based on their modes of action, which could be modified with changes in parameters, such as temperature or pH, as per requirements. The toxicity of NPs may show potential problems for use as medicine. On the other hand, plant mediated synthesis of NPs is considered environmentally friendly. Metallic NPs from plant extracts could be synthesized on a large scale for use in various applications. Plant-based NPs have several advantages and applications in the fields of medicine and pharmaceuticals. Many studies have confirmed their antimicrobial properties alone or in combination with antibiotics to diminish the issue of drug resistance. However, assessment of their safety and risks associated with human health and environmental concerns need to be explored and discussed. More detailed studies of action mechanisms that mediate the antibacterial effect of NPs are also required before plant-based NPs can be used in hospitals or other health care settings. In conclusion, plant-mediated synthesized metallic NPs could be exploited for the treatment of patients; however, in vivo assays are needed to evaluate the therapeutic efficacy against infections to analyze all principles associated with their use.

REFERENCES

Abubakar, A.R. and Haque, M. (2020) Preparation of medicinal plants: Basic extraction and fractionation procedures for experimental purposes. J Pharm Bioallied Sci. 12(1), 1–10. doi: 10.4103/jpbs.JPBS_175_19.

Bainton, N.J., Stead, P., Chhabra, S.R., Bycroft, B.W., Salmond, G.P.C., Stewart, G.S.A.B. and Williams, P. (1992) N-(3-oxohexanoyl)-L-homoserine lactone regulates carbapenem antibiotic production in *Erwinia carotovora*. Biochem J. 288, 997–1004.

Eberhard, A. (1972) Inhibition and activation of bacterial luciferase synthesis. J Bacteriol. 109, 1101–1105.

Engebrecht, J., Nealson, K. and Silverman, M. (1983) Bacterial bioluminescence: Isolation and genetic analysis of functions from *Vibrio fischeri*. Cell. 32, 773–781.

Fernando, S.I.D. and Judan Cruz, K.G. (2020) Ethnobotanical biosynthesis of gold nanoparticles and its downregulation of quorum sensing-linked AhyR gene in *Aeromonas hydrophila*. SN Appl. Sci. 2, 570.

Fong, J., Zhang, C., Yang, R., Boo, Z.Z., Tan, S.K., Nielsen, T.E., Givskov, M., Liu, X., Bin, W., Su, H. and Yang, L. (2018) Combination therapy strategy of quorum quenching enzyme and quorum sensing inhibitor in suppressing multiple quorum sensing pathways of *P. aeruginosa*. Sci Rep. 8, 1155.

Gambello, M.J. and Iglewski, B.H. (1991) Cloning and characterisation of the *Pseudomonas aeruginosa* lasR gene, a transcriptional activator of elastase expression. J. Bacteriol. 173, 3000–3009.

González, J.E. and Marketon, M.M. (2003) Quorum sensing in nitrogen-fixing rhizobia. Microbiol Mol Biol Rev. 67(4), 574–592.

Greenberg, E.P. (1997) Quorum sensing in Gram-negative bacteria. ASM News. 63, 371–377.

Iravani, S. (2011) Green synthesis of metal nanoparticles using plants. Green Chem. 13(10), 2638–2650.

Kempner, E.S. and Hanson, F.E. (1968) Aspects of light production by *Photobacterium fischeri*. J Bacteriol. 95, 975–979.

Latifi, A., Winson, M.K., Foglino, M., Bycroft, B.W., Stewart, G.S.A.B. and Williams, P. (1995) Multiple homologues of LuxR and LuxI control expression of virulence determinants and secondary metabolites through quorum sensing in *Pseudomonas aeruginosa* PAO1. Mol. Microbiol. 17, 333–343.

Lu, L., Li, M., Yi, G., Liao, L., Cheng, Q., Zhu, J., Zhang, B., Wang, Y., Chen, Y. and Zeng, M. (2022) Screening strategies for quorum sensing inhibitors in combating bacterial infections. J Pharm Anal. 12(1), 1–14.

Nealson, K.H., Platt, T. and Hastings JW. (1970) Cellular control of the synthesis and activity of the bacterial luminescent system. J Bacteriol. 104(1), 313–322.

Ochsner, U.A., Koch, A.K., Fiechter, A. and Reiser, J. (1994) Isolation and characterization of a regulatory gene affecting rhamnolipid biosurfactant synthesis in *Pseudomonas aeruginosa*. J Bacteriol. 176, 2044–2054.

Park, J., Jagasia, R., Kaufmann, G. F., Mathison, J. C., Ruiz, D. I., Moss, J. A., Meijler, M. M., Ulevitch, R. J. and Janda, K. D. (2007) Infection control by antibody disruption of bacterial quorum sensing signaling. Chem Biol. 14(10), 1119–1127.

Pearson, J.P., Gray, K.M., Passador, L., Tucker, K.D., Eberhard, A., Iglewski, B.H. and Greenberg, E.P. (1994) Structure of the autoinducer required for expression of *Pseudomonas aeruginosa* virulence genes. Proc Natl Acad Sci USA. 91, 197–201.

Pena, R.T., Blasco, L., Ambroa, A., González-Pedrajo, B., Fernández-García, L., López, M., Bleriot, I., Bou, G., García-Contreras, R. and Wood, T.K. (2019) Tomás M. Relationship Between Quorum Sensing and Secretion Systems. Front Microbiol. 10, 1100.

Qais, F.A, Ahmad, I., Altaf, M., Manoharadas, S., Al-Rayes, B.F., Abuhasil, M.S.A. and Almaroai, Y.A. (2021) Biofabricated silver nanoparticles exhibit broad-spectrum antibiofilm and antiquorum sensing activity against gram-negative bacteria. RSC Adv. 11, 13700–13710.

Sawai, J. and Yoshikawa, T.J. (2004) Quantitative evaluation of antifungal activity of metallic oxide powders (MgO, CaO and ZnO) by an indirect conductimetric assay. Appl Microbiol. 96, 803.

Silpe, J.E. and Bassler, B.L. (2019) A host-produced quorum-sensing autoinducer controls a phage lysis-lysogeny decision. Cell. 176(1–2), 268–280.e13.

Swift, S., Winson, M.K., Chan, P.F., Bainton, N.J., Birdsall, M., Reeves, P.J., Rees, C.E.D., Chhabra, S.R., Hill, P.J., Throup, J.P., Bycroft, B.W., Salmond, G.P.C., Williams, P. and Stewart, G.S.A.B. (1993) A novel strategy for the isolation of luxI homologues: Evidence for the widespread distribution of a LuxR:LuxI superfamily in enteric bacteria. Mol Microbiol. 10, 511–520.

Winson, M.K., Camara, M., Latifi, A., Foglino, M., Chhabra, S.R., Daykin, M., Bally, M., Chapon, V., Salmond, G.P.C., Bycroft, B.W., Lazdunski, A., Stewart, G.S.A.B. and Williams, P. (1995) Multiple N-acyl-L-homoserine lactone signal molecules regulate production of virulence determinants and secondary metabolites in *Pseudomonas aeruginosa*. Proc Natl Acad Sci USA. 92, 9427–9431.

Winson, M.K., Swift, S., Fish, L., Throup, J.P., Jørgensen, F., Chhabra, S.R., Bycroft, B.W., Williams, P. and Stewart, G.S.A.B. (1998) Construction and analysis of luxCDABE-based plasmid sensors for investigating N-acyl homoserine lactone-mediated quorum sensing. FEMS Microbiol Lett. 163, 185–192.

Youk, H. and Lim, W.A. (2014) Secreting and sensing the same molecule allows cells to achieve versatile social behaviors. Science. 343(6171), 628–628.

Zhou, L., Zhang, Y., Ge, Y., Zhu, X. and Pan, J. (2020) Regulatory mechanisms and promising applications of quorum sensing-inhibiting agents in control of bacterial biofilm formation. Front Microbiol. 11, 589640.

8 Fungal Compounds in Down-Regulating Quorum Sensing

Barun Kumar, Dibyajit Lahiri, Soumya Pandit, Ankit Kumar, Kalpana Sharma, Prakriti Negi, and Moupriya Nag

8.1 INTRODUCTION

Communication is key in life. This seems to be a clear statement when referring to evolutionary organisms. For a long time, microorganisms were thought to work as individual creatures that sought nutrition and multiplied. However, the unexpected finding of density-dependent intermicrobial interaction more than 20 years ago resulted in the realization that microbes are capable of complex interaction (De Sordi and Mühlschlegel 2009). Earlier, it was assumed that quorum sensing was present only in bacteria, although an investigation into cell intensity morphological transition in *Candida albicans* showed that farnesol also serves as a quorum-sensing molecule (QSM) inside this eukaryote. The extensive array of microbial species causes various connections across various organisms. Interaction soil microbes are among the fundamental elements determining soil fertility, which contributes significantly to the health of ecology and crop production. To meet the demands of the world's growing population and effectively use the opportunities afforded by soil biota, a better comprehension of biological components is therefore critical for fertilizer application (Albarracín Orio et al. 2020). In pathogenicity and the adaptability of microbes to host tissues, regulating the activity of pathogenicity is an important step. Several harmful bacteria have effectively controlled their toxicity by a technique known as quorum sensing (QS), which coordinates the production of virulent proteins around population density (Albuquerque and Casadevall 2012). Use of traditional anthelmintics causes production of enzymes that may affect the drug-efflux mechanisms thereby lowering the density of the antibiotic inside the cells (Thomas Bovbjerg Rasmussen et al. 2005a).

Originally the QS hypothesis was thought to apply only to a select few microorganisms, but it is now found in the regulatory mechanism of many bacterial species. QS compensatory signalling pathways exist within species, across species, and even among representatives of both prokaryotes and eukaryotes (Vicente-Soler

DOI: 10.1201/9781003300557-8

et al. 2016). This chapter's primary goal is to describe the QS strategies found in fungi and their possible biotechnology uses, including optimizing industrial applications or minimizing fungal diseases (Barriuso and Centro de Investigaciones Biológicas, Consejo Superior de Investigaciones Científicas 2015). Demand for microbiological QS has risen sharply during the past two decades. The hypothesis of QS blocker synthesis was triggered by detecting the QS-mediated pathogenicity expression in several important clinical diseases. It is understood that inhibiting QS is a good strategy for creating new medicines (Wynendaele et al. 2013). It is possible to build different synthetic therapeutic chemicals by utilizing the biochemical diversity seen inside this range of QS signalling molecules (Maillé et al. 2017). Several quorum-sensing molecules and blockers have recently been observed and described in a variety of fungal species. This chapter will largely concentrate on fungus quorum-sensing compounds and the development of inhibitors by fungi with possible uses in the therapy of fungi illnesses (Mehmood et al. 2019).

8.2 QUORUM SENSING—FUNGI

Small diffusible signalling molecules that collect inside the extracellular environment regulate quorum sensing across bacteria and fungi (Hogan 2006). The signalling molecules are known as autoinducers. They accumulate in solution as urbanization increases; colonies from microbes may connect with others in the atmosphere (Wongsuk et al. 2016). New genes are triggered or inhibited, and actions like viral replication, virulence, or cooperation begin when all these chemicals' levels cross the threshold concentration. These systems, referred to as quorum sensing (QS), were initially detected in bacteria and, more recently, in fungus. The dimorphic harmful yeast *C. albicans* functions as the prototype eukaryote organism for these projects. The QS systems of several fungi with clinical or industrial importance have recently been characterized. Consequently, further investigation into these systems may be important for their possible use in preventing illness (Barriuso and Barriuso 2015). Mostly in the genera *Aspergillus, Candida,* and *Saccharomyces*, QS activates several numbers of fungi signalling pathways which serve as sensing mechanisms for biofilm development and tissue culture frequency. In past years, several observations of QS-like activities in fungi, strangely, have shown a morphological change from filamentous mycelial form to yeast (Padder et al. 2018). Farnesol (Fig.8.1) has been the most extensively researched of the QSMs reported through the fungus. It hinders *C. albicans* from transforming from yeast to the mycelial phase and does not eliminate hyphae that have formed. Farnesol is synthesized in larger quantities at 37–40°C and prevents the growth of mycelial colonies throughout 1–50 mm. (Sharma and Jangid 2015) Numerous pathogens, especially *Staphylococcus aureus, Saccharomyces cerevisiae*, *Aspergillus* species, *Paracoccidioides brasiliensis,* and *Mycobacterium smegmatis*, have been proven to be adversely impacted by farnesol. (Albuquerque and Casadevall 2012)

In addition, after exposing *C. albicans* skin to hydrogen peroxide, Westwater and colleagues found no proof of a survival benefit of exogenous tyrosol

(Westwater et al. 2005). Another QSM created by *C. albicans* is tyrosol. It was noted that there is a long lag phase before increment begins when a strong culture of *C. albicans* is diluted into the bacterial culture (Han et al. 2011). Since biofilms comprise cells with different morphologies, farnesol likely seems responsible for influencing cellular morphology and, ultimately, in establishing matured biofilms. This is worthwhile to know as biofilms developed over 24 hours begin to react again with farnesol. The data support the idea that emerging biofilm suffers commitment or decommitment in a similar way to planktonic cells in cell medium.

Farnesol also can influence monocyte progression into dendritic cells. The microorganism's LuxR-type receptor protein, which would be a part of the LuxI-LuxR-type QS system found in many Gram-negative microbes, has proven to be highly controlled by QSM AHL adhesion: in the absence of its ligand, just a tiny portion of LuxR receptor proteins are steady, whereas a large number of the specific receptors have a substantial adverse effect.

8.2.1 Regulation and Mechanism

Quorum sensing is only loosely associated with cell numbers. After a cell produces an autoinducer, the molecule rapidly diffuses out of the cell. The more cells in a given space, the faster the autoinducer builds up, and the more likely it will reenter the cell, triggering the luminescence response. Under laboratory conditions, the artificial addition of the autoinducer to the appropriate concentrations will even cause cells grown at low density to glow.

8.2.2 *Vibrio fischeri*

A discovery that fundamentally changed the thought about microbes was made while studying a peculiar marine bacterium called *V. fischeri*. In the dark, colonies of *V. fischeri* are bioluminescent. Although these bacteria can survive as planktonic forms in marine environments, *V. fischeri* predominantly lives in symbiosis with certain marine animals, such as the Hawaiian bobtail squid (*Euprymna scolopes*). The bacteria colonize and receive nutrients inside the light organ, a structure located in the squid's mantle cavity. The bacteria glow under only certain conditions. The genes needed to make light are 'off' when bacterial cell densities

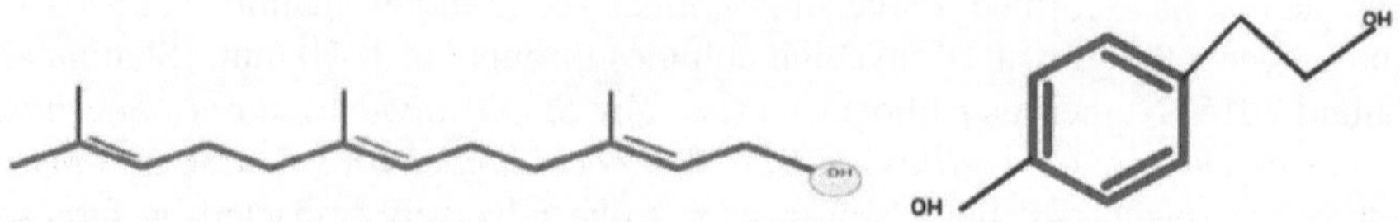

(A) (B)

FIGURE 8.1 General structure of farnesol and tyrosol.

are low, such as in seawater, but turn on under crowded conditions, such as in the squid's light organ when enough cells are together to make a visual impact.

8.2.3 Benefit of Quorum Sensing to *V. fischeri*

The bacteria provide luminescence for the squid after sunset when the animal emerges from its hiding place in the sand to search for food. As the squid swims in the moonlit night, its light organ projects light downwards to hide the squid from any predators below. Looking up, a fish would perceive the light of the night sky instead of the dark silhouette of a squid.

8.3 ORIGIN AND DEVELOPMENT OF FUNGAL QUORUM SENSING

The quorum sensing mechanism was very well proven in several organisms by the end of the 20th century. Its abundance in *E.* coli is presumably what inspired investigators to extend their studies into eukaryotes. If eukaryotes genuinely have a QS mechanism, it will provide fresh insights into the relationships between mycelial and yeast forms, the pathogenesis and toxicity of fungi, and other functions of these compounds in regulatory systems (Thomas B. Rasmussen and Givskov 2006). Phenotypic differences were once thought to be an evolved trait in some fungi: from soil-specific saprophytic mycelial form to human– or animal-specific disease-causing yeast. Dimorphism can be referred to as a 'dual personality' of fungi, i.e., the same organism trying to behave differently with a change in conditions, such as heat, carbon, nitrogen, blood glucose, and many other natural features (Thomas Bovbjerg Rasmussen et al. 2005b). There are two main theories regarding how cell-signaling management has developed in fungi. First, fungi have been residing in various environments where they are required to communicate with neighboring cells. For example, a dolipore septum is a small structure in some forms of fungus that enables multiple cellular types to connect with one another inwardly. Second, fungi must interact with one another outwardly to live. For example, expression of pathogenicity by fungi or bacteria is found to be successful only by developing a quorum-sensing mechanism (Sharma and Jangid 2015).

8.3.1 Bioengineering of Fungal QS

QS mechanisms from various microorganisms have been deliberately manipulated to create interaction across diverse species using synthesized and computational science methods. Systems biology uses technology and standardization principles to create new regulation circuit designs with novel characteristics and various bioactivities, permitting behaviour and labour division within such a community.

C. albicans is a common fungi detected in Membrane bioreactor (MBR) – activated sludge. Recently, some microorganisms and fungi have been shown to generate

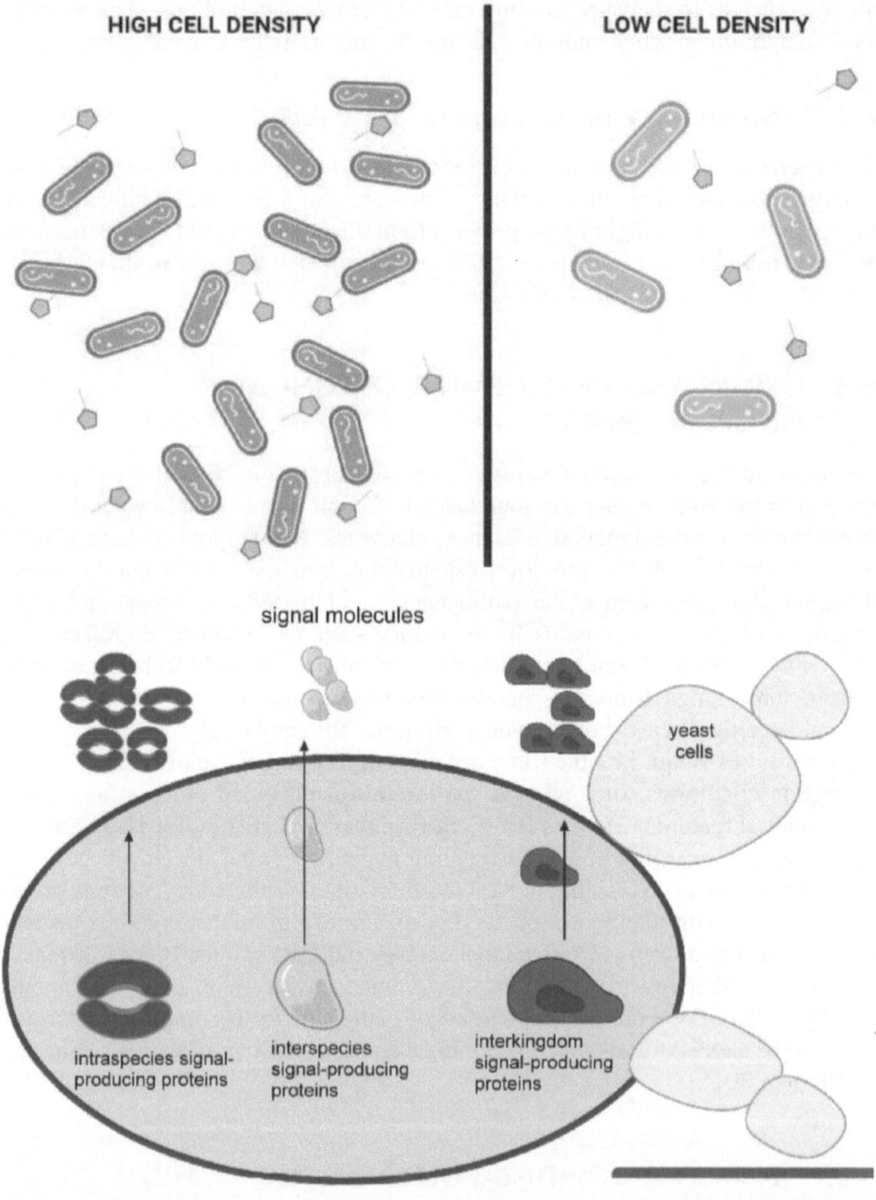

FIGURE 8.2 Modulation of mediated signaling in fungi: (a) specific (Left panel) and (b) non-specific (Right panel).

farnesol, which can conflict with fungal growth by limiting the growth of cells or affecting metabolic pathways (Choo et al. 2020). QS has been demonstrated to represent significant roles throughout many crucial bioreactors during the past few decades, such as microbial activity (both primary and secondary metabolism), pathogenesis (host colonization, invasion, antibiotic-resistant, and so on), plasmid conjugation, motility, and biofilm formation. This is despite the reality that many QS signal transduction or gene networks have been found in bacteria extracted via pollution treatment technologies (Yong and Zhong 2012).

8.4 DOWN-REGULATION OF QS

QS is based on interactions that include signalling generation, recognition, or gene activation/inactivation. Any of these stages could be disrupted, causing the QS mechanism to fail and possibly having a devastating affect on the survival and pathogenicity of bacteria and fungi (Pan and Ren 2009). One of the major downsides of using correspondent varieties to recognize quorum-sensing inhibitors is that even the reporter phenotypic expression may be caused by various factors in addition to quorum-sensing disruption (Bulman et al. 2011). The easiest way to rule out this possibility is to confirm that the complexation does not affect the investigative journalist subtype when it is separate from quorum sensing (for example, by placing GFP, lacZ, or fluorescence emission alleles under the cis-regulatory element) (Defoirdt et al. 2013). The samples without farnesol or Amph-B had never shown the establishment of pseudohyphae, which are chains of extended, irregular blastospores. Farnesol could also regulate pathogenicity activation and biofilm growth. The surface structure of the cells did not seem as uniform after the injection of farnesol as it did earlier, indicating a potential degradation of cytosolic content. At great amounts, farnesol substantially changes the bacterial cell surface. (Décanis et al. 2011). In a multitude of cancerous cells, farnesol has already been claimed to limit the growth of cells and/or trigger apoptosis; unfortunately, the published IC50 values differed greatly for different tumour types and cell differentiation lines (Scheper et al. 2008).

8.4.1 Autoinducer

Induction of a QS gene system requires accumulation of a molecule called an autoinducer. Autoinducers mediate quorum sensing; autoinducer-receptor complexes direct the expression of QS-dependent target genes. Autoinducer-receptor complexes activate expression of the gene ciphering the autoinducer synthase, which elevates the extracellular autoinducer concentration as the bacteria enter into quorum-sensing mode.

After the cell produces an autoinducer, the molecule rapidly diffuses out of the cell. The more cells there are in a given space, the faster the autoinducer builds up and the more likely it will reenter cells and trigger the response. N-acyl homoserine lactones (AHLs) are used as autoinducers by many Gram-negative pathogens, such as *P. aeruginosa*.

Gram-negative quorum-sensing bacteria utilize small molecules as autoinducers, and two types of cognate receptor detection determine these autoinducers: cytoplasmic transcription factors or transmembrane two-component histidine sensor kinases.

Gram-positive bacteria generally use oligopeptides as autoinducers, and the partner receptors are transmembrane two-component histidine sensor kinases.

8.4.1.1 *Ophiostoma piceae*

O. piceae is a wood-staining fungus that causes discoloration on pine and spruce. Ignoring that it was not a pathogenic species but associated with nonaggressive bark beetles, the disease results in significant economic losses in the forestry industry. Its ability to produce a structural shift from yeast into hyphal aspects dependent upon fungal growth factors most likely helps its adaptation to various environmental conditions (de Salas et al. 2015). Even during the exponential development stage, farnesol development in the fermentation media of this dimorphic fungi organizes the morphological change between yeast and hyphae, possibly controlled by transcribed inhibitor (Barriuso et al. 2018).

8.4.1.2 *P. aeruginosa*

Many research organizations have now employed reliable control strains. It is necessary to examine how QS causes pathology in *P. Aeruginosa* since QS influences a huge variety of elements in this bacterium's activity. In parallel to jungle strains of *P. aeruginosa*, several publications investigating QS early illness have employed organisms that had ablation of one or more of six QS-related genes. Analyses on the function of *P. aeruginosa* QS in transmission have shown that reductions of one or more QS domains diminish *P. aeruginosa*'s lethality relative to wild-type *P. aeruginosa* in the decayed model, a murine model of rapid pneumonia, and a rat prototype of chronic lung infections. Recently, researchers have worked to clarify the physical process by which bacterial QS is regulated. Several routinely used drugs are this often to *P. aeruginosa*. Even though several strains are sensitive to gentamicin, tobramycin, colistin, and amikacin, tolerant variants have arisen. Gentamicin and carbenicillin are often used to treat severe *Pseudomonas* infections. Several vaccinations are currently being investigated. However, none are presently ready for broad use ("What Is Pseudomonas Aeruginosa?" n.d.). The manufacture of stimuli, some of which are produced by mass transportation or propagation outside the cell, is the common method across all QSMs. High QSM contents cause QSMs to attach to the transmitter, which allows them to connect with promoter region repeats and stimulate the transcriptional controllers of the gene. QS can be hindered in 3 stages:

(1) Slowing down the manufacturing of QSMs;
(2) Deteriorating via enzymatic reactions the above QSMs; and
(3) Blocking senses via homologues of QSMs.

("Fungal Quorum-Sensing Molecules and Inhibitors with Potential Antifungal Activity: A Review—PMC" n.d.)

8.4.2 The Lux System

The autoinducer quickly diffuses out of the cell, typically without a chance to interact further with the lux system inside the cell; however, when the autoinducer is at a high concentration outside of the cell (as in crowded conditions), a number of these molecules are likely to reenter the cells and bind to the regulatory molecule lux r. The lux r-autoinducer complex binds to the operon and activates transcription;

hence, in the expression of the target genes, more lux I and more autoinducers are produced, which ultimately further enhances the expression of this operon. The enzyme responsible for light production, luciferase, consists of the product of lux a and lux b genes. Luciferase catalyzes a redox reaction that produces oxidized and reduced chemical products and blue-green light because the lux proteins, like other proteins, require energy. Cells turn this system on only when appropriate, such as when they are crowded together in the Hawaiian bobtail squid's light organ.

QS regulation can be employed as a strategy to invade hosts successfully: a single bacterium expressing virulence genes could be determined and dealt with by the host's immune response, but a coordinated attack by a population of bacteria may swamp a host before it has a chance to defend itself. Many bacteria control several aspects of life using QS, including biofilm formation, bioluminescence, virulence, and DNA exchange.

8.4.3 Biofilm

Biofilm is a distinct structure formed by bacteria adsorbed on the surface of inert or active materials to adapt to the living environment. It is made up of an extracellular matrix that includes polysaccharides and proteins. It shows new biological traits and stronger environmental adaptability.

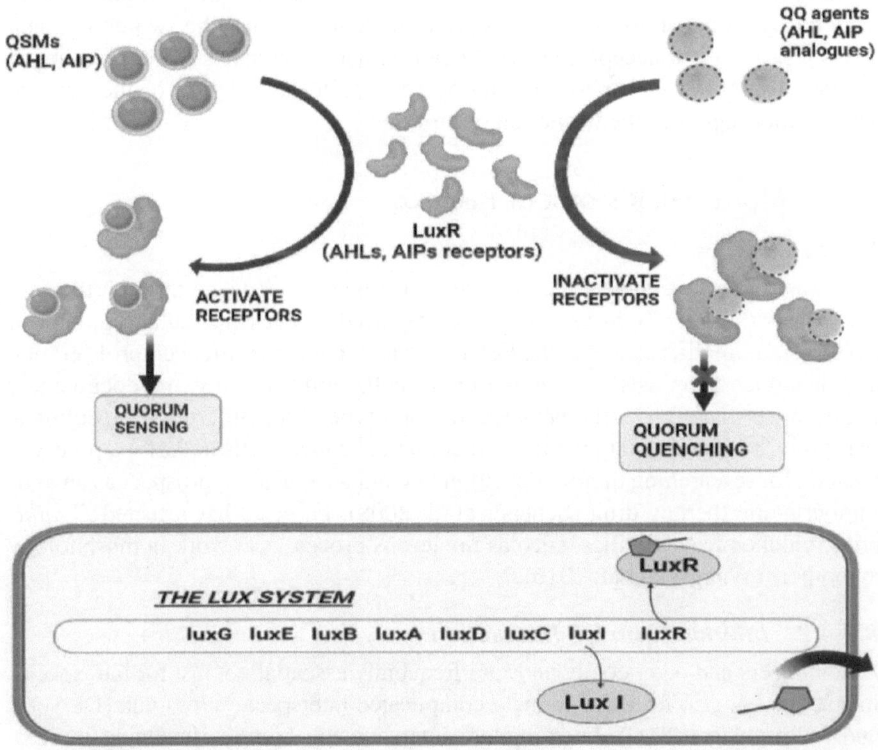

FIGURE 8.3A, B Regulation and antiregulation of QS.

TABLE 8.1
Fungi and Their Effect on Down-Regulation of QS

QSMs	Effects	References
Farnesol	Limits development as well as morphogenesis; involvement in the susceptibility to oxidative stress hampers hyphal development	(Nickerson et al. 2006)
Tyrosol	Causes morphogenesis; antifungal properties	(Chang et al. 2019)
Cryptococcus neoformans QSP1	Supports the development and generation of virulence genes	(Bulman et al. 2011)

The formation of bacterial biofilms is parallel to the development of drug resistance. It is generally believed that bacterial biofilms can result in bacterial drug resistance through penetration restriction mechanisms, nutrition restriction mechanisms, and drug resistance phenotypic mechanisms.

When a signal molecule reaches a certain threshold, the signal molecule attaches to the corresponding signal molecule receptor and activates the receptor. The activated receptor then triggers the relevant transcriptional regulators to make extracellular polysaccharides, toxic factors, alginates, and so on, thereby causing bacteria to form biofilms. The QS system controls the biofilm of Gram-positive bacteria by using oligopeptides as signal molecules, which the two-component sensing protein can accept after modification, and it regulates the expression of the target gene through phosphorylation and dephosphorylation of the protein. This further regulates the formation of biofilms.

8.4.4 Molecular Response of Farnesol

8.4.4.1 Farnesol

Farnesol, an extracellularly secreted quorum-sensing substance through the fungal species *C. albicans*, has recently been reported as an extracellular regulator of death in mammalian cancer cells. Farnesol was shown to restrict cell proliferation and/or induce apoptosis in various tumour cells; unfortunately, the documented IC50 values varied greatly between tumour types and different cell cultures. Farnesol's potential to target sites in mammalian cancer cells makes it a potential weapon for researching tumour development and an appealing prospect as an antitumourigenic therapy drug (Scheper et al. 2008). Farnesol has affected *C. albicans*' biochemical activities, such as numerous processes at work in morphology and others (Wongsuk et al. 2016).

8.4.4.2 Interkingdom QS Interaction

Autoinducers and other compounds are frequently essential not just for intraspecies interaction but also for the extremely complicated interspecies cross-talk (De Sordi and Mühlschlegel 2009). Interkingdom interaction in biofilms involving the caries-promoting microbes *Streptococcus mutans* and *C. albicans* has been documented using QS technologies (Barriuso et al. 2018). The morphological yeast *C. albicans* best describes QS systems (Fig. 8.2a and 8.2b). This variety is the causative agent of the bulk of frequent mycosis and a component of human and animal flora (Barriuso et al. 2018). *C. albicans* tends to change among budding yeast and polarized filament production, and this capacity is likely to be critical for developing sickness. One of the first fungi discovered to have quorum-sensing machinery was a dimorphic fungus recognized for its relevance as an invasive human pathogen (Hogan 2006). Indeed, a full growth medium and a GlcNAc-induced differentiated test for GTF showed a decrease in the percentage of mycelia. Morphogenesis is viewed as a crucial pathogenicity factor for *C. albicans* and some other dimorphic fungi. QSM does not just prevent GTF but also produces a morphological shift to rapidly expanding yeast cells in both situations (Hornby et al. 2001). The management of morphology in *C. albicans* regulated by E,E-farnesol is complex but has been properly examined.

Eukaryotes can disrupt microbial interaction by generating molecular signals that may interact against bacterial QS.

According to Vicente-Soler et al. (2016), Whenever bacteria and fungus co-culture, they usually produce antifungal substances. For example, *Bacillus thuringiensis and Pseudomonas putida* synthesize undiscovered antifungal pharmaceuticals that lower urediniospore germination and disease severity generated by the leaf rust pathogens *Hemileia vastatrix* (Khalid and Keller 2021). Moreover, numerous QS indicators, primarily those belonging to the same family, have similar features, implying the probability of cross-species signal disruption (Zhang and Dong 2004). Its capacity for signalling molecules to traverse species and subspecies is a vital organizational and tremendously productive research area (Clinton and Rumbaugh 2016).

8.4.4.2.1 C. albicans *with Bacterial Association*

C. albicans is a human body commensal as well as an opportunistic pathogen that is commonly blamed for nosocomial infections. Most of these infections are probably linked to biofilm growth in or on implanted medical equipment. *C. albicans* cells might interact with bacteria within biofilms, primarily through biochemical and metabolic indirect connections and/or direct interactions, including both yeast, hyphal version of the fungal cell, or, relatively rarely, combination forms (Bernard et al. 2020).

C. albicans has been connected to oral streptococci in denture- and mucosal-related disorders. Oral streptococci are the main colonizers of oral cavity surfaces, generating biofilms on which oral bacteria such as *C. albicans* cling, supporting the development and persistence of polymicrobial ecosystems (Dutton et al. 2016). The overall significance of fungus-bacterium relationships is rapidly gaining recognition in biomedical work. *Fusarium* species frequently generate fusaric acid, which is dangerous for both eukaryotes and prokaryotes and prevents the bacterial formation of antifungal compounds (Tarkka et al. 2009). Although *C. albicans* easily develops biofilms upon abiotic materials and mucosal surfaces of something like the oral cavity, there is currently moderate evidence concerning biofilm formation on the vaginal mucosa (Ponde et al. 2021). Microbial QS activity extends to partnerships involving bacteria and eukaryotes, including host-pathogen relationships in medical and commercial settings that seem to be of particular interest (Lowery et al. 2008). Furthermore, although the initial contact between *C. albicans* and *S. aureus* looks to just be synergistic, it is plausible that the interaction turns antagonistic or adversarial at a certain point, even during biofilm's growth. Farnesol, a candidal QS molecule, has been demonstrated to influence biofilm formation by *S. aureus*, together with affecting cellular membrane integrity, viability, and sensitivity to several clinically important antibiotics (Shirtliff et al. 2009). Ultimately, research analysis revealed that perhaps the denture microbiome/mycobiome is largely resistant to oral hygiene challenges, yet *Candida* species may interact with critical oral taxa. When the chance occurs, these interactions could affect the network structure and mark a change between health and illness. (Delaney et al. 2019)

8.4.5 Antimicrobial Resistance

It was discovered that microorganisms could develop resistance to any of the drugs used. Most pathogenic microorganisms are capable of developing resistance to at least some antimicrobial agents.

The discovery of microbial quorum-sensing systems has provided new hope for studying the regulatory mechanism of drug resistance mechanisms and overcoming drug resistance. The results show that the quorum-sensing system regulates various cellular processes of microorganisms, such as pathogenic gene expression, toxin production, and extracellular polysaccharide synthesis, and plays an important regulatory role in the process of drug efflux pumps and the formation of microbial biofilms.

Multiple-drug resistance has made it difficult to solve the issue of microbial resistance. More and more microorganisms have evolved multiple resistance mechanisms and thus have become "superbacteria."

There are three types of antimicrobial resistance. First, intrinsic resistance is said to be the resistance exhibited due to the inherent properties of the bacterium. Second, acquired resistance is defined as the resistance shown when a previously sensitive bacterium acquires a resistance mechanism by either a mutation or acquiring new genetic material from an exogenous source (horizontal gene transfer, of which there are three types: the AMR gene being associated with mobile genetic elements, including mobile introns; loss of synteny [genetic loci] of the insertion site in the host; and obtaining an AMR gene through gene transfer. Multiple copies of resistant genes give bacteria the best chance to steer clear of antibiotics). Third, adaptive resistance is the resistance to one or more antibiotics caused by a specific environmental signal (e.g., stress, growth state, pH, concentrations of ions, nutrient conditions, or subinhibitory levels of antibiotics).

Antimicrobial resistance mechanisms (the main methods for resistance) can be categorized as:

- limiting uptake of a drug
- modifying a drug target

- inactivating a drug
- active drug efflux

Intrinsic resistance may entail limiting uptake, drug inactivation, and drug efflux; acquired resistance mechanisms utilize drug target modification, drug inactivation, and drug efflux. Because of differences in structure, there is a variation in the methods used by Gram-negative and Gram-positive bacteria. Gram-negative bacteria utilize all four main mechanisms, whereas Gram-positive bacteria less commonly limit the uptake of a drug (no lipopolysaccharide in the outer membrane) and do not have the capacity for some drug efflux mechanisms.

8.5 CONCLUSION

QSMs are tiny compounds released by cells that function as signalling molecules, which govern the network based on cell population concentration. The relevance of quorum-sensing processes as therapeutical targets may be tested if the relevance of quorum sensing in pathogenic fungus is set up and the precise mechanistic details are understood. Diverse intraspecies and interspecies quorum quenching methods have been discovered in bacteria and fungi, as stated earlier. It has become apparent that such microbial signalling molecules have an impact on the host organisms as well. In this section, we explained the mechanisms of fungal quorum sensing and its related components. Table 8.1 demonstrates the critical roles of QSMs in yeast and dimorphic fungal cells. Antimicrobial resistance emergence may be unavoidable in the evolutionary process, whereas the mechanisms that safeguard its perseverance, even without antibiotic selection pressure, are not fully explained properly. There is still a lot to learn regarding the dynamics of antimicrobial-resistant determinants within microbial communities.

REFERENCES

Albarracín Orio, Andrea G., Daniel Petras, Romina A. Tobares, Alexander A. Aksenov, Mingxun Wang, Florencia Juncosa, Pamela Sayago, Alejandro J. Moyano, Pieter C. Dorrestein, and Andrea M. Smania. 2020. "Fungal–Bacterial Interaction Selects for Quorum Sensing Mutants with Increased Production of Natural Antifungal Compounds." *Communications Biology* 3 (1): 1–9. https://doi.org/10.1038/s42003-020-01342-0.

Albuquerque, Patrícia, and Arturo Casadevall. 2012. "Quorum Sensing in Fungi – A Review." *Medical Mycology* 50 (4): 337–45. https://doi.org/10.3109/13693786.2011.652201.

Barriuso, Jorge, and Jorge Barriuso. 2015. "*Quorum Sensing* Mechanisms in Fungi." *AIMS Microbiology* 1 (1): 37–47. https://doi.org/10.3934/microbiol.2015.1.37.

Barriuso, Jorge, and Centro de Investigaciones Biológicas, Consejo Superior de Investigaciones Científicas. 2015. "*Quorum Sensing* Mechanisms in Fungi." *AIMS Microbiology* 1 (1): 37–47. https://doi.org/10.3934/microbiol.2015.1.37.

Barriuso, Jorge, Deborah A. Hogan, Tajalli Keshavarz, and María Jesús Martínez. 2018. "Role of Quorum Sensing and Chemical Communication in Fungal Biotechnology and Pathogenesis." *FEMS Microbiology Reviews* 42 (5): 627–38. https://doi.org/10.1093/femsre/fuy022.

Bernard, C., M. Girardot, and C. Imbert. 2020. "Candida Albicans Interaction with Gram-Positive Bacteria within Interkingdom Biofilms." *Journal de Mycologie Médicale* 30 (1): 100909. https://doi.org/10.1016/j.mycmed.2019.100909.

Bulman, Zackery, Phuong Le, André O. Hudson, and Michael A. Savka. 2011. "A Novel Property of Propolis (Bee Glue): Anti-Pathogenic Activity by Inhibition of N-Acyl-Homoserine Lactone Mediated Signaling in Bacteria." *Journal of Ethnopharmacology* 138 (3): 788–97. https://doi.org/10.1016/j.jep.2011.10.029.

Chang, Aiping, Shiwei Sun, Li Li, Xiaoyun Dai, Hui Li, Qiaomei He, and Hu Zhu. 2019. "Tyrosol from Marine Fungi, a Novel Quorum Sensing Inhibitor against Chromobacterium Violaceum and Pseudomonas Aeruginosa." *Bioorganic Chemistry* 91: 103140. https://doi.org/10.1016/j.bioorg.2019.103140.

Choo, Kwang-Ho, Pyung-Kyu Park, and Hyun-Suk Oh. 2020. "Quorum Sensing and Quorum Quenching in Membrane Bioreactors." In *Current Developments in Biotechnology and Bioengineering*, 245–74. Elsevier. https://doi.org/10.1016/B978-0-12-819809-4.00012-7.

Clinton, Allie, and Kendra P. Rumbaugh. 2016. "Interspecies and Interkingdom Signaling via Quorum Signals." *Israel Journal of Chemistry* 56 (5): 265–72. https://doi.org/10.1002/ijch.201400132.

Décanis, Nadège, Neftaha Tazi, Alexandra Correia, Manuel Vilanova, and Mahmoud Rouabhia. 2011. "Farnesol, a Fungal Quorum-Sensing Molecule Triggers Candida Albicans Morphological Changes by Downregulating the Expression of Different Secreted Aspartyl Proteinase Genes." *The Open Microbiology Journal* 5: 119–26. https://doi.org/10.2174/1874285801105010119.

Defoirdt, Tom, Gilles Brackman, and Tom Coenye. 2013. "Quorum Sensing Inhibitors: How Strong Is the Evidence?" *Trends in Microbiology* 21 (12): 619–24. https://doi.org/10.1016/j.tim.2013.09.006.

Delaney, Christopher, Lindsay E. O'Donnell, Ryan Kean, Leighann Sherry, Jason L. Brown, Gareth Calvert, Christopher J. Nile, et al. 2019. "Interkingdom Interactions on the Denture Surface: Implications for Oral Hygiene." *Biofilm* 1: 100002. https://doi.org/10.1016/j.bioflm.2019.100002.

De Sordi, Luisa, and Fritz A. Mühlschlegel. 2009. "Quorum Sensing and Fungal–Bacterial Interactions in Candida Albicans: A Communicative Network Regulating Microbial Coexistence and Virulence." *FEMS Yeast Research* 9 (7): 990–99. https://doi.org/10.1111/j.1567-1364.2009.00573.x.

Dutton, Lindsay C., Howard F. Jenkinson, Richard J. Lamont, and Angela H. Nobbs. 2016. "Role of Candida Albicans Secreted Aspartyl Protease Sap9 in Interkingdom Biofilm Formation." *Pathogens and Disease* 74 (3): ftw005. https://doi.org/10.1093/femspd/ftw005.

"Fungal Quorum-Sensing Molecules and Inhibitors with Potential Antifungal Activity: A Review—PMC." n.d. Accessed July 17, 2022. https://www.ncbi.nlm.nih.gov/pmc/articles/PMC6571750/.

Han, Ting-Li, Richard D. Cannon, and Silas G. Villas-Bôas. 2011. "The Metabolic Basis of Candida Albicans Morphogenesis and Quorum Sensing." *Fungal Genetics and Biology* 48 (8): 747–63. https://doi.org/10.1016/j.fgb.2011.04.002.

Hogan, Deborah A. 2006. "Talking to Themselves: Autoregulation and Quorum Sensing in Fungi." *Eukaryotic Cell* 5 (4): 613–19. https://doi.org/10.1128/EC.5.4.613-619.2006.

Hornby, Jacob M., Ellen C. Jensen, Amber D. Lisec, Joseph J. Tasto, Brandon Jahnke, Richard Shoemaker, Patrick Dussault, and Kenneth W. Nickerson. 2001. "Quorum Sensing in the Dimorphic Fungus Candida Albicans Is Mediated by Farnesol." *Applied and Environmental Microbiology* 67 (7): 2982–92. https://doi.org/10.1128/AEM.67.7.2982-2992.2001.

Khalid, Saima, and Nancy P. Keller. 2021. "Chemical Signals Driving Bacterial–Fungal Interactions." *Environmental Microbiology* 23 (3): 1334–47. https://doi.org/10.1111/1462-2920.15410.

Lowery, Colin A., Tobin J. Dickerson, and Kim D. Janda. 2008. "Interspecies and Interkingdom Communication Mediated by Bacterial Quorum Sensing." *Chemical Society Reviews* 37 (7): 1337–46. https://doi.org/10.1039/B702781H.

Maillé, Émilie, Manon Ruffin, Damien Adam, Hatem Messaoud, Shantelle L. Lafayette, Geoffrey McKay, Dao Nguyen, and Emmanuelle Brochiero. 2017. "Quorum Sensing Down-Regulation Counteracts the Negative Impact of Pseudomonas Aeruginosa on CFTR Channel Expression, Function and Rescue in Human Airway Epithelial Cells." *Frontiers in Cellular and Infection Microbiology* 7. https://www.frontiersin.org/articles/10.3389/fcimb.2017.00470.

Mehmood, Arshad, Guorong Liu, Xin Wang, Guannan Meng, Chengtao Wang, and Ya Liu. 2019. "Fungal Quorum-Sensing Molecules and Inhibitors with Potential Antifungal Activity: A Review." *Molecules* 24 (10): 1950. https://doi.org/10.3390/molecules24101950.

Nickerson, Kenneth W., Audrey L. Atkin, and Jacob M. Hornby. 2006. "Quorum Sensing in Dimorphic Fungi: Farnesol and Beyond." *Applied and Environmental Microbiology* 72 (6): 3805–13. https://doi.org/10.1128/AEM.02765-05.

Padder, Sajad Ahmad, Rajendra Prasad, and Abdul Haseeb Shah. 2018. "Quorum Sensing: A Less Known Mode of Communication among Fungi." *Microbiological Research* 210: 51–58. https://doi.org/10.1016/j.micres.2018.03.007.

Pan, Jiachuan, and Dacheng Ren. 2009. "Quorum Sensing Inhibitors: A Patent Overview." *Expert Opinion on Therapeutic Patents* 19 (11): 1581–1601. https://doi.org/10.1517/13543770903222293.

Ponde, Nicole O., Léa Lortal, Gordon Ramage, Julian R. Naglik, and Jonathan P. Richardson. 2021. "Candida Albicans Biofilms and Polymicrobial Interactions." *Critical Reviews in Microbiology* 47 (1): 91–111. https://doi.org/10.1080/1040841X.2020.1843400.

Rasmussen, Thomas B., and Michael Givskov. 2006. "Quorum-Sensing Inhibitors as Anti-Pathogenic Drugs." *International Journal of Medical Microbiology* 296 (2–3): 149–61. https://doi.org/10.1016/j.ijmm.2006.02.005.

Rasmussen, Thomas Bovbjerg, Mette E. Skindersoe, Thomas Bjarnsholt, Richard K. Phipps, Kathrine Bisgaard Christensen, Peter Ostrup Jensen, Jens Bo Andersen, et al. 2005a. "Identity and Effects of Quorum-Sensing Inhibitors Produced by Penicillium Species." *Microbiology* 151 (5): 1325–40. https://doi.org/10.1099/mic.0.27715-0.

Rasmussen, Thomas Bovbjerg, Mette E. Skindersoe, Thomas Bjarnsholt, Richard K. Phipps, Kathrine Bisgaard Christensen, Peter Ostrup Jensen, Jens Bo Andersen, et al. 2005b. "Identity and Effects of Quorum-Sensing Inhibitors Produced by Penicillium Species." *Microbiology* 151 (5): 1325–40. https://doi.org/10.1099/mic.0.27715-0.

Salas, Felipe de, María Jesús Martínez, and Jorge Barriuso. 2015. "Quorum-Sensing Mechanisms Mediated by Farnesol in Ophiostoma Piceae: Effect on Secretion of Sterol Esterase." *Applied and Environmental Microbiology* 81 (13): 4351–57. https://doi.org/10.1128/AEM.00079-15.

Scheper, Mark A., Mark E. Shirtliff, Timothy F. Meiller, Brian M. Peters, and Mary Ann Jabra-Rizk. 2008. "Farnesol, a Fungal Quorum-Sensing Molecule Triggers Apoptosis in Human Oral Squamous Carcinoma Cells." *Neoplasia* 10 (9): 954–63. https://doi.org/10.1593/neo.08444.

Sharma, Rohit, and Kamlesh Jangid. 2015. "Fungal Quorum Sensing Inhibitors." In *Quorum Sensing vs Quorum Quenching: A Battle with No End in Sight*, edited by Vipin Chandra Kalia, 237–57. Springer. https://doi.org/10.1007/978-81-322-1982-8_20.

Shirtliff, Mark E., Brian M. Peters, and Mary Ann Jabra-Rizk. 2009. "Cross-Kingdom Interactions: Candida Albicans and Bacteria." *FEMS Microbiology Letters* 299 (1): 1–8. https://doi.org/10.1111/j.1574-6968.2009.01668.x.

Tarkka, Mika T., Alain Sarniguet, and Pascale Frey-Klett. 2009. "Inter-Kingdom Encounters: Recent Advances in Molecular Bacterium–Fungus Interactions." *Current Genetics* 55 (3): 233–43. https://doi.org/10.1007/s00294-009-0241-2.

Vicente-Soler, Jerónima, Marisa Madrid, Alejandro Franco, Teresa Soto, José Cansado, and Mariano Gacto. 2016. "Quorum Sensing as Target for Antimicrobial Chemotherapy." In *New Weapons to Control Bacterial Growth*, edited by Tomas G. Villa and Miguel Vinas, 161–84. Springer International Publishing. https://doi.org/10.1007/978-3-319-28368-5_8.

Westwater, Caroline, Edward Balish, and David A. Schofield. 2005. "*Candida Albicans*-Conditioned Medium Protects Yeast Cells from Oxidative Stress: A Possible Link between Quorum Sensing and Oxidative Stress Resistance." *Eukaryotic Cell* 4 (10): 1654–61. https://doi.org/10.1128/EC.4.10.1654-1661.2005.

"What Is Pseudomonas Aeruginosa?" n.d. Accessed August 1, 2022. https://www.ehagroup.com/resources/pathogens/pseudomonas-aeruginosa/.

Wongsuk, Thanwa, Potjaman Pumeesat, and Natthanej Luplertlop. 2016. "Fungal Quorum Sensing Molecules: Role in Fungal Morphogenesis and Pathogenicity: Quorum Sensing in Fungi." *Journal of Basic Microbiology* 56 (5): 440–47. https://doi.org/10.1002/jobm.201500759.

Wynendaele, Evelien, Antoon Bronselaer, Joachim Nielandt, Matthias D'Hondt, Sofie Stalmans, Nathalie Bracke, Frederick Verbeke, Christophe Van De Wiele, Guy De Tré, and Bart De Spiegeleer. 2013. "Quorumpeps Database: Chemical Space, Microbial Origin and Functionality of Quorum Sensing Peptides." *Nucleic Acids Research* 41 (D1): D655–59. https://doi.org/10.1093/nar/gks1137.

Yong, Yang-Chun, and Jian-Jiang Zhong. 2012. "Impacts of Quorum Sensing on Microbial Metabolism and Human Health." In *Future Trends in Biotechnology*, edited by Jian-Jiang Zhong, Vol. 131, 25–61. Advances in Biochemical Engineering/Biotechnology. Springer. https://doi.org/10.1007/10_2012_138.

Zhang, Lian-Hui, and Yi-Hu Dong. 2004. "Quorum Sensing and Signal Interference: Diverse Implications." *Molecular Microbiology* 53 (6): 1563–71. https://doi.org/10.1111/j.1365-2958.2004.04234.x.

9 Antimicrobial Peptides
Sources, Types, and Antibiofilm Activity

Bhavini K, Abigail Fernandes, Vinothkannan Ravichandran, Dibyajit Lahiri, and Renitta Jobby

9.1 INTRODUCTION

In this era, when we are dependent on antibiotics, increasing incidences of antibiotic-resistant and multidrug-resistant microorganisms pose a significant threat. Microorganisms form biofilms by colonizing different surfaces and their cells communicate with each other via quorum sensing. Biofilms are difficult to treat as they are microbial communities living inside matrices such as exopolysaccharides. Novel strategies to suppress the growing antimicrobial resistance and biofilm formation should be explored for managing the resistant pathogens. Antimicrobial peptides (AMPs) have been found in all major classes of organisms ranging from prokaryotes to eukaryotes (Zhang & Gallo, 2016), and have a wide range of structural and functional variety. AMPs are promising candidates for therapeutics against drug-resistant pathogens. Some of AMPs are already available on the market worldwide, with many more being investigated and in clinical trial stages (Koo & Seo, 2019; Dijksteel et al., 2021).

Basically, AMPs are short peptides or oligopeptides that target a broad spectrum of microorganisms, including bacteria, viruses, and even parasites. Also, eukaryotes and prokaryotes both produce AMPs as a part of their innate immune response (Rossi et al., 2008). AMPs are distinguished by their size (ranging from 15–30 amino acids), charge (usually positively charged), and ability to target pathogen membranes. Cell membranes of the pathogen are negatively charged and together with the biofilm surface, they attract the positively charged AMPs. Antimicrobial activities of AMPs can be further improved with modifications in their amino acid sequences.

Antimicrobial peptides have a varied range of activity against bacteria, fungi, mycobacteria, and viruses. Furthermore, cytotoxic properties possessed by AMPs against cancerous cells have been inferred (Roudi et al., 2017). Most AMPs have varied antimicrobial and antibiofilm potency. For example, it was deduced that when a synthetic cathelicidin (a type of AMP), NA-CATH:ATRA1-ATRA1 was employed at doses well below the MIC point,

DOI: 10.1201/9781003300557-9

it could suppress *Staphylococcus aureus* biofilm development. It was also found that the peptide LL-37 regulates *Pseudomonas aeruginosa* biofilm formation by inhibiting expression of genes involved for the process (Dean et al., 2011). The down-regulated genes in *P. aeruginosa* were involved in pili formation, flagella assembling, and quorum sensing (Overhage et al., 2008). Certain AMPs have also been seen to have antibacterial properties; for example, lactoferrin produced from milk inhibits the growth of *P. aeruginosa* biofilms. Furthermore, biofilm breakdown may be aided by AMPs binding to extracellular DNA (Algburi et al., 2017).

It is also claimed that using an antibiotic in combination with an antimicrobial peptide is preferable to using antibiotics separately. For example, a study suggests that using vancomycin with a lower MIC value in combination with lipopeptides such Pal-Lys-Lys-NH2 and Pal-Lys-Lys has resulted in a decline in the biofilm formation rate of *S. aureus* on vascular grafts (Cirioni et al., 2007). Another study, revealed that *P. aeruginosa* biofilms are eradicated by using a combination of specially produced broad-spectrum D-enantiomeric AMPs and antibiotics (de La Fuente-Núñez et al., 2015).

The first antimicrobial protein to be discovered (in 1922, by Alexander Fleming) was lyzozyme. With the discovery of penicillin, the focus shifted away from antimicrobial peptides. However, due to a spike in the number of drug-resistant bacteria in the postantibiotic era, AMPs are again in the spotlight. Nisin was one among the earliest AMPs discovered, around the same time as the discovery of penicillin (Abee & Delves-Broughton, n.d.), in milk. Other AMPs, like gramicidin (Dubos, 1939a, 1939b; Dubos & Cattaneo, 1939), tyrocidine (Dubos & Hotchkiss, 1941), alamethicin, and purothionin (Ohtani et al., 1977), were isolated after nisin discovery. AMPs are not limited to bacteria or fungi, but are also found in plants, chordates. and nonchordates (Bahar & Ren, 2013). Defensin is the first reported animal-originated AMP, isolated in 1956 from rabbit leukocytes (Zeya & Spitznagel, 1963). Subsequently, other AMPs, like bombinin and lactoferrin, from epithelium cells (Kiss & Michl, 1962) and cow milk (González-Chávez et al., 2009), respectively, were also discovered.

More than 5,000 AMPs have been successfully identified or synthesized to date. The Antimicrobial Peptide Database contains data related to all of the AMPs, and is updated regularly. There are currently 3,791 AMPs in the database, with 431 from bacteria, 4 from archaea, 7 from protozoa, 6 from fungi, 824 from plants, and 2519 from animal species (Kang et al., 2019). Cationic peptides and anionic peptides are both found in vertebrates, invertebrates, and plants, but cationic AMPs are the most common.

9.2 SOURCES OF ANTIMICROBIAL PROTEINS

As already mentioned, AMPs can be naturally derived from various sources like prokaryotes (bacteria) and eukaryotes (protozoa, fungi, insects, plants, and animals). Some of the most common sources of AMPs are mentioned here.

9.2.1 AMPs from Mammals

In vertebrates, more than 1,770 species of AMPs have been discovered to date. The most common sources of mammalian AMPs are humans and cattle. These AMPs support the innate immune system by interacting with inorganic molecules like hydrogen peroxide and nitric oxide, as well as antibacterial proteins like lysozyme, cathepsin G, and lactoferrin (Zhang et al., 2017).

The primary families of AMPs identified in animals are human host defence peptides (HDPs), such as cathelicidins and defensins. They were found to regulate immune response, apoptosis, and wound healing, in addition to antibacterial action (Wang, 2014). Many such AMPs or HDPs can be found on the skin, ears, eyes, mouth, lungs, intestine, and other areas of the mammal's body. Cathelicidin LL-37, a mammalian AMP, is found in the infant skin, while human beta-defensin 2 (hBD-2) is found in the elderly (Gschwandtner et al., 2014). Dairy is a vital source of AMPs; alpha, beta-lactoglobulin, lactoferrin, casein fractions, LfcinB, and other AMPs have been found in dairy products (Huan et al., 2020). AMPs are also found in breast milk, and they serve a vital function (Field, 2005).

9.2.2 AMPs from Plants

AMPs derived from plants were inferred to protect the plants from diseases caused by pathogens from air and soil. Plant-derived AMPs are classified into various groups such as thionins, defensins, and cyclotides (Zhang et al., 2021). Thionins are present in abundance in plant stems, roots, and leaves (Höng et al., 2021), and Gram-negative bacteria (Li et al., 2021) and fungi (Li et al., 2021). Plant AMPs have cysteine residues, which involves the formation of numerous disulphide bonds, critical for structural stability of the AMPs (Tang et al., 2018).

9.2.3 AMPs from Amphibians

Amphibians produce AMPs to help them defend themselves or endure pathogen attacks. Amphibian skin is regarded as one of the most valuable sources of AMPs. Maganins are one of the most extensively identified classes of AMPs in amphibians, particularly frog species. Skin secretions of *Xenopus*, *Silurana*, *Hymenochirus*, and *Pseudhymenochirus* species of frogs have been found to contain it (Michael Conlon & Mechkarska, 2014). Cancrin is another AMP found in marine amphibians like *Rana cancrivora* (Lu et al., 2008). Amphibians also produce cathelicidins, which are another class of AMPs. Cathelicidin-PV, for example, an AMP found in the skin of *Paa yunnanensis*, shows potent antibacterial activity and against a variety of fungi. Its efficacy against clinically isolated drug-resistant and conventional pathogen strains has also been studied, but was shown to have a poor haemolytic activity (Zhang et al., 2021).

9.2.4 AMPs from Insects

Insects produce AMPs in their bodies and blood cells which help them survive and adapt to harsh conditions. The amount of AMPs produced in an insect is different for each species. For example, insects like *Harmonia axyridis* (ladybird variety) and *Hermetia illucens* (black soldier fly) produce around 50 AMPs, whereas certain other insect species, like *Acyrthosiphon pisum* (pea aphids), have no AMPs (Huan et al., 2020). Cecropin is the most well-known and most studied kind of insect AMP having anti-inflammatory and anticancer properties, and it is present in such insects as silkworms, bees, and *Drosophila melanogaster*. Another AMP, derived from bees' royal jelly, is Jellein-1, whose synthesized form by lauric acid conjugation can work against *Leishmania major* parasites (Zahedifard et al., 2020).

9.2.5 AMPs from Microorganisms

Microorganisms such as bacteria and fungus are considered the most prevalent source of AMPs since chemical synthesis is very expensive and biological synthesis is comparatively easy. Nisin was the first AMP discovered in bacteria. Because of its antibacterial properties, nisin plays a significant role as a natural food preservative. Copsin is an AMP found in *Coprinopsis cinerea* (a mushroom species) that inhibits membrane synthesis in Gram-positive bacteria including *Enterococcus faecalis* and *Listeria monocytogenes*, resulting in bactericidal effects (Zhang et al., 2021).

9.3 STRUCTURE OF AMPs

AMPs have been divided into four groups based on their secondary structures. The α-helical and β-sheet are the two most common structures, while loop and extended peptide structures have also been studied (Di Somma et al., 2020).

9.3.1 α-Helical

These AMPs maintain a linear form when in aqueous solutions; however, when present in organic solvents or during interactions with bacterial membranes they appear as amphipathic structures (Di Somma et al., 2020). There is a distance of 0.15 nm and an angle of 100 degrees between adjacent amino acids in this α-helical conformation of AMPs (Bahar & Ren, 2013). The peptides isolated from *Xenopus laevis*, Magainin-2 and human LL-37, are amphipathic peptides belonging to this group (see Figure 9.1) (Kuroda et al., 2015).

9.3.2 β-Sheet

The peptides belonging to this group consist of two or more β-strands linked by stabilizing disulphide bonds. The disulphide bridges lead to the creation of amphipathic structures which are rigid and do not change conformation. This group of AMPs consists of the cathelicidin protegrins, tachyplesin, and defensins which are the dominant β-sheet members (see Figure 9.1) (Di Somma et al., 2020).

9.3.3 Loop

Peptides that are stabilized by disulphide, amide, or isopeptide bonds are classified as loop AMPs. Thanatin is an insect-derived peptide that contains β-hairpin structure and a single disulphide bond between the Cys11 and Cys18 (Ma et al., 2019). Lactoferricin B is another important member of this group. It was isolated from bovine sources and showed antifungal properties (Hwang et al., 1998).

9.3.4 Extended Peptides

This class consists of peptides that show irregular structures and are rich in arginine and tryptophan residues. Tritrpticin, of the cathelicidin family, and indolicidin, a tridecapeptide from bovine leukocytes, belong to the extended peptides. These peptides are short and contain transformable residues that can lead to enhanced antimicrobial and anticancer properties (Di Somma et al., 2020).

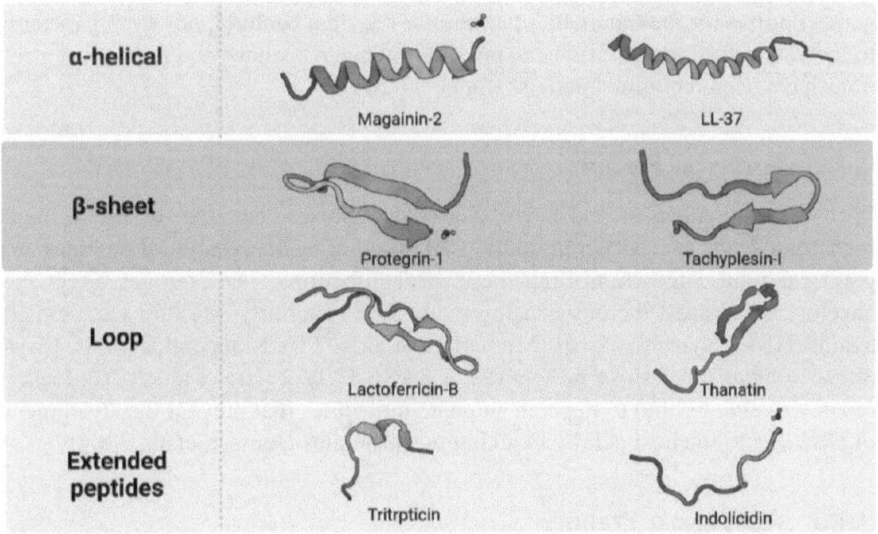

FIGURE 9.1 Structures of the antimicrobial peptides based on their secondary structures. PDB codes; 2MAG, Magainin-2; 2K6O, LL-37; 1PG1, Protegrin-1; 1MA2, Tachyplesin-I; 1LFC, Lactoferricin-B; 8TFV, Thanatin; 1D6X, Tritrpticin; 1G89, Indolicidin. Image created with BioRender.com.

9.4 TYPES OF AMPs

These peptides can be further divided based on their mode of action: antibacterial, antiviral, antifungal, or antiparasitic (Huan et al., 2020). We discuss these different categories in the following sections.

9.4.1 Antibacterial Peptides

These are the most investigated AMPs. Chiefly these peptides are cationic, targeting the bacterial cell membrane which leads to disintegration of the membrane. Many of the peptides have hydrophilic as well as hydrophobic domains, making them amphipathic. This also improves their ability to bind to lipid components. AMPs like buforin II inhibit cellular pathways by diffusing into the cells without changing membrane integrity (Park et al., 1998). Bacteriocins like nisin have been tested against antibiotic-resistant Gram-positive as well as Gram-negative bacteria (Bahar & Ren, 2013; Huan et al., 2020). Albopeptide 6 is a nonribosomal peptide isolated from *Streptomyces albodaciens* that is active against *Enterococcus faecium* (Wang et al., 2021). Mutacin-1140, produced by *Streptococcus mutans*, is known to inhibit the peptidoglycan synthesis in multidrug-resistant *S. aureus* and *Enteroccocus* (Hasper et al., 2006). Lysostaphin, produced by *Staphyloccocus simulans*, is a low-toxic peptide that targets the pentaglycine cross-links and hydrolyzes the cell wall (Chen et al., 2014). Antibacterial AMPs do not have specific target molecules or pathways and hence there are fewer chances for the development of resistance. AMPs individually or as combination are novel alternative options for the treatment of antibiotic-resistant bacteria; however, exposure to low doses for long duration can possibly induce resistance and high concentrations have been recommended (Zhang et al., 2021).

9.4.2 Antiviral Peptides

With the emergence of and havoc caused by viruses like SARS-COV-2, there is increased resistance to commonly used antiviral agents. Antiviral peptides are good candidates for controlling these viral infections. The peptides target the envelope of viruses, thereby causing membrane instability. Melittin was studied against HSV-1 syncytial virus (Memariani et al., 2020). Lipopeptide EK1C4 was effective against the spike protein (S) of SARS-COV-2 (Xia et al., 2020). Lactoferrin was able to bind to heparan sulphate molecules that prevent the attachment of HS virus to the host cell by blocking receptor sites (Jenssen et al., 2004).

9.4.3 Antifungal Peptides

Antifungal peptides have been grouped into membrane-traversing and non-membrane-traversing peptides. The antifungal peptides interfere with synthesis of DNA and RNA and can also interfere with protein synthesis. They can inhibit the growth of fungi by inducing apoptotic mechanisms, disrupting membranes, and metabolic turnovers (Di Somma et al., 2020). A few of the synthetic peptides are also found to have antifungal potential, for example, a concoction of 37 different peptides isolated from *Lactobacillus plantarum* TE10 was effective against *Aspergillus flavus* MD3 (Muhialdin et al., 2020).

9.4.4 Antiparasitic Peptides

There are several peptides that show antiparasitic activity. Magainins and cecropins were amongst the earliest identified peptides. These peptides have shown good activity against parasites causing deadly diseases like malaria and leishmania (Rhaiem & Houimel, 2016). Although protozoan parasites are multicellular organisms, the inhibiting mechanisms of antiparasitic peptides remain fairly similar in most cases. A synthetic peptide, Epi-1, was able to destroy the membrane of *Trichomonas vaginalis* (Neshani et al., 2019). However, a few peptides do not possess the same mechanism; for instance, Jelleine-1 obtained from bee jelly had specific protein targets affecting the *Leishmania* parasite (Rivas & Rojas, 2019).

9.5 MECHANISM OF ACTION OF AMPs

As described earlier, most of the AMPs interact with the cell membrane to impart their antimicrobial activities. Nonetheless there are a few peptides that have intracellular targets. In both cases a preliminary interaction with the membrane is involved (Bahar & Ren, 2013). The AMPs can be divided depending upon the region of action into membrane-active and non-membrane-active. As they are amphipathic, the cationic faces are involved during the interactions with the negative cell membrane and insertion into the cell. While the hydrophobic faces are used during the insertion of AMP in cell membranes, Huan et al. 2020 provide the classification of the mechanism of action based on the target. The pole and carpet models are two major membrane-targeting mechanisms, and the pole model is further classified into toroidal pore model and barrel-stave model. In the toroidal pore model, peptides insert into the membrane in a vertical alignment where the hydrophobic faces interact with the center section of lipid bilayer while the hydrophilic faces the pore. This is often referred to as the wormhole model, since the AMPs bend and form a ring (see Figure 9.2) (Huan et al., 2020). Bacterial antimicrobial peptide lacticin Q is an example of the toroidal pore model (Omardien et al., 2018). In the barrel-stave model, as the name suggests, staves are formed parallel to the membrane, followed by barrel formation, wherein the peptides are vertically inserted (see Figure 9.2). An aggregation of the peptides allows perpendicular insertion into the lipid bilayer and forms channels that lead to cytoplasmic outflow, sometimes leading to cell membrane rupture and death. Alamethicin and protegrin-1 can generate barrels and half-barrels that form pores (Lipkin & Lazaridis, 2015). The carpet model is also termed the detergent-like model, wherein molecules for micelles and small regions of membrane are covered by the AMPs (see Figure 9.2). The hydrophobic sides, positioned parallel like

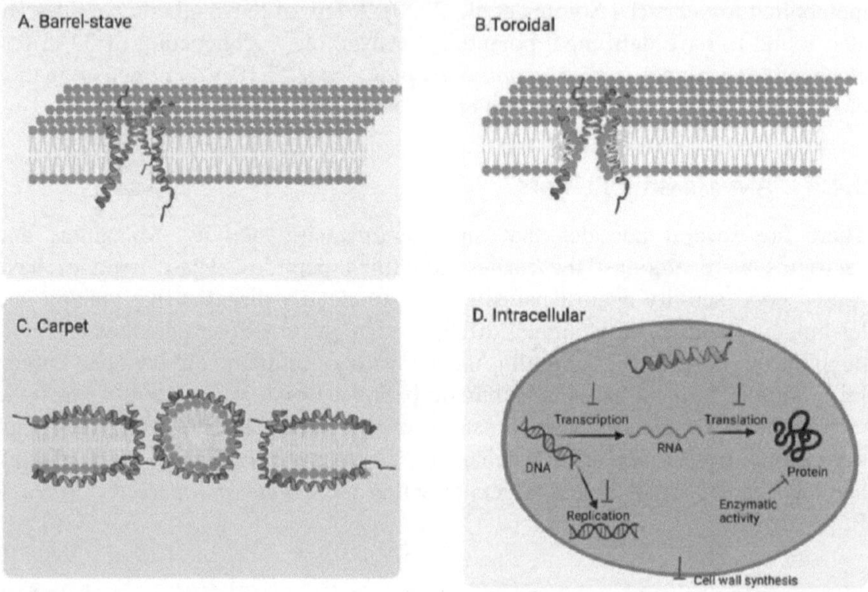

FIGURE 9.2 Mechanisms used by antimicrobial peptides. A, B, C are modes of membrane-targeting peptides. D is the mode of action used by non-membrane-targeting peptides.

a carpet, curl inwards, forming a pore in the bilayer in detergent-like behaviour (Bahar & Ren, 2013; Huan et al., 2020). β-sheet AMPs and cathelicidin LL-37 work via this mechanism (Shenkarev et al., 2011; Correa et al., 2019).

Further, AMPs also penetrate into the cells and act on certain targets inside the cytoplasm. The non-membrane-targeting mechanisms are categorized into protein biosynthesis inhibition, protease activity, nucleic acid synthesis, and division of cells (Huan et al., 2020). In the category of protein synthesis inhibitors, the peptides often inhibit the transcription, translation, and assembly. Bac7 targets the ribosomes and thereby inhibits protein translation (Mardirossian et al., 2014). AMPs like TFP1–1TC24 affect enzymes that can degrade or inhibit nucleic acid synthesis (He et al., 2017). Histatin 5 inhibits proteases, and AMPs including indolicidin are known to inhibit serine proteases and chymotrypsin (Le et al., 2017). The peptides that inhibit cell division work by hindering DNA replication and DNA damage response, interrupting the cell cycle, and inhibiting chromosome separation (Huan et al., 2020). The AMP APP penetrates into the membrane of *Candida albicans*, binds to the DNA and causes S-phase arrest (Li et al., 2016).

9.6 AMPs AND BIOFILMS

Biofilms are a complex structure made of several microbes. It is a self-secreting matrix that makes a scaffold on which other nutrients may adhere. Biofilms have created a nuisance in almost all sectors including environment, food, and biomedical and healthcare systems, and caused a severe loss to the economy (Di Somma et al., 2020). The formation and development of biofilms occurs in four stages: attachment, adhesion, development and maturity, and dispersion and aging (Crouzet et al., 2014). It is rather difficult to discern whether an AMP is showing antibiofilm activity. The minimum concentration of AMP for biofilm inhibition should be below the minimum inhibitory concentration, as disrupting biofilms that have been already formed is much more difficult than inhibiting the formation. AMPs have been shown to affect biofilm formation by inhibiting different stages of the process, like formation, adhesion, disrupting preformed biofilms, and down-regulating quorum sensing (Di Somma et al., 2020). Protegrin-1 is an AMP isolated from porcine leukocytes that was able to disrupt the biofilms created against *Acenitobacter baumannii* strains isolated from surgical wounds (Morroni et al., 2019). Pleurocidin was able to inhibit and eradicate biofilms of antibiotic-resistant *S. aureus* (Ko et al., 2019). *S. mutans* is widely known to form dental caries leading to oral biofilm formation. Human β-defensin 3 and the synthetic form of this peptide were studied as a potential oral antibiofilm agent (Ahn et al., 2017). Nisin A disrupts the biofilms of MRSA, while LL-37 peptide affects the cell signalling mechanism of *P. aeruginosa* biofilms (Okuda et al., 2013; Overhage et al., 2008).

9.7 AMPs AS QUORUM QUENCHERS

Quorum sensing (QS) is a communication mechanism used by bacterial cells in a population to collectively adapt to the cell density and environment. This mechanism makes the bacteria switch to a community stage at high cell density in order to perform functions which they cannot perform at an individual level – for instance, biofilm formation. It works on the basis of autoinducers, which act as the signal molecules. Gram-negative bacteria produce short peptides as an autoinducer, whereas Gram-negative bacteria involve N-acyl homoserine lactones as their QS signal molecules (Miller & Bassler, 2003).

The QS system can control gene expression, biofilm formation, and extracellular polysaccharide production, allowing bacteria to work together as a unit. Pathogenicity and virulence are expressed through the QS system by producing and releasing QS signal molecules, detecting and binding to the membrane receptors at the high-cell-density stage, and by binding of the receptor-signal complex from the cell to the promoter region to further initiate transcription of virulence genes. The QS system can up-regulate virulence genes, while QS inhibition down-regulates the pathogenicity, which aids the immune system in eradicating infected bacteria (Chen et al., 2002). Mechanisms that can be used to disrupt this bacterial communication include interfering with the production of autoinducers using small-molecule QS inhibitors, or by quenching or scavenging the autoinducers using quorum-quenching antibodies so that QS does not happen effectively (Zhong & He, 2021; Park et al., 2007)

Quorum quenching (QQ) is the act of interfering with the QS system. QQ-based QS disruption methods are increasingly gaining the research focus as QQ does not affect microbial growth, nor is it involved in development of resistance. Quorum quenching can be attained in three major ways: inhibition of signal molecule production, inhibition of the binding of signal molecules to target receptors, and degradation of the signal molecule.

Several antimicrobial peptides have been identified that quench bacterial quorum sensing and target them. Such peptides can be derived from natural compounds, microorganisms, animals, plants, or fungi, or can even be synthesized chemically. For example, down-regulation of Las and Rhl QS systems of *P. aeruginosa* by human antimicrobial peptides like cathelicidic LL-37 and indolicidin is found to hinder biofilm formation (Overhage et al., 2008). According to scientists Tung and Quoc (2021), a range of β-turn mimetic-based peptides were also found to likewise act as potent quorum sensing inhibitors. Peptides P1, P4, and P5, among others, showed excellent anti-quorum sensing activity on the *P. aeruginosa* lasB-gfp reporter strain without compromising its growth (Tung & Quoc, 2021). TNRHNPHHLHHV, a synthetic peptide, was found to inhibit LuxS enzyme activity and therefore inhibit QS synthesizing autoinducer molecules (Han & Lu, 2009). LIVRHK and LIVRRK, two novel synthetic peptides, were also discovered to block QS-regulated virulence factors and biofilm formation in *P. aeruginosa* POA1. In this case, transcriptional variations linked to QS, as well as considerable down-regulation of gene expression of lasI, lasR, rhlI, and rhlR, were identified (Taha et al., 2019).

9.8 BIOENGINEERING ANTIMICROBIAL PEPTIDES

There are very few studies targeting the interrelation of AMP structure and the antimicrobial activity range of AMPs, as even AMPs that are very similar have different and vague modes of action and act on a different organism range. Various factors, among them size and charge, amphipathicity, and hydrophobicity, are important for determining the structure and the required antimicrobial spectrum specificity.

The length of a typical AMP should be around 8 amino acids and it should be amphipathic. The length of the AMP affects the mechanism of action and determines the cytotoxicity. The net charge of AMPs also plays an important role as it affects the interaction between the negatively charged cell membranes of the target cell. Hydrophobic residues in the AMPs also tend to influence the properties and thereby the activities too. Increased hydrophobic residues result in enhanced antimicrobial activity, whereas a lesser number of hydrophobic residues subsequently tend to decrease the antimicrobial efficacy. Therefore, bioengineering of AMPs should involve selection of hydrophobic residues within the optimal range for the desired effect. Another important property of AMPs, amphipathicity, ensures antimicrobial efficacy, and AMPs should be engineered with an amphipathic structure for targeting cells of a specific range (Bahar & Ren, 2013).

All the aforementioned properties must be satisfied for efficient bioengineering of any AMP. Most AMPs are bioengineered or synthesized in their active form, but certain ones require post-translational modifications, like methylation, phosphorylation, disulphide bridges, amidation, and the like for effective function.

Covalent alteration of an AMP can have a significant impact on its structure and function. Amino acid content alteration is one of the most researched AMP modification and bioengineering techniques. Changing the amino acid composition can potentially have an impact on cytotoxicity. A bioengineered AMP, P60.4, effective against MRSA in nasal applications, was found to have lesser cytotoxic effects on eukaryotic cells as neutral amino acids like Asn and Gln were removed from the main sequence of the AMP LL37 and two more positively charged Arg residues were added (Nell et al., 2006). Bioengineering and modification of AMPs by use of unnatural amino acids into the core sequence is also a significantly growing area. One of the unnatural amino acids, β-didehydrophenylalanine, when incorporated into the sequence of an AMP called VS1 resulted in a new variety which had more potential and higher stability against proteases (Bahar & Ren, 2013).

Bioengineering of AMPs has the potential to produce novel peptides to combat antibiotic-resistant superbugs. Another method for easily producing AMPs with specific features is to design synthetic AMPs using homology modelling. Three synthetic peptides, AMP72, AMP126, and AMP2041, were created using existing AMPs such as arenicin, protegrin, and thanatin as templates. These novel synthesized AMPs had reduced cytotoxicity compared to the original AMPs and antibacterial activity against Gram-negative bacteria that was dose dependent (Romani et al., 2013). In the case of nisin bioengineering to create novel AMPs, the nisin Z derivatives N20K, M21K, N27K, and H31K were found to have improved solubility qualities when compared to nisin, an AMP (Rollema et al., 1995; Yuan et al., 2004). Single amino acid residue changes in nisin have also resulted in new forms of AMPs with improved antimicrobial action against drug-resistant diseases such as VRE, MRSA, and others, as well as a broader range of activity against Gram-negative organisms. Variants of nisin have resulted from further substitutions and changes at various locations in the protein (D. Field et al., 2015).

9.9 CHALLENGES IN MANIPULATING QS

Microorganisms are developing methods to evade or resist the impact of AMPs. However, given the diverse architecture of bacterial membranes, establishing a foolproof, effective resistance mechanism against AMPs is difficult. Resistance to AMPs is either inducible or constitutive (Yeaman & Yount, 2003). Various mechanisms of inducible resistance include modifications, acylation, proteolytic enzyme activation, efflux pumps, and others. Electrostatic shielding, membrane potential variations, and increased biofilm growth are all examples of constitutive resistance.

The interaction between the AMPs and the EPS is hypothesized to be connected to biofilm resistance to AMPs. *S. aureus* biofilms are defended against the action of LL-37 and defensins AMPs because of the presence of positively charged PIA exopolymer, in contrast to the typical EPS. This happens via electrostatic repulsion as the peptides are also positively charged (Otto, 2006; Vuong et al., 2004). Alginate, a Gram-negative bacteria-secreted EPS, can resist positively charged AMPs by trapping them in hydrophobic environments, forming a complex (AMP-alginate) that then neutralizes the two species (Chan et al., 2004). MprF membrane protein confers AMP resistance in Gram-positive bacteria, which is involved in the addition of alanine or lysine to phosphatidylglycerol (PG) to form alanyl-phosphatidylglycerol (APG) and lysyl-phosphatidylglycerol (LPG). It was shown that adding lysine or arginine residues to the membrane made *S. aureus* MprF mutants resistant to the action of AMP. LL-37, polymyxin B, and polymyxin E AMPs have been demonstrated to be resistant to *P. aeruginosa*'s two-component regulatory machinery, pmrA-pmrB. The efficacy of the interaction of AMPs with the bacterial membrane is reduced when the outer membrane lipopolysaccharide is modified. Further, addition of an acyl group to lipid A was also considered to result in the peptide resistance (Di Somma et al., 2020). Resistance to antimicrobial peptides is not as specific as antibiotic resistance as it only covers a few antimicrobial peptides.

9.10 CONCLUSIONS AND FUTURE PERSPECTIVES

After several years of rigorous research, new therapeutic strategies to combat biofilms are still a challenge for the scientific and medical community. Antimicrobial peptides can potentially combat the antimicrobial resistant microbes and can also disrupt the biofilms and could therefore be an alternative and promising strategy. The activity of AMPs could be hampered due to the production of exopolysaccharides; however, bioengineered peptides can alter the signalling pathways involved. Since high concentrations of antibiotics are related to increased chances of toxicity, the combinational approach should be considered while conceptualizing these strategies. Furthermore, standard *in vivo* model systems should be determined for better understanding and predictions of the activity. A lot of the peptides that have demonstrated significant activity should be directed towards preclinical trials to effectively control the emergence and spread of pathogens.

9.10.1 Acknowledgement

All images were created using BioRender.com.

REFERENCES

Abee, T., & Delves-Broughton, J. (n.d.). *Bacteriocins—nisin*. Springer. Retrieved February 26, 2022, from https://link.springer.com/chapter/10.1007/978-0-387-30042-9_8

Ahn, K. B., Kim, A. R., Kum, K. Y., Yun, C. H., & Han, S. H. (2017). The synthetic human beta-defensin-3 C15 peptide exhibits antimicrobial activity against *Streptococcus mutans*, both alone and in combination with dental disinfectants. *Journal of Microbiology (Seoul, Korea)*, *55*(10), 830–836. https://doi.org/10.1007/s12275-017-7362-y

Algburi, A., Comito, N., Kashtanov, D., Dicks, L. M. T., & Chikindas, M. L. (2017). Control of biofilm formation: Antibiotics and beyond. *Applied and Environmental Microbiology*, *83*(3). https://doi.org/10.1128/AEM.02508-16/ASSET/E9B1F33C-B8D3-4317-BDB4-71CBA1750489/ASSETS/GRAPHIC/ZAM9991176320001.JPEG

Bahar, A. A., & Ren, D. (2013). Antimicrobial peptides. *Pharmaceuticals*, *6*(12), 1543–1575. https://doi.org/10.3390/PH6121543

Chan, C., Burrows, L. L., & Deber, C. M. (2004). Helix induction in antimicrobial peptides by alginate in biofilms*. *Journal of Biological Chemistry*, *279*(37), 38749–38754. https://doi.org/10.1074/JBC.M406044200

Chen, C., Fan, H., Huang, Y., Peng, F., Fan, H., Yuan, S., & Tong, Y. (2014). Recombinant lysostaphin protects mice from methicillin-resistant *Staphylococcus aureus* pneumonia. *BioMed Research International*, *2014*, 602185.

Chen, X., Schauder, S., Potier, N., van Dorsselaer, A., Pelczer, I., Bassler, B. L., & Hughson, F. M. (2002). Structural identification of a bacterial quorum-sensing signal containing boron. *Nature*, *415*(6871), 545–549. https://doi.org/10.1038/415545a

Cirioni, O., Giacometti, A., Ghiselli, R., Kamysz, W., Silvestri, C., Orlando, F., Mocchegiani, F., Vittoria, A. della, Kamysz, E., Saba, V., & Scalise, G. (2007). The lipopeptides Pal-Lys-Lys-NH2 and Pal-Lys-Lys soaking alone and in combination with intraperitoneal vancomycin prevent vascular graft biofilm in a subcutaneous rat pouch model of staphylococcal infection. *Peptides*, *28*(6), 1299–1303. https://doi.org/10.1016/J.PEPTIDES.2007.03.017

Corrêa, J. A. F., Evangelista, A. G., de Melo Nazareth, T., & Luciano, F. B. (2019). Fundamentals on the molecular mechanism of action of antimicrobial peptides. *Materialia*, *8*, 100494. https://doi.org/10.1016/j.mtla.2019.100494

Crouzet, M., Le Senechal, C., Brözel, V. S., Costaglioli, P., Barthe, C., Bonneu, M., Garbay, B., & Vilain, S. (2014). Exploring early steps in biofilm formation: Set-up of an experimental system for molecular studies. *BMC Microbiology*, *14*, 253. https://doi.org/10.1186/s12866-014-0253-z

Dean, S. N., Bishop, B. M., & van Hoek, M. L. (2011). Natural and synthetic cathelicidin peptides with anti-microbial and anti-biofilm activity against Staphylococcus aureus. *BMC Microbiology*, *11*(1), 114. https://doi.org/10.1186/1471-2180-11-114/FIGURES/4

de La Fuente-Núñez, C., Reffuveille, F., Mansour, S. C., Reckseidler-Zenteno, S. L., Hernández, D., Brackman, G., Coenye, T., & Hancock, R. E. W. (2015). D-enantiomeric peptides that eradicate wild-type and multidrug-resistant biofilms and protect against lethal Pseudomonas aeruginosa infections. *Chemistry and Biology*, *22*(2), 196–205. https://doi.org/10.1016/J.CHEMBIOL.2015.01.002

Dijksteel, G. S., Ulrich, M. M. W., Middelkoop, E., & Boekema, B. K. H. L. (2021). Review: Lessons learned from clinical trials using antimicrobial peptides (AMPs). *Frontiers in Microbiology*, *12*, 287. https://doi.org/10.3389/FMICB.2021.616979/BIBTEX

Di Somma, A., Moretta, A., Canè, C., Cirillo, A., & Duilio, A. (2020). Antimicrobial and antibiofilm peptides. *Biomolecules*, *10*(4), 652. https://doi.org/10.3390/biom10040652

Dubos, R. J. (1939a). Studies on a bactericidal agent extracted from a soil Bacillus : I. Preparation of the agent its activity in vitro. *The Journal of Experimental Medicine*, *70*(1), 1–10. https://doi.org/10.1084/JEM.70.1.1

Dubos, R. J. (1939b). Studies on a bactericidal agent extracted from a soil bacillus: II. Protective effect of the bactericidal agent against experimental pieuococcus infections in mice. *Journal of Experimental Medicine*, *70*(1), 11–18. https://doi.org/10.1084/jem.70.1.11

Dubos, R. J., & Cattaneo, C. (1939). Studies on a bactericidal agent extracted from a soil Bacillus: III. Preparation and activity of a protein-free fraction. *The Journal of Experimental Medicine*, *70*(3), 249–256. https://doi.org/10.1084/jem.70.3.249

Dubos, R. J., & Hotchkiss, R. D. (1941). The production of bactericidal substances by aerobic sporulating Bacilli. *The Journal of Experimental Medicine*, *73*(5), 629. https://doi.org/10.1084/JEM.73.5.629

Field, C. J. (2005). The immunological components of human milk and their effect on immune development in infants. *The Journal of Nutrition, 135*(1), 1–4. https://doi.org/10.1093/JN/135.1.1

Field, D., Cotter, P. D., Hill, C., & Ross, R. P. (2015). Bioengineering lantibiotics for therapeutic success. *Frontiers in Microbiology, 6*, 1363. https://doi.org/10.3389/FMICB.2015.01363/BIBTEX

González-Chávez, S. A., Arévalo-Gallegos, S., & Rascón-Cruz, Q. (2009). Lactoferrin: Structure, function and applications. *International Journal of Antimicrobial Agents, 33*(4), 301.e1–301.e8. https://doi.org/10.1016/J.IJANTIMICAG.2008.07.020

Gschwandtner, M., Zhong, S., Tschachler, A., Mlitz, V., Karner, S., Elbe-Bürger, A., & Mildner, M. (2014). Fetal human keratinocytes produce large amounts of antimicrobial peptides: Involvement of histone-methylation processes. *The Journal of Investigative Dermatology, 134*(8), 2192–2201. https://doi.org/10.1038/JID.2014.165

Han, X., & Lu, C. (2009). Biological activity and identification of a peptide inhibitor of LuxS from Streptococcus suis serotype 2. *FEMS Microbiology Letters, 294*(1), 16–23. https://doi.org/10.1111/J.1574-6968.2009.01534.X

Hasper, H. E., Kramer, N. E., Smith, J. L., Hillman, J. D., Zachariah, C., Kuipers, O. P., de Kruijff, B., & Breukink, E. (2006). An alternative bactericidal mechanism of action for lantibiotic peptides that target lipid II. *Science, 313*, 1636–1637.

He, S.-W., Zhang, J., Li, N.-Q., Zhou, S., Yue, B., & Zhang, M. (2017). A TFPI-1 peptide that induces degradation of bacterial nucleic acids, and inhibits bacterial and viral infection in half-smooth tongue sole, *Cynoglossus semilaevis. Fish and Shellfish Immunology, 60*, 466–473. https://doi.org/10.1016/j.fsi.2016.11.029

Höng, K., Austerlitz, T., Bohlmann, T., & Bohlmann, H. (2021). The thionin family of antimicrobial peptides. *PloS One, 16*(7). https://doi.org/10.1371/JOURNAL.PONE.0254549

Huan, Y., Kong, Q., Mou, H., & Yi, H. (2020) Antimicrobial peptides: Classification, design, application and research progress in multiple fields. *Frontiers in Microbiology, 11*, 582779. https://doi.org/10.3389/fmicb.2020.582779

Hwang, P. M., Zhou, N., Shan, X., Arrowsmith, C. H., & Vogel, H. J. (1998). Three-dimensional solution structure of lactoferricin B, an antimicrobial peptide derived from bovine lactoferrin. *Biochemistry, 37*(12), 4288–4298. https://doi.org/10.1021/bi972323m

Jenssen, H., Andersen, J. H., Uhlin-Hansen, L., Gutteberg, T. J., & Rekdal, Ø. (2004). Anti-HSV activity of lactoferricin analogues is only partly related to their affinity for heparan sulfate. *Antiviral Research, 61*, 101–109.

Kang, X., Dong, F., Shi, C., Liu, S., Sun, J., Chen, J., Li, H., Xu, H., Lao, X., & Zheng, H. (2019). DRAMP 2.0, an updated data repository of antimicrobial peptides. *Scientific Data, 6*(1). https://doi.org/10.1038/S41597-019-0154-Y

Kiss, G., & Michl, H. (1962). Uber das Giftsekret der Gelbbauchunke, Bombina variegata L. *Toxicon, 1*(1), 33–34. https://doi.org/10.1016/0041-0101(62)90006-5

Ko, S. J., Kang, N. H., Kim, M. K., Park, J., Park, E., Park, G. H., Kang, T. W., Na, D. E., Park, J. B., Yi, Y. E., Jeon, S. H., & Park, Y. (2019). Antibacterial and anti-biofilm activity, and mechanism of action of pleurocidin against drug resistant *Staphylococcus aureus. Microbial Pathogenesis, 127*, 70–78. https://doi.org/10.1016/j.micpath.2018.11.052

Koo, H. B., & Seo, J. (2019). Antimicrobial peptides under clinical investigation. *Peptide Science, 111*(5), e24122. https://doi.org/10.1002/PEP2.24122

Kuroda, K., Okumura, K., Isogai, H., & Isogai, E. (2015). The human cathelicidin antimicrobial peptide LL-37 and mimics are potential anticancer drugs. *Frontiers in Oncology, 5*, 144. https://doi.org/10.3389/fonc.2015.00144

Le, C.-F., Fang, C.-M., & Sekaran, S. D. (2017). Intracellular targeting mechanisms by antimicrobial peptides. *Antimicrobial Agents and Chemotherapy*, *61*, e02340-16. https://doi.org/10.1128/AAC.02340-16

Li, J., Hu, S., Jian, W., Xie, C., & Yang, X. (2021). Plant antimicrobial peptides: Structures, functions, and applications. *Botanical Studies*, *62*(1). https://doi.org/10.1186/S40529-021-00312-X

Li, L., Sun, J., Xia, S., Tian, X., Cheserek, M. J., & Le, G. (2016). Mechanism of antifungal activity of antimicrobial peptide APP, a cell-penetrating peptide derivative, against *Candida albicans*: Intracellular DNA binding and cell cycle arrest. *Applied Microbiology and Biotechnology*, *100*, 3245–3253. https://doi.org/10.1007/s00253-015-7265-y

Li, T., Li, L., Du, F., Sun, L., Shi, J., Long, M., & Chen, Z. (2021). Activity and mechanism of action of antifungal peptides from microorganisms: A review. *Molecules*, *26*, 3438. https://doi.org/10.3390/molecules26113438

Lipkin, R. B., & Lazaridis, T. (2015). Implicit membrane investigation of the stability of antimicrobial peptide β-barrels and ARCS. *The Journal of Membrane Biology*, *248*, 469–486. https://doi.org/10.1007/s00232-014-9759-4

Lu, Y., Ma, Y., Wang, X., Liang, J., Zhang, C., Zhang, K., Lin, G., & Lai, R. (2008). The first antimicrobial peptide from sea amphibian. *Molecular Immunology*, *45*(3), 678–681. https://doi.org/10.1016/J.MOLIMM.2007.07.004

Ma, B., Fang, C., Lu, L., Wang, M., Xue, X., Zhou, Y., Li, M., Hu, Y., Luo, X., & Hou, Z. (2019). The antimicrobial peptide thanatin disrupts the bacterial outer membrane and inactivates the NDM-1 metallo-β-lactamase. *Nature Communications*, *10*, 3517. https://doi.org/10.1038/s41467-019-11503-3

Mardirossian, M., Grzela, R., Giglione, C., Meinnel, T., Gennaro, R., Mergaert, P., & Scocchi, M. (2014). The host antimicrobial peptide Bac71–35 binds to bacterial ribosomal proteins and inhibits protein synthesis. *Chemistry & Biology*, *21*, 1639–1647. https://doi.org/10.1016/j.chembiol.2014.10.009

Memariani, H., Memariani, M., Moravvej, H., & Shahidi-Dadras, M. (2020). Melittin: A venom-derived peptide with promising anti-viral properties. *European Journal of Clinical Microbiology & Infectious Diseases: Official Publication of the European Society of Clinical Microbiology*, *39*(1), 5–17. https://doi.org/10.1007/s10096-019-03674-0

Michael Conlon, J., & Mechkarska, M. (2014). Host-defense peptides with therapeutic potential from skin secretions of frogs from the family pipidae. *Pharmaceuticals*, *7*(1), 58. https://doi.org/10.3390/PH7010058

Miller, M. B., & Bassler, B. L. (2003). Quorum sensing in bacteria. *Annual Reviews*, *55*, 165–199. https://doi.org/10.1146/ANNUREV.MICRO.55.1.165

Morroni, G., Simonetti, O., Brenciani, A., Brescini, L., Kamysz, W., Kamysz, E., Neubauer, D., Caffarini, M., Orciani, M., Giovanetti, E., Offidani, A., Giacometti, A., & Cirioni, O. (2019). In vitro activity of Protegrin-1, alone and in combination with clinically useful antibiotics, against *Acinetobacter baumannii* strains isolated from surgical wounds. *Medical Microbiology and Immunology*, *208*(6), 877–883. https://doi.org/10.1007/s00430-019-00624-7

Muhialdin, B. J., Algboory, H. L., Kadum, H., Mohammed, N. K., Saari, N., Hassan, Z., & Hussin, A. S. M. (2020). Antifungal activity determination for the peptides generated by *Lactobacillus plantarum* TE10 against Aspergillus flflavus in maize seeds. *Food Control*, *109*, 106898. https://doi.org/10.1016/j.foodcont.2019.106898

Nell, M. J., Tjabringa, G. S., Wafelman, A. R., Verrijk, R., Hiemstra, P. S., Drijfhout, J. W., & Grote, J. J. (2006). Development of novel LL-37 derived antimicrobial peptides with LPS and LTA neutralizing and antimicrobial activities for therapeutic application. *Peptides*, *27*(4), 649–660. https://doi.org/10.1016/J.PEPTIDES.2005.09.016

Neshani, A., Zare, H., Akbari Eidgahi, M. R., Khaledi, A., & Ghazvini, K. (2019). Epinecidin-1, a highly potent marine antimicrobial peptide with anticancer and immunomodulatory activities. *BMC Pharmacology and Toxicology*, *20*, 33. https://doi.org/10.1186/s40360-019-0309-7

Ohtani, K., Okada, T., Yoshizumi, H., & Kagamiyama, H. (1977). Complete primary structures of two subunits of purothionin A, a lethal protein for brewer's yeast from wheat flour. *Journal of Biochemistry*, *82*(3), 753–767. https://doi.org/10.1093/OXFORDJOURNALS.JBCHEM.A131752

Okuda, K. I., Zendo, T., Sugimoto, S., Iwase, T., Tajima, A., Yamada, S., Sonomoto, K., & Mizunoe, Y. (2013). Effects of bacteriocins on methicillin-resistant *Staphylococcus aureus* biofifilm. *Antimicrobial Agents and Chemotherapy*, *57*, 5572–5579.

Omardien, S., Drijfhout, J. W., Vaz, F. M., Wenzel, M., Hamoen, L. W., Zaat, S. A. J., & Brul, S. (2018). Bactericidal activity of amphipathic cationic antimicrobial peptides involves altering the membrane fluidity when interacting with the phospholipid bilayer. *Biochimica et biophysica acta. Biomembranes*, *1860*(11), 2404–2415. https://doi.org/10.1016/j.bbamem.2018.06.004

Otto, M. (2006). Bacterial evasion of antimicrobial peptides by biofilm formation. *Current Topics in Microbiology and Immunology*, *306*, 251–258. https://doi.org/10.1007/3-540-29916-5_10

Overhage, J., Campisano, A., Bains, M., Torfs, E. C. W., Rehm, B. H. A., & Hancock, R. E. W. (2008). Human host defense peptide LL-37 prevents bacterial biofilm formation. *Infection and Immunity*, *76*(9), 4176–4182. https://doi.org/10.1128/IAI.00318-08

Park, C. B., Kim, H. S., & Kim, S. C. (1998). Mechanism of action of the antimicrobial peptide buforin ii: Buforin ii kills microorganisms by penetrating the cell membrane and inhibiting cellular functions. *Biochemical and Biophysical Research Communications*, *244*, 253–257.

Park, J., Jagasia, R., Kaufmann, G. F., Mathison, J. C., Ruiz, D. I., Moss, J. A., Meijler, M. M., Ulevitch, R. J., & Janda, K. D. (2007). Infection control by antibody disruption of bacterial quorum sensing signaling. *Chemistry & Biology*, *14*(10), 1119–1127. https://doi.org/10.1016/J.CHEMBIOL.2007.08.013

Rhaiem, R. B., & Houimel, M. (2016). Targeting *Leishmania* major parasite with peptides derived from a combinatorial phage display library. *Acta Tropica*, *159*, 11–19. https://doi.org/10.1016/j.actatropica.2016.03.018

Rivas, L., & Rojas, V. (2019). Cyanobacterial peptides as a tour de force in the chemical space of antiparasitic agents. *Archives of Biochemistry and Biophysics*, *664*, 24–39. https://doi.org/10.1016/j.abb.2019.01.030

Rollema, H. S., Kuipers, O. P., Both, P., de Vos, W. M., & Siezen, R. J. (1995). Improvement of solubility and stability of the antimicrobial peptide nisin by protein engineering. *Applied and Environmental Microbiology*, *61*(8), 2873–2878. https://doi.org/10.1128/AEM.61.8.2873-2878.1995

Romani, A. A., Baroni, M. C., Taddei, S., Ghidini, F., Sansoni, P., Cavirani, S., & Cabassi, C. S. (2013). In vitro activity of novel in silico-developed antimicrobial peptides against a panel of bacterial pathogens. *Journal of Peptide Science : An Official Publication of the European Peptide Society*, *19*(9), 554–565. https://doi.org/10.1002/PSC.2532

Rossi, L. M., Rangasamy, P., Zhang, J., Qiu, X. Q., & Wu, G. Y. (2008). Research advances in the development of peptide antibiotics. *Journal of Pharmaceutical Sciences*, *97*(3), 1060–1070. https://doi.org/10.1002/JPS.21053

Roudi, R., Syn, N. L., & Roudbary, M. (2017). Antimicrobial peptides as biologic and immunotherapeutic agents against cancer: A comprehensive overview. *Frontiers in Immunology*, *8*, 1320. https://doi.org/10.3389/FIMMU.2017.01320/BIBTEX

Shenkarev, Z. O., Balandin, S. V., Trunov, K. I., Paramonov, A. S., Sukhanov, S. V., Barsukov, L. I., Arseniev, A. S., & Ovchinnikova, T. V. (2011). Molecular mechanism of action of β-hairpin antimicrobial peptide arenicin: Oligomeric structure in dodecylphosphocholine micelles and pore formation in planar lipid bilayers. *Biochemistry*, *50*(28), 6255–6265. https://doi.org/10.1021/bi200746t

Taha, M. N., Saafan, A. E., Ahmedy, A., el Gebaly, E., & Khairalla, A. S. (2019). Two novel synthetic peptides inhibit quorum sensing-dependent biofilm formation and some virulence factors in Pseudomonas aeruginosa PAO1. *Journal of Microbiology (Seoul, Korea)*, *57*(7), 618–625. https://doi.org/10.1007/S12275-019-8548-2

Tang, S. S., Prodhan, Z. H., Biswas, S. K., Le, C. F., & Sekaran, S. D. (2018). Antimicrobial peptides from different plant sources: Isolation, characterisation, and purification. *Phytochemistry*, *154*, 94–105. https://doi.org/10.1016/J.PHYTOCHEM.2018.07.002

Tung, T. T., & Quoc, T. N. (2021). Discovery of novel β-turn mimetic-based peptides as novel quorum sensing inhibitors of gram-negative bacteria. *Bioorganic & Medicinal Chemistry Letters*, *46*. https://doi.org/10.1016/J.BMCL.2021.128170

Vuong, C., Voyich, J. M., Fischer, E. R., Braughton, K. R., Whitney, A. R., DeLeo, F. R., & Otto, M. (2004). Polysaccharide intercellular adhesin (PIA) protects Staphylococcus epidermidis against major components of the human innate immune system. *Cellular Microbiology*, *6*(3), 269–275. https://doi.org/10.1046/J.1462-5822.2004.00367.X

Wang, G. (2014). Human antimicrobial peptides and proteins. *Pharmaceuticals (Basel, Switzerland)*, *7*(5), 545–594. https://doi.org/10.3390/PH7050545

Wang, S., Fang, Q., Lu, Z., Gao, Y., Trembleau, L., Ebel, R., Andersen, J. H., Philips, C., Law, S., & Deng, H. (2021). Discovery and biosynthetic investigation of a new antibacterial dehydrated non ribosomal tripeptide. *Angewandte Chemie*, *60*(6), 3229–3237.

Xia, S., Liu, M., Wang, C., Xu, W., Lan, Q., Feng, S., Qi, F., Bao, L., Du, L., Liu, S., Qin, C., Sun, F., Shi, Z., Zhu, Y., Jiang, S., & Lu, L. (2020). Inhibition of SARS-CoV-2 (previously 2019-nCoV) infection by a highly potent pan coronavirus fusion inhibitor targeting its spike protein that harbors a high capacity to mediate membrane fusion. *Cell Research*, *30*, 343–355. https://doi.org/10.1038/s41422-020-0305-x

Yeaman, M. R., & Yount, N. Y. (2003). Mechanisms of antimicrobial peptide action and resistance. *Pharmacological Reviews*, *55*(1), 27–55. https://doi.org/10.1124/PR.55.1.2

Yuan, J., Zhang, Z. Z., Chen, X. Z., Yang, W., & Huan, L. D. (2004). Site-directed mutagenesis of the hinge region of nisinZ and properties of nisinZ mutants. *Applied Microbiology and Biotechnology*, *64*(6), 806–815. https://doi.org/10.1007/S00253-004-1599-1

Zahedifard, F., Lee, H., No, J. H., Salimi, M., Seyed, N., Asoodeh, A., & Rafati, S. (2020). Comparative study of different forms of Jellein antimicrobial peptide on Leishmania parasite. *Experimental Parasitology*, *209*. https://doi.org/10.1016/J.EXPPARA.2019.107823

Zeya, H. I., & Spitznagel, J. K. (1963). Antibacterial and enzymic basic proteins from leukocyte lysosomes: Separation and identification. *Science (New York, N.Y.)*, *142*(3595), 1085–1087. https://doi.org/10.1126/SCIENCE.142.3595.1085

Zhang, F., Cui, X., Fu, Y., Zhang, J., Zhou, Y., Sun, Y., Wang, X., Li, Y., Liu, Q., & Chen, T. (2017). Antimicrobial activity and mechanism of the human milk-sourced peptide Casein201. *Biochemical and Biophysical Research Communications*, *485*(3), 698–704. https://doi.org/10.1016/J.BBRC.2017.02.108

Zhang, L. J., & Gallo, R. L. (2016). Antimicrobial peptides. *Current Biology*, *26*(1), R14–R19. https://doi.org/10.1016/J.CUB.2015.11.017

Zhang, Q. Y., Yan, Z. bin, Meng, Y. M., Hong, X. Y., Shao, G., Ma, J. J., Cheng, X. R., Liu, J., Kang, J., & Fu, C. Y. (2021). Antimicrobial peptides: Mechanism of action, activity and clinical potential. *Military Medical Research*, *8*(1), 1–25. https://doi.org/10.1186/S40779-021-00343-2

Zhong, S., & He, S. (2021). Quorum sensing inhibition or quenching in acinetobacter baumannii: The novel therapeutic strategies for new drug development. *Frontiers in Microbiology*, *12*, 55. https://doi.org/10.3389/FMICB.2021.558003/BIBTEX

10 Algal Compounds in Down-Regulating Quorum Sensing of Biofilm

Ashmita Samanta, Drisha Roy, Moupriya Nag, Dibyajit Lahiri, Debasmita Bhattacharya, and Rina Rani Ray

10.1 INTRODUCTION

"Biofilm" refers to sessile communities immersed in an autogenous extracellular matrix where the bacterial cells are blanketed from stressful environmental conditions and the effects of antimicrobials (Solano et al., 2014). The 3D matrix in which the microbial cells are embedded is composed of a diverse array of extracellular polymeric substances (EPSs). The EPS is responsible for the integrity and structural uniqueness of the biofilm community (Karygianni et al., 2020). The multicellular biofilm community has an evolved lifestyle that has provided them with ways to formulate antimicrobial-resistant techniques. The bacterial communication system within the biofilm involves cell-to-cell communication in response to their environment. This process is called quorum sensing (QS). Cells coordinate their behavior by producing and detecting signal molecules that are collectively termed autoinducers. The QS system provides a switch to the community members to alter to a biofilm lifestyle when the cellular consistency surpasses a threshold. Increase in the population density results in the aggregation of the signal in the matrix, which then becomes adequate to trigger the reaction mechanism. There are three major QS systems: (i) the acyl-homoserine lactone (AHL) QS system in Gram-negative bacteria; (ii) the autoinducing peptide (AIP) QS system in Gram-positive bacteria; and (iii) the autoinducer-2 (AI-2) QS system present in both Gram-negative and Gram-positive bacteria (Brackman & Coenye, 2014). The biofilm matrix furnishes the resistance activity of bacteria with myriad virulence factors, protecting them from stressful conditions and compelling them to become multidrug-resistant (Sharma et al., 2019). Various inhibitor compounds can disrupt the QS signalling process and transmute their gene expression. These compounds are referred to as quorum sensing inhibitors. The molecules act primarily through the

DOI: 10.1201/9781003300557-10

quenching, or to say it simply, inhibition of the QS algorithm (Kalia, 2013). The process of quorum sensing inhibition may also prove to be useful in marine biofilm formation. The International Maritime Organization in September 2008 prohibited the use of nonbiodegradable organotin as antifouling paint due to its high toxicity levels (Qian et al., 2009). This indicates that immediate action must be taken to inhibit – and in future, possibly prevent – biofouling natural or chemically synthesized compounds. Marine microalgae are a good source of biogenic compounds applicable to antimicrobial uses and antifouling techniques. The current focus of phycological studies is the antifouling and antibiofilm properties of algal-derived compounds. The availability of algal resources with their diverse chemical composition in green, red, and brown algae opens a new avenue for us to explore in search of answers to our long unanswered questions of treating biofilm and marine fouling fields.

10.2 THE TIE-IN BETWEEN BIOFILM AND QUORUM SENSING

Biofilm organizes microcolonies of bacteria embedded in a self-secreted organic polymer matrix formulated with polysaccharides, proteins, and extracellular DNAs (Tremblay et al., 2013). The extracellular matrix acts as one of the integral components in maintaining the architectural integrity of the biofilm community. The analytical study of biofilm by infrared spectroscopy provides us with data about the chemical content of the matrix and its involvement in the architectural integrity of the structure. EPS provides an adequate interpretation of the structural and functional relationship in the extracellular matrix and thus paints a vivid picture of the phenotypical characteristics of biofilm (Di Martino et al., 2018). The microcolonies subsisting within the EPS communicate via quorum sensing (QS), which particularly helps in biofilm growth, expression of virulence factors, and physiological information of the biofilm association (Pircalabioru & Chifiriuc, 2020). EPS plays a significant part in the dissemination of the antibiotic resistance property of biofilm which in due course ensures the synthesis of persister cells (Mah & O'Toole, 2001). The complex process of bacterial biofilm formation comprises five stages: (i) reversible attachment, when bacteria affix to the substrate surface; (ii) irreversible attachment, which encompasses interaction between bacterial cells and the substratum with the help of bacterial adhesins including flagella, pili/fimbriae, and nonfimbrial adhesions (Berne et al., 2015); (iii) synthesis of EPS; (iv) development of mature biofilm, wherein the bacterial cells continue to form microcolonies and biofilm continues to grow; (v) dispersal/detachment, the terminal process of biofilm development when the bacterial cells emigrate from the biofilm community and continue to form other microcolonies in other biofilm communities (Muhammad et al., 2020; Singh et al., 2017).

10.3 MODES OF BACTERIAL QUORUM SENSING

The intercellular communication of the bacterial cells with their neighbours helps them to develop a "molecular circuit" called quorum sensing (QS) (Lee & Zhang, 2015). QS requires a pair of genes: one or more than one synthase gene and one or more than one receptor (see Figure 10.1). The bacteria present in the biofilm interact with each other by synthesizing and reacting to signal molecules called autoinducers (Waters & Bassler, 2005). The receptor protein controls the transcriptional activity of the signalling molecule it perceives which in turn regulates the synthesis of the receptor protein itself; hence, they are correctly referred to as autoinducers (Lee & Zhang, 2015). When the population achieves a required cell density, the chemical signals (autoinducers) interact with a receptor protein to change their unicellular mode of living into a multicellular one (Li & Tian, 2012). There is a wide range of autoinducers that differ in chemical structure and mode of mechanism. QS signal molecules are characterized by low molecular weight and include a variety of chemicals such as acyl-homoserine lactones (AHLs), ribose derivative AI-2 molecules, diffusible signal factor family signals, and peptides (Solano et al., 2014). Quorum sensing was discovered in 1970 by Nealson et al. when they found out that luminescence occurs in the marine bacterium *Vibrio fischeri* at high cell density. The autoinducer involved in the luminescence process was later found out to be 3-oxohexanoyl-homoserine lactone (von Bodman et al., 2008). Bioluminescence materializes only when *V. fischeri* is at an elevated-cell-concentration-regulated process of quorum sensing. *V. fischeri* is able to emit bioluminescence due to the accumulation of threshold concentration of AHL molecule-regulated light production by the members of the community (Schauder & Bassler, 2001). There are three main QS systems: (i) the AHL QS system in Gram-negative bacteria; (ii) the "oligopeptide-two-component" autoinducing peptide system in Gram-positive bacteria (Dunny & Leonard, 1997); and (iii) the autoinducer-2 system, present in both Gram-negative and Gram-positive bacteria. Although a lot of the signalling molecules exhibit species specificity, some molecules take part in interspecies communication.

10.3.1 Mechanism of Quorum Sensing in Gram-Positive and Gram-Negative Bacteria

In Gram-negative bacteria, the QS system uses small diffusible AHL as signalling molecules whose yield is based on a LuxI-type synthase. AHLs attach and trigger the LuxR transcriptional activator (Li & Tian, 2012; Waters & Bassler, 2005). The AHL signalling molecules penetrate the cell membrane and accumulate in the surrounding extracellular environment regulated by cell density. When the quantity of AHLs reaches a threshold, the cells respond to the change in the environment and consequently manipulate the manifestation of quorum-sensing genes. Cognate LuxI enzymes produce a specific type of AHL when fatty acyl derivative is assigned to the homocysteine moiety of S-adenosyl-methionine (SAM) (Hanzelka & Greenberg, 1996; Parsek & Greenberg, 2000). An example of Gram-negative bacteria where the process of quorum sensing is controlled by the LuxI/LuxR-type is *Pseudomonas aeruginosa* which consists of two pairs of LuxI/LuxR homologues—LasI/LasR and RhlI/RhlR. In *V. fischeri* the coding of luciferase is done by lux AB genes, and the lux CDE genes encode enzymes

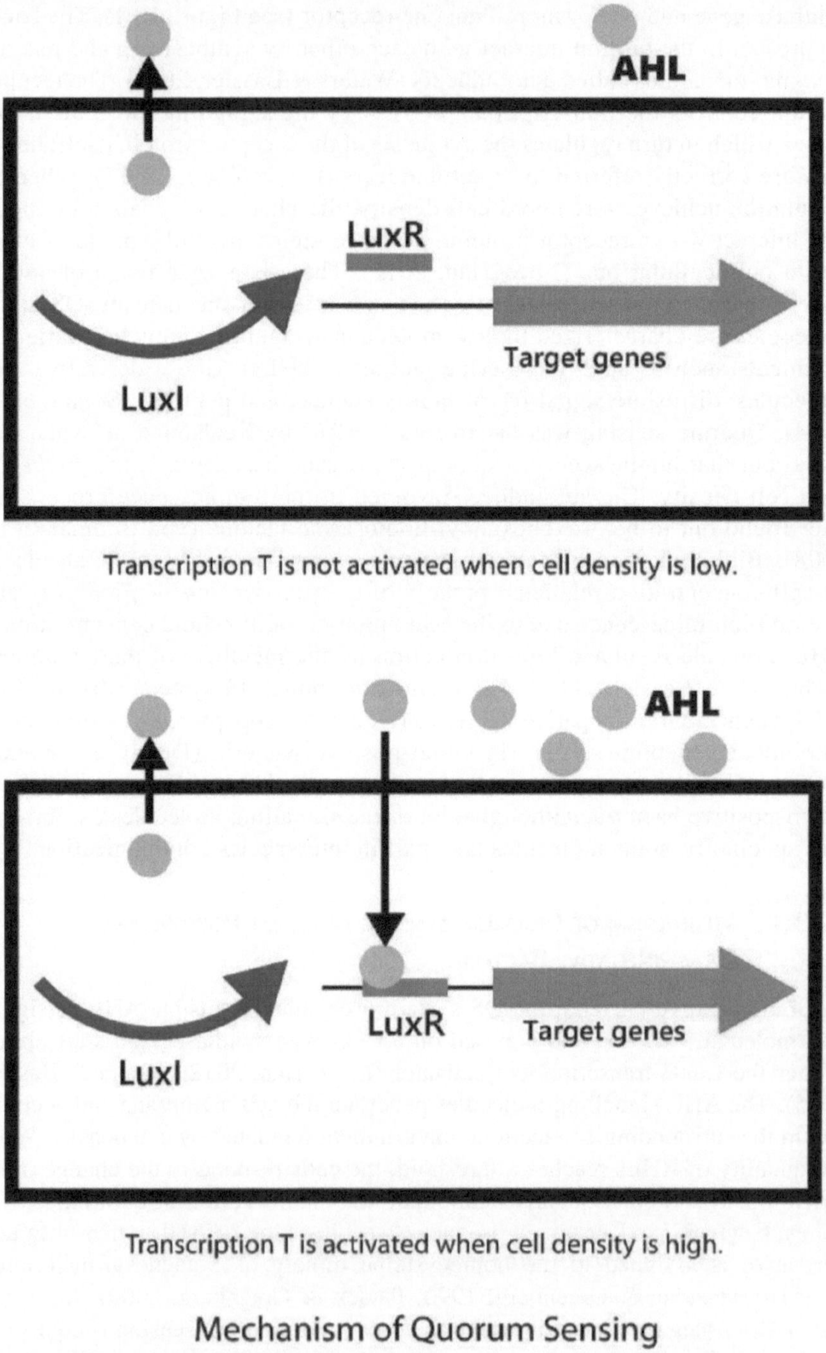

FIGURE 10.1 Quorum sensing involving a complex circuit of genes.

responsible for the production of substrates for luciferase, which ultimately results in the process of bioluminescence (Nazzaro et al., 2013; Paluch et al., 2020).

Another type of QS system is allocated for both Gram-positive and Gram-negative bacteria and is based on AI-2 molecules (Vendeville et al., 2005; Waters & Bassler, 2005). This type of QS system paves the way for interspecies communication and can be considered as a universal dialect for interspecies communication (Xavier & Bassler, 2005). The production of AI-2 is based on a metabolic enzyme *luxS* encoded synthase which catalyzes the separation of S-ribosylhomocysteine to homocysteine and 4, 5-dihydroxy-2, 3-pentanedione (DPD) (X. Chen et al., 2002). The AI-2 QS system was initially identified in a species of marine bacteria, *Vibrio harveyei*. The process of bioluminescence of *V. harveyei* is regulated by furanosyl borate by the mechanism of monitoring cell density (X. Chen et al., 2002; Vendeville et al., 2005). The process of AI-2 receptors and a two-component signal regulator in *Vibrio* species and signal transduction system using an ABC transporter in *Salmonella enterica serovar Typhimurium* and *Escherichia coli* induces the mechanism of quorum sensing (Vendeville et al., 2005).

10.4 QUORUM QUENCHING

The alarming rise of antibiotic-resistant microorganisms has become a global threat. We must now take desperate measures, acting quickly to come up with new tactics to combat the antibiotic-resistant microorganisms. The process of quorum quenching (QQ) represents a new key to decipher the dilemma of microbial resistance. Biofilm community members communicate via the intricate QS system, so interference with the process might lead to the inhibition of biofilm production. Quorum quenching explains the various ways we can inhibit or quench the synthesis of signalling molecules to diminish the dissemination of biofilm synthesis. The enzymatic deterioration of signalling molecules of the QS system halts the accumulation in the extracellular environment and makes drastic changes in gene expression. The inhibition (quenching) of the production of autoinducer molecules prevents the expression of virulency, including biofilm production by drug-averting bacteria (Paluch et al., 2020). The enzymes which take part in the inactivation of the QS signal are called quorum-quenching enzymes. QS inhibitors are defined as a class of chemicals that interfere with the QS pathway. The active substances which possess quorum quenching ability are collectively termed quorum sensing inhibitors (QSIs) (Carradori et al., 2020). Enzymes that are known to degrade signalling molecules and peptides include AHL lactonases and decarboxylases, AHL-acylase, and deaminases.

Four strategies accompany the process of quorum quenching: (i) suppressing the production of the signal molecule by creating hindrance of Lux-I synthase proteins (Lade et al., 2014); (ii) enzymatic degradation of signal molecules by AHL acylase and AHL lactonase (Lade et al., 2014); (iii) meddling with signal receptor analogues (Koch et al., 2005); and (iv) jamming the AHL-LuxR complex (G. Chen et al., 2011; Rampioni et al., 2014). These strategies can be used for

TABLE 10.1
Enzymes Having Quorum-Quenching Ability

QS signal molecules	Mechanism of action	Reference
AHL acylase	Cuts the acyl side chain and hydrolyzes amide bond	(Lin et al., 2003)
AHL lactonase	Hydrolyzes the lactone moiety and opens the HSL ring	(Sakr et al., 2013)
Oxidoreductase	Catalyzes the redox reaction of the acyl side chain	(Jawahir & Zolman, 2021)

achieving the required inhibition algorithm of the AHL-regulated QS network in Gram-positive bacteria as well as AIP-regulated Gram-negative bacteria.

Analogues of the AHL signal molecules act as adversaries and interfere with the signal-receptor binding by modifying the acyl side chain. Analogue molecules in which the acyl side chain of *V. fischeri*, *P. aeruginosa,* and *Agrobacterium tumefacient* was modified exhibited inhibition of the AHL molecular binding (Hanzelka & Greenberg, 1996; Passador et al., 1996; Zhu et al., 1998; Hirakawa & Tomita, 2013). Any alteration in the amide function takes a toll on the binding of the receptor proteins since the amide is responsible for producing hydrogen bonds with tyrosine and aspartic acid in the AHL binding process (Hirakawa & Tomita, 2013; Vannini, 2002). When the amide function was replaced by a triazolyl dihydro furanone, it showed diminished biofilm synthesis along with its inhibition against *Burkholderia cenocepacia* and *P. aeruginosa* (Brackman et al., 2012). Analogue molecules such as C8-HSL antagonize AHL-*TraR* binding in *A. tumefaciens* and C10-HSL antagonizes *LuxR* in *V. fischeri* (Geske et al., 2005). Few diols containing compounds which include pyrogallol and boronic acids demonstrate antagonistic activity against AI-2-LuxP binding (Ni et al., 2008). The most effective and natural way of quorum quenching is the process of enzymic deterioration of AHL molecules (see Table 10.1), categorized under three unique categories of enzymes: (i) lactonases hydrolyze the lactone moiety of the acyl side chain; (ii) amylases hydrolyze the amide bond between the lactone ring and side chain; and (iii) oxidoreductases catalyze the redox reaction of the acyl side chain of the AHL molecule (Rehman & Leiknes, 2018). The hydrolysis of the AHL signal molecule accounts for the total loss of their functional activity, whereas the redox reaction takes part in down-regulating its effects. This implies the advantage of hydrolyzing enzymes over the redox ones in the halting process of the QS pathway (Chowdhary et al., 2007). Several AHL-degrading enzymes are found in nature. These include: AHL acylase *Aiid* from *Ralstonia* sp. XJ12b (Lin et al., 2003), *Hac A* and *Hac B* from *Pseudomonas syringae* strain B728a (Shepherd & Lindow, 2009); AHL lactonase *AttM* from *A. tumifaciens* c58 (Lee & Zhang, 2015), *AiiA* from *Bacillus cereus* and *Bacillus mycoides* (Dong et al., 2002), *QIcA* from *Acidobacteria* sp. (Riaz et al., 2008); and oxidoreductase P450BM3 from *Bacillus megaterium* (Chowdhary et al., 2007). The final method is blockage of

TABLE 10.2
Quorum-Quenching Genes Obtained from Different Bacterial Species

Source	Genes obtained	AHL degradation	Mechanism of action	Reference
Ralstonia sp. XJ12B	*AiiD*	3-oxo-C8-HSL, 3-oxo-C10-HSL, 3-oxo-C12-HSL	Hydrolyzes AHL amide group	Lin et al, 2003
P. syringae strain B728a	*HacA*	C8-HSL, C10-HSL, C12-HSL	Hydrolyzes inactive AHLs	(Shepherd & Lindow, 2009)
P. syringae strain B728a	*HacB*	C6-HSL, C8-HSL, C10-HSL, C12-HSL	by cleaving the amide linkages	(Shepherd & Lindow, 2009)
A. tumifaciens c58	*AttM*	3-oxo-C8-HSL	Hydrolyzes lactone ring of 3OC8HSL	(Zhang et al., 2002)
B. cereus and *B. mycoides*	*AiiA*	C6-HSL, C8-HSL, C10-HSL	Hydrolyzes lactone ring	(Dong et al., 2002)
Acidobacteria sp.	*QlcA*	3-oxo-C6-HSL, C7-HSL, 3-oxo-C8-HSL, 3-oxo-C10-HSL	Hydrolyzes lactone ring	(Riaz et al., 2008)
B. megaterium CYP102A1	*P450BM3*	3-oxo-C12-HSL, 3-oxo-C14-HSL, C16-HSL, C18-HSL, C20-HSL	Oxidation of AHLs and their products	(Chowdhary et al., 2007)

the consecutive reaction of signal transduction. Savrin, a small molecule inhibitor, infringes upon the transcriptional regulator *AgrA*, attaches to DNA resulting in the inhibition of the production of RNAIII which, when combined with *AgrA*, is accountable for the synthesis of virulence factors (Paluch et al., 2020; Sully et al., 2014). The techniques and methods which are currently in use to analyze quorum quenching and quorum sensing molecules comprise methods such as colorimetry, bioluminescence, fluorescence, and electrochemistry (Paluch et al., 2020; van der Meer & Belkin, 2010) (see Table 10.2).

10.5 BIOFOULING

Accumulation of microorganisms, plants, and algal components on the surface of submerged underwater objects like ships or submarines has become a problem. These accumulations result in the primary degradation of the object. The process of such accumulation of microorganisms, small plants, and algae is termed biofouling. This is considered a natural process that results from biofilm arrangement on surfaces that are immersed or submerged and has become a major issue for multiple industries. The formation of biofilm is essential to biofouling. Biofilm progresses through numerous stages, including pathways like conditioning substratum, followed by the adhesion of pioneer bacteria and the secretion of extracellular polymeric essences and low macrofouling, a temporary attachment, to hard macrofouling, the permanent attachment (Alemán-Vega et al., 2020).

Because excessive growth of biofouling components has become a problem for the marine industry, preventive measures have been taken to combat algal and microorganism growth on submerged surfaces. Among these are the use of biosurfactants, which is regarded as a promising way to prevent biofouling. Biosurfactants are amphipathic surface-active compounds that are synthesized by means of several microorganisms and substantially decrease the surface tension. They are considered anti-adhesives and antimicrobial/antibiofilm agents and are comparatively less toxic, which allow for bioremediation properties (Alemán-Vega et al., 2020). Along with biosurfactants, the use of toxic biocides, like copper, that kill organisms and accumulate in the environment are also used widely in the market. But the use of toxic antifouling methods has shown several disadvantages, whereas the nontoxic methods available in the market are costly and not as effective (Dahms & Dobretsov, 2017). Control of biofouling compounds has become more important in the past few decades as significant problems are observed related to health risks, environmental impact, and financial losses. The issues are also accompanied by reduction in speed and increasing cost of repairing biofouled equipment, as well as environmental concerns, corrosion, and safety hazards. Some common marine foulers are the green algae *Ulva australis* and brown algae *Ectocarpus* (Bixler & Bhushan, 2012).

10.6 MECHANISM OF BIOFOULING

The formation of biofoulers is observed when a microorganism transitions to a stationary sessile lifestyle from a free-floating planktonic one. This results in the formation of a biofilm. The microorganisms adhere to each other along a hard surface which has an adhesive property called the extracellular polymeric substance (EPS). All unprotected artificial things in the aquatic environment first come to be colonized via biota, a manner referred to as biofouling. The process of biofouling includes three important procedures. The first procedure follows the adsorption of dissolved natural molecules and proceeds to colonization by prokaryotes as well as unicellular eukaryotes (microfouling). Subsequent recruitment of invertebrate larvae and algal spores (macrofouling) is then observed (Dobretsov et al., 2009). The biofilm then grows and becomes more diverse, enticing more microorganisms by sending out chemical 'messages' (Bixler & Bhushan, 2012). The process of biofouling where the organisms collect and grow on the surface is called colonization (Bixler & Bhushan, 2012). Marine biofilm formation proceeds with the development of a biofilm layer, followed by the signalling that attracts larger macrofoulers. The previously affected layer of biofilm, when attracting the other microorganisms, secretes EPS, which makes the bonds among the organisms much stronger. The establishment of the purposeful and structural integrity of biofilms is considered the process of EPS. They are considered the fundamental component for the determination of the physicochemical properties of a biofilm and have an essential part in the cell attachment process between the organisms. The preliminary attachment and the formation of the EPS leads to permanent attachment, followed by final growth and dispersion (Bixler & Bhushan, 2012). Moreover, underwater environments are ideal for the growth of biofoulers as they receive an adequate amount of nutrients. The rate of growth for a biofilm depends on several factors, including the substrate, flow, velocity, shear stress, and temperature, along with bio-adhesion. These factors have an essential role in the tenacity of biofoulers on the surface. Bio-adhesion is a two-step process consisting of initial attachment and irretrievable attachment. This process of attachment occurs through frail, revocable van der Waals bonds, which are comparatively sturdier than electrostatic repulsive forces along with the secretion of EPS (Bixler & Bhushan, 2012).

10.7 PREVENTION OF BIOFOULING: ANTIFOULING

With the rising concern of biofouling in marine industries, scientists are developing antifouling processes which will be both economical and effective. Over the decades, several methods have been introduced to prevent biofouling. Biosurfactants are one such method, but their toxicity and other disadvantageous factors make them less appealing. In general, currently available procedures for antifouling are expensive and not very effective. Thus, research into antifouling is still ongoing.

Antifouling is generally achieved by coating, painting, or treating the surface with a solid that will control or prevent the attachment of undesirable organisms. For example, tributyltin (TBT) coatings were used in the past, but TBT, a biocide, was found to have excessive drawbacks such as toxicity, which led to its early ban. Similarly, other biocide preparations, such as irgarol, chlorothalonil, dichlofluanid, and diuron, were used as marine antifouling coatings in what was hoped to be an environmentally friendly manner, but later they were found to be lethal, collecting in marinas and harbors (Saha et al., 2018). Studies suggested these biocides leak into the seawater and kill barnacles and other marine life that attach to a solid surface. Furthermore, studies have concluded that these compounds persist in the water, threaten sea life, harm the environment, and possibly disrupt the food chain (Telegdi et al., 2016).

One of the latest strategies, biomimicry, is believed to be an improvement in environmentally friendly antifouling coatings. To reduce or prevent toxicity, it is suggested to incorporate natural antifouling compounds instead of synthetic compounds into marine paints. However, for this to be feasible at the large-scale industry level, such compounds need to be cost efficient in terms of manufacturing, readily available, easy to use, and nonpoisonous to marine biota (Saha et al., 2018).

Among other new strategies, scientists have devised methods which involve the marine ecosystem. It has been observed that many marine organisms have the ability to prevent and repel biofouling. Algae, sponges, and corals have the ability to prevent the colonization of hosts by other fouling organisms (Dahms & Dobretsov,

2017). These marine organisms do not rely on one specific mechanism; rather, they depend on several factors, including the combined use of physical, chemical, and biological strategies to prevent biofouling (Dahms & Dobretsov, 2017).

10.8 MARINE ANTIFOULING FROM ALGAL COMPOUNDS

Over the past decade, several types of research have been conducted on marine algae as an antifouling compound. Although it is still ongoing, scientists could conclude that marine algal organisms are effective in antifouling. Algal compounds are the primary producers of a temperate ecosystem as well as the largest biomass producers in the environment. Moreover, marine algal compounds are largely responsible for producing a compound that protects the marine environment (Saha et al., 2018). Studies have shown evidence that marine organisms like algal compounds have anti-QS abilities which may be able to temper economic losses of marine industries from the major issue of biofouling. Quorum sensing is a population density–dependent and structured genetic circuitry involving extracellular signalling molecules produced by the bacteria for their own survival under extreme conditions (Jha et al., 2013). It is the biggest factor responsible for the formation of biofilm layers on surfaces. Communication through the process of cell signalling QS has been both a biological and economic issue because biofouling is a major cause of economic loss in the maritime industries (Saha et al., 2018). The potential ability of algal compounds to prevent the growth of fouling organisms without adverse effects on the environment marks a major advance in science (see Table 10.3).

10.9 QUORUM-SENSING INHIBITION IN THE MARINE ENVIRONMENT

Bacteria in a community may also signal their existence to each other via production, detection, and responding to minor diffusible sign molecules called autoinducers. This process was first described in the marine bioluminescent bacterium *V. fischeri* and the intercellular communication came to be known as quorum sensing (QS) (Li & Tian, 2012). A study suggested that the light organ of the sepiolid squid, *Euprymna scolopes*, is colonized by *V. fischeri*. Within the light organs of squid, bacteria can construct very significant population density. The light emission via microorganisms is regulated with the aid of AHLs and occurs best at an excessive population density (Dobretsov et al., 2009). This discovery led to the assumption that free-living and invertebrate-associated marine bacteria both can produce QS signals. *Phaeobacter gallaeciencis* is a bacteria and member of the Roseobacter clade that releases compounds throughout microalgae decomposition which up-regulates roseobacticides biosynthesis in the bacteria. This mechanism was called 'turning on an algicidal switch' in the bacterium (Dow, 2021). Further studies on this bacterium substantiated the biosynthesis of both roseobacticides and tropodithietic acid. These composites are correspondingly controlled by quorum sensing, particularly OH-C10-AHL (Dow, 2021).

TABLE 10.3
List of Algal Compounds in Down-Regulation of QS

Algal source	Name of compound	Mechanism of action	Reference
Delisea pulchra	Halogenated furanones	Competitive binding to Lux-R proteins resulting in heightened LuxR turnover	(Paluch et al., 2020)
Ahnfeltiopsis flabelliformis	Betonicine, floridoside, isetheonic acid	Combined effect of the mixture consisting of the three compounds exhibited QS inhibition moderated by N-OHL (octanoyl-DL-homoserine lactone) and transcriptional activator protein TraR	(Liu et al., 2008)
Asparogopsis taxiformis	Methanolic extract, mahorone, 5-bromomahorone	Cinnamaldehyde used as positive control showed QS inhibition of violacein which is produced as QS-regulated behavior	(Jha et al., 2013)
Laurencia viridis	28-Hydroxysaiyacenol B Saiyacenol C 15,16-epoxythyrsiferol A 15,16-epoxythyrsiferol B	Shows inhibitory effects on diatom growth and macroalgal zoospores	(Cen-Pacheco et al., 2015)
Cystoseira tamarascifolia	Cystophloroketals A-E monocyclic meroditerpenoid	Shows possibility for inhibition of foulers such as *Balanus amphitrite* and *Mytilus edulis*	(El Hattab et al., 2015)
Bonnemaisonia hamifera	1,1,3,3-Tetrabromo-2-heptanone	Alterations in community composition were perceived on an intricate level	(Persson et al., 2011)

Because marine algal organisms complete their life cycle in the seawater, it was previously believed they could provide a surface for biofilm formation. However, it has been observed to be only a rare occurrence. This led to the idea of algal compounds possessing quorum sensing inhibition properties and a defence mechanism for the production of biofilm on the surface (Jha et al., 2013). This property of marine algal organisms was added to antifouling compounds as an important agent. Recently, several investigators have reported substitute antifouling compounds from macroalgal extracts like the cystophloroketals and chromanols, along with additional reports of conditioning from crude extracts of numerous polarities (Saha et al., 2018).

The first algal species from which quorum sensing inhibitory compounds were isolated was red macroalgae, *Delisea pulchra*, and its role in AHL regulatory systems and quorum sensing inhibition was studied (Jha et al., 2013). More than 30 species have been studied to understand the quorum sensing inhibition properties of algal compounds. The algal compounds considered to be active antifouling compounds are genera like the red algae *Asparagopsis* and *Laurencia*, and the brown alga *Sargassum*. Recent discoveries relating to the red alga *Asparagopsis taxiformis*, found exclusively in the Indian Ocean, has led to the detection of two new highly brominated cyclopentenones, namely mahorone and 5-bromomahorone, that possess antimicrobial activity that can be used against biofouling compounds (Saha et al., 2018). The study of these 30 different algal species was divided into three categories according to their quorum sensing inhibition capabilities identified using *Chromobacterium violaceum* CV026. Through the study, it was seen that only *A. taxiformis* showed positive quorum sensing inhibition properties (Saha et al., 2018).

10.10 CONTROL OF BIOFOULING BY QUORUM QUENCHERS

The Gram-negative aquatic bacteria are chiefly responsible for biofouling and have been found to possess AHL-mediated quorum sensing features (Lade et al., 2014). Most of them are characterized by their LuxR-N-acyl-homoserine lactones (AHLs) mechanism. However, some specific QS signalling systems are present in Gram-negative bacteria and are obtained through these bacteria. The AHLs are commonly produced via homologues of the AHL synthase LuxI from S-adenosyl methionine with the interaction of intracellular pool action of acyl-carrier proteins (Dobretsov et al., 2009). Several quorum quenching strategies are in use to prevent the signalling interaction and further prevent the growth of the biofouling compounds. The quorum sensing process can be interrupted by the following methods: i) blocking of the LuxI-type synthase proteins by inhibition of AHL synthesis; ii) destruction of enzymes of AHLs molecules by AHL-acylase and AHL-lactonase to prevent them from accumulating; and iii) formation of the AHL/LuxR complex through the intrusion of signal receptors or blockage (Lade et al., 2014).

The LuxR-type complexes are found to bind to specific regions within the exact promoter sequences. They can activate transcription of QS genes, which is the prime reason for luminescence, as well as trigger the formation of biofilm, virulence, and other pertinent procedures. LuxR proteins are tainted rather quickly in the absence of AHLs. The further stabilization of LuxR occurs when it binds to the cognate AHL signal (Dobretsov et al., 2009). The inhibition of AHL-mediated quorum sensing methods to deduce Gram-negative bacteria is applied whereas, for Gram-positive bacteria, the application of AIPs-mediated quorum sensing is observed (Lade et al., 2014). The traditional process of antifouling involved killing or inhibiting the growth of bacteria by the use of antimicrobial compounds or antibiotics. However, this method is inefficient as bacteria are important for decomposition of contaminants, so indiscriminate killing of bacteria became another major issue. The use of antimicrobial compounds will likely lead to more intense research involving compounds that have nonantibiotic capabilities and can specifically block AHL-mediated biofouling traits without interfering with microbial growth (Lade et al., 2014).

10.11 CONCLUSION

In the past few years, numerous reports concluded that marine algal compounds were effective against the formation of biofilm. Biofilm is the unwanted formation of biological organisms (specifically, bacteria) on the surfaces of marine equipment like ships and submarines. The formation of such compounds has become a rising problem and is affecting marine industries economically as well as threatening the health of the environment. The process of preventing biofilm formation is called antifouling. Traditionally, various surfactants and paints were used as antifouling components. However, the toxic effect of these compounds on the marine ecosystem led to a decrease in use and increased interest in more eco-friendly materials. Studies suggested algal compounds could prevent the growth of bacteria on the biofilm layer, and further, that they could be the biomass producer in the environment.

Marine algal organisms are composed of quorum-sensing inhibition components that prevent the cell signalling system between the bacteria and stop the interaction and more accumulations of it on the surface. Several marine algae were studied for quorum sensing inhibition and among all, *A. taxiformis* showed a positive response in building quorum sensing inhibitory complexes as well as growth inhibition complexes. The formation of these complexes gave evidence of preventing the formation of biofilm without any adverse toxic effects on the environment. Studies also showed the utility of algal compounds in green synthesis of nanoparticles for antifouling applications. However, isolation of the essential biogenic compounds that can be obtained from marine algae is at present very costly and time-consuming as researchers must have considerable biological and chemical expertise to identify these biogenic compounds. Moreover, it is also necessary to understand the ecological factors of the algal compound for it to be good for the environment. Progress in isolating these antifouling compounds is intensifying in hopes of immediate application and preventing further damage in the marine industries.

REFERENCES

Alemán-Vega, M., Sánchez-Lozano, I., Hernández-Guerrero, C. J., Hellio, C., & Quintana, E. T. (2020). Exploring antifouling activity of biosurfactants producing marine bacteria isolated from Gulf of California. *International Journal of Molecular Sciences*, *21*(17), 6068. https://doi.org/10.3390/ijms21176068

Berne, C., Ducret, A., Hardy, G. G., & Brun, Y. V. (2015). Adhesins involved in attachment to abiotic surfaces by gram-negative bacteria. *Microbiology Spectrum*, *3*(4). https://doi.org/10.1128/microbiolspec.MB-0018-2015

Bixler, G. D., & Bhushan, B. (2012). Biofouling: Lessons from nature. *Philosophical Transactions of the Royal Society A: Mathematical, Physical and Engineering Sciences*, *370*(1967), 2381–2417. https://doi.org/10.1098/rsta.2011.0502

Brackman, G., & Coenye, T. (2014). Quorum sensing inhibitors as anti-biofilm agents. *Current Pharmaceutical Design*, *21*(1), 5–11. https://doi.org/10.2174/1381612820666140905114627

Brackman, G., Risseeuw, M., Celen, S., Cos, P., Maes, L., Nelis, H. J., Van Calenbergh, S., & Coenye, T. (2012). Synthesis and evaluation of the quorum sensing inhibitory effect of substituted triazolyldihydrofuranones. *Bioorganic & Medicinal Chemistry*, *20*(15), 4737–4743. https://doi.org/10.1016/j.bmc.2012.06.009

Carradori, S., Di Giacomo, N., Lobefalo, M., Luisi, G., Campestre, C., & Sisto, F. (2020). Biofilm and quorum sensing inhibitors: The road so far. *Expert Opinion on Therapeutic Patents*, *30*(12), 917–930. https://doi.org/10.1080/13543776.2020.1830059

Cen-Pacheco, F., Santiago-Benítez, A. J., García, C., Álvarez-Méndez, S. J., Martín-Rodríguez, A. J., Norte, M., Martín, V. S., Gavín, J. A., Fernández, J. J., & Daranas, A. H. (2015). Oxasqualenoids from *Laurencia viridis*: Combined spectroscopic–computational analysis and antifouling potential. *Journal of Natural Products*, *78*(4), 712–721. https://doi.org/10.1021/np5008922

Chen, G., Swem, L. R., Swem, D. L., Stauff, D. L., O'Loughlin, C. T., Jeffrey, P. D., Bassler, B. L., & Hughson, F. M. (2011). A strategy for antagonizing quorum sensing. *Molecular Cell*, *42*(2), 199–209. https://doi.org/10.1016/j.molcel.2011.04.003

Chen, X., Schauder, S., Potier, N., Dorsselaer, A. V., Pelczer, I., Bassler, B. L., & Hughson, F. M. (2002). Structural identification of a bacterial quorum-sensing signal containing boron. *Nature*, *415*, 545–549.

Chowdhary, P. K., Keshavan, N., Nguyen, H. Q., Peterson, J. A., González, J. E., & Haines, D. C. (2007). *Bacillus megaterium* CYP102A1 oxidation of acyl homoserine lactones and acyl homoserines. *Biochemistry*, *46*(50), 14429–14437. https://doi.org/10.1021/bi701945j

Dahms, H., & Dobretsov, S. (2017). Antifouling compounds from marine macroalgae. *Marine Drugs*, *15*(9), 265. https://doi.org/10.3390/md15090265

Di Martino, P. & Groupe Biofilm et Comportement Microbien aux Interfaces, Laboratoire ERRMECe-EA1391, Université de Cergy-Pontoise, rue Descartes site de Neuville-sur-Oise 95031 Cergy-Pontoise, cedex France. (2018). Extracellular polymeric substances, a key element in understanding biofilm phenotype. *AIMS Microbiology*, *4*(2), 274–288. https://doi.org/10.3934/microbiol.2018.2.274

Dobretsov, S., Teplitski, M., & Paul, V. (2009). Mini-review: Quorum sensing in the marine environment and its relationship to biofouling. *Biofouling*, *25*(5), 413–427. https://doi.org/10.1080/08927010902853516

Dong, Y.-H., Gusti, A. R., Zhang, Q., Xu, J.-L., & Zhang, L.-H. (2002). Identification of quorum-quenching *N*-acyl homoserine lactonases from *Bacillus* species. *Applied and Environmental Microbiology*, *68*(4), 1754–1759. https://doi.org/10.1128/AEM.68.4.1754-1759.2002

Dow, L. (2021). How do quorum-sensing signals mediate algae–bacteria interactions? *Microorganisms*, *9*(7), 1391. https://doi.org/10.3390/microorganisms9071391

Dunny, G. M., & Leonard, B. A. B. (1997). Cell-cell communication in gram-positive bacteria. *Annual Review of Microbiology*, *51*(1), 527–564. https://doi.org/10.1146/annurev.micro.51.1.527

El Hattab, M., Genta-Jouve, G., Bouzidi, N., Ortalo-Magné, A., Hellio, C., Maréchal, J.-P., Piovetti, L., Thomas, O. P., & Culioli, G. (2015). Cystophloroketals A–E, unusual phloroglucinol–meroterpenoid hybrids from the brown alga *Cystoseira tamariscifolia*. *Journal of Natural Products*, *78*(7), 1663–1670. https://doi.org/10.1021/acs.jnatprod.5b00264

Geske, G. D., Wezeman, R. J., Siegel, A. P., & Blackwell, H. E. (2005). Small molecule inhibitors of bacterial quorum sensing and biofilm formation. *Journal of the American Chemical Society*, *127*(37), 12762–12763. https://doi.org/10.1021/ja0530321

Hanzelka, B. L., & Greenberg, E. P. (1996). Quorum sensing in Vibrio fischeri: Evidence that S-adenosylmethionine is the amino acid substrate for autoinducer synthesis. *Journal of Bacteriology*, *178*(17), 5291–5294. https://doi.org/10.1128/jb.178.17.5291-5294.1996

Hirakawa, H., & Tomita, H. (2013). Interference of bacterial cell-to-cell communication: A new concept of antimicrobial chemotherapy breaks antibiotic. *Frontiers in Microbiology, 4*. https://doi.org/10.3389/fmicb.2013.00114

Jawahir, V., & Zolman, B. K. (2021). Long chain acyl CoA synthetase 4 catalyzes the first step in peroxisomal indole-3-butyric acid to IAA conversion. *Plant Physiology, 185*(1), 120–136. https://doi.org/10.1093/plphys/kiaa002. PMID: 33631795; PMCID: PMC8133310.

Jha, B., Kavita, K., Westphal, J., Hartmann, A., & Schmitt-Kopplin, P. (2013). Quorum sensing inhibition by asparagopsis taxiformis, a marine macro alga: Separation of the compound that interrupts bacterial communication. *Marine Drugs, 11*(12), 253–265. https://doi.org/10.3390/md11010253

Kalia, V. C. (2013). Quorum sensing inhibitors: An overview. *Biotechnology Advances, 31*(2), 224–245. https://doi.org/10.1016/j.biotechadv.2012.10.004

Karygianni, L., Ren, Z., Koo, H., & Thurnheer, T. (2020). Biofilm matrixome: Extracellular components in structured microbial communities. *Trends in Microbiology, 28*(8), 668–681. https://doi.org/10.1016/j.tim.2020.03.016

Koch, B., Liljefors, T., Persson, T., Nielsen, J., Kjelleberg, S., & Givskov, M. (2005). The LuxR receptor: The sites of interaction with quorum-sensing signals and inhibitors. *Microbiology, 151*(11), 3589–3602. https://doi.org/10.1099/mic.0.27954-0

Lade, H., Paul, D., & Kweon, J. H. (2014). Quorum quenching mediated approaches for control of membrane biofouling. *International Journal of Biological Sciences, 10*(5), 550–565. https://doi.org/10.7150/ijbs.9028

Lee, J., & Zhang, L. (2015). The hierarchy quorum sensing network in Pseudomonas aeruginosa. *Protein & Cell, 6*(1), 26–41. https://doi.org/10.1007/s13238-014-0100-x

Li, Y.-H., & Tian, X. (2012). Quorum sensing and bacterial social interactions in biofilms. *Sensors, 12*(3), 2519–2538. https://doi.org/10.3390/s120302519

Lin, Y.-H., Xu, J.-L., Hu, J., Wang, L.-H., Ong, S. L., Leadbetter, J. R., & Zhang, L.-H. (2003). Acyl-homoserine lactone acylase from Ralstonia strain XJ12B represents a novel and potent class of quorum-quenching enzymes: Quorum-quenching AHL-acylase. *Molecular Microbiology, 47*(3), 849–860. https://doi.org/10.1046/j.1365-2958.2003.03351.x

Liu, H. B., Koh, K. P., Kim, J. S., Seo, Y., & Park, S. (2008). The effects of betonicine, floridoside, and isethionic acid from the red alga Ahnfeltiopsis flabelliformis on quorum-sensing activity. *Biotechnology and Bioprocess Engineering, 13*(4), 458–463. https://doi.org/10.1007/s12257-008-0145-x

Mah, T.-F. C., & O'Toole, G. A. (2001). Mechanisms of biofilm resistance to antimicrobial agents. *Trends in Microbiology, 9*(1), 34–39. https://doi.org/10.1016/S0966-842X(00)01913-2

Muhammad, M. H., Idris, A. L., Fan, X., Guo, Y., Yu, Y., Jin, X., Qiu, J., Guan, X., & Huang, T. (2020). Beyond risk: Bacterial biofilms and their regulating approaches. *Frontiers in Microbiology, 11*, 928. https://doi.org/10.3389/fmicb.2020.00928

Nazzaro, F., Fratianni, F., & Coppola, R. (2013). Quorum sensing and phytochemicals. *International Journal of Molecular Sciences, 14*(6), 12607–12619. https://doi.org/10.3390/ijms140612607

Ni, N., Choudhary, G., Li, M., & Wang, B. (2008). Pyrogallol and its analogs can antagonize bacterial quorum sensing in Vibrio harveyi. *Bioorganic & Medicinal Chemistry Letters, 18*(5), 1567–1572. https://doi.org/10.1016/j.bmcl.2008.01.081

Paluch, E., Rewak-Soroczyńska, J., Jędrusik, I., Mazurkiewicz, E., & Jermakow, K. (2020). Prevention of biofilm formation by quorum quenching. *Applied Microbiology and Biotechnology, 104*(5), 1871–1881. https://doi.org/10.1007/s00253-020-10349-w

Parsek, M. R., & Greenberg, E. P. (2000). Acyl-homoserine lactone quorum sensing in Gram-negative bacteria: A signaling mechanism involved in associations with higher organisms. *Proceedings of the National Academy of Sciences*, *97*(16), 8789–8793. https://doi.org/10.1073/pnas.97.16.8789

Passador, L., Tucker, K. D., Guertin, K. R., Journet, M. P., Kende, A. S., & Iglewski, B. H. (1996). Functional analysis of the Pseudomonas aeruginosa autoinducer PAI. *Journal of Bacteriology*, *178*(20), 5995–6000. https://doi.org/10.1128/jb.178.20.5995-6000.1996

Persson, F., Svensson, R., Nylund, G. M., Fredriksson, N. J., Pavia, H., & Hermansson, M. (2011). Ecological role of a seaweed secondary metabolite for a colonizing bacterial community. *Biofouling*, *27*(6), 579–588. https://doi.org/10.1080/08927014.2011.589001

Pircalabioru, G. G., & Chifiriuc, M.-C. (2020). Nanoparticulate drug-delivery systems for fighting microbial biofilms: From bench to bedside. *Future Microbiology*, *15*(8), 679–698. https://doi.org/10.2217/fmb-2019-0251

Qian, P.-Y., Xu, Y., & Fusetani, N. (2009). Natural products as antifouling compounds: Recent progress and future perspectives. *Biofouling*, *26*(2), 223–234. https://doi.org/10.1080/08927010903470815

Rampioni, G., Leoni, L., & Williams, P. (2014). The art of antibacterial warfare: Deception through interference with quorum sensing–mediated communication. *Bioorganic Chemistry*, *55*, 60–68. https://doi.org/10.1016/j.bioorg.2014.04.005

Rehman, Z. U., & Leiknes, T. (2018). Quorum-quenching bacteria isolated from red sea sediments reduce biofilm formation by Pseudomonas aeruginosa. *Frontiers in Microbiology*, *9*, 1354. https://doi.org/10.3389/fmicb.2018.01354

Riaz, K., Elmerich, C., Moreira, D., Raffoux, A., Dessaux, Y., & Faure, D. (2008). A metagenomic analysis of soil bacteria extends the diversity of quorum-quenching lactonases. *Environmental Microbiology*, *10*(3), 560–570. https://doi.org/10.1111/j.1462-2920.2007.01475.x

Saha, M., Goecke, F., & Bhadury, P. (2018). Minireview: Algal natural compounds and extracts as antifoulants. *Journal of Applied Phycology*, *30*(3), 1859–1874. https://doi.org/10.1007/s10811-017-1322-0

Sakr, M. M., Aboshanab, K. M., Aboulwafa, M. M., & Hassouna, N. A. (2013). Characterization and complete sequence of lactonase enzyme from *Bacillus weihenstephanensis* isolate P65 with potential activity against acyl homoserine lactone signal molecules. *Biomed Research International*, *2013*, 192589. https://doi.org/10.1155/2013/192589. Epub 2013 Jul 9. PMID: 23936779; PMCID: PMC3722983.

Schauder, S., & Bassler, B. L. (2001). The languages of bacteria. *Genes & Development*, *15*(12), 1468–1480. https://doi.org/10.1101/gad.899601

Sharma, D., Misba, L., & Khan, A. U. (2019). Antibiotics versus biofilm: An emerging battleground in microbial communities. *Antimicrobial Resistance & Infection Control*, *8*(1), 76. https://doi.org/10.1186/s13756-019-0533-3

Shepherd, R. W., & Lindow, S. E. (2009). Two dissimilar *N*-acyl-homoserine lactone acylases of *Pseudomonas syringae* influence colony and biofilm morphology. *Applied and Environmental Microbiology*, *75*(1), 45–53. https://doi.org/10.1128/AEM.01723-08

Singh, P. K., Bartalomej, S., Hartmann, R., Jeckel, H., Vidakovic, L., Nadell, C. D., & Drescher, K. (2017). Vibrio cholerae combines individual and collective sensing to trigger biofilm dispersal. *Current Biology*, *27*(21), 3359–3366.e7. https://doi.org/10.1016/j.cub.2017.09.041

Solano, C., Echeverz, M., & Lasa, I. (2014). Biofilm dispersion and quorum sensing. *Current Opinion in Microbiology*, *18*, 96–104. https://doi.org/10.1016/j.mib.2014.02.008

Sully, E. K., Malachowa, N., Elmore, B. O., Alexander, S. M., Femling, J. K., Gray, B. M., DeLeo, F. R., Otto, M., Cheung, A. L., Edwards, B. S., Sklar, L. A., Horswill, A. R., Hall, P. R., & Gresham, H. D. (2014). Selective chemical inhibition of AGR quorum sensing in staphylococcus aureus promotes host defense with minimal impact on resistance. *PLoS Pathogens*, *10*(6), e1004174. https://doi.org/10.1371/journal.ppat.1004174

Telegdi, J., Trif, L., & Románszki, L. (2016). Smart anti-biofouling composite coatings for naval applications. In *Smart Composite Coatings and Membranes* (pp. 123–155). Elsevier. https://doi.org/10.1016/B978-1-78242-283-9.00005-1

Tremblay, Y. D., Lévesque, C., Segers, R. P., & Jacques, M. (2013). Method to grow Actinobacillus pleuropneumoniaebiofilm on a biotic surface. *BMC Veterinary Research*, *9*(1), 213. https://doi.org/10.1186/1746-6148-9-213

van der Meer, J. R., & Belkin, S. (2010). Where microbiology meets microengineering: Design and applications of reporter bacteria. *Nature Reviews Microbiology*, *8*(7), 511–522. https://doi.org/10.1038/nrmicro2392

Vannini, A. (2002). The crystal structure of the quorum-sensing protein TraR bound to its autoinducer and target DNA. *The EMBO Journal*, *21*(17), 4393–4401. https://doi.org/10.1093/emboj/cdf459

Vendeville, A., Winzer, K., Heurlier, K., Tang, C. M., & Hardie, K. R. (2005). Making "sense" of metabolism: Autoinducer-2, LUXS and pathogenic bacteria. *Nature Reviews Microbiology*, *3*(5), 383–396. https://doi.org/10.1038/nrmicro1146

von Bodman, S. B., Willey, J. M., & Diggle, S. P. (2008). Cell-cell communication in bacteria: United we stand. *Journal of Bacteriology*, *190*(13), 4377–4391. https://doi.org/10.1128/JB.00486-08

Waters, C. M., & Bassler, B. L. (2005). Quorum sensing: Cell-to-cell communication in bacteria. *Annual Review of Cell and Developmental Biology*, *21*(1), 319–346. https://doi.org/10.1146/annurev.cellbio.21.012704.131001

Xavier, K. B., & Bassler, B. L. (2005). Interference with AI-2-mediated bacterial cell–cell communication. *Nature*, *437*(7059), 750–753. https://doi.org/10.1038/nature03960

Zhu, Z., Labbé, S., Peña, Ma. M. O., & Thiele, D. J. (1998). Copper differentially regulates the activity and degradation of yeast Macl transcription factor. *Journal of Biological Chemistry*, *273*(3), 1277–1280. https://doi.org/10.1074/jbc.273.3.1277

11 Microbial Volatile Organic Compounds as Quorum Quenchers

Ashmita Samanta, Drisha Roy, Dibyajit Lahiri, Rina Rani Ray, Debasmita Bhattacharya, and Moupriya Nag

11.1 INTRODUCTION

The broad biochemical and phylogenetic diversity of microorganisms is responsible for producing volatile gases that affect the air's composition in the environment. The density variation of microbes present in the air can differ drastically; for instance, indoor air contains 102 to 109 microbes per dm^3 (Misztal et al., 2018). Microbes need energy to grow and reproduce, thus they use chemicals to transform available substrates into energy and create microbial biomass. In the presence of microbial rivals, metabolic energy can switch to secondary metabolism, much like the biosynthesis of poisons or antibiotics. Nonvolatile, semivolatile, and volatile chemicals are released from microbial cells during the process of metabolism. The metabolism of microorganisms like bacteria and fungi results in the production of microbial volatile organic compounds (mVOCs). They are produced as waste products during both primary metabolism (from the synthesis of DNA, amino acids, and fatty acids, for example) and secondary metabolism, especially in the metabolic oxidation of glucose from a number of intermediates of the primary metabolism. As a result, the synthesis of mVOCs is significantly influenced by microbial species, growth phases, and environmental conditions such as nutrition, pH, humidity, and temperature (Korpi et al., 2009).

The produced mVOCs could undergo additional chemical transformations into different chemicals in the environment. Unsaturated hydrocarbons (isoprene/terpenes) and ozone (and other oxidants) reactions have recently been experimentally examined as potential sources of mVOCs in the atmosphere. Although it has been believed that the intermediate molecules produced during the reactions are significantly more irritating than the corresponding original reactants and end-products, aldehydes, ketones, and organic acids are the main products of these reactions. It is crucial to select compounds that can only be of microbial origin and use these as indicators of microbial effect in order to separate microbial volatiles from other

DOI: 10.1201/9781003300557-11

sources. As a result, the 'real' microbial influence will constantly be underestimated because so few volatiles will be the original microbial volatile organic molecules (mVOCs). Other, more prevalent volatile organic compounds (VOCs) present in higher quantities in indoor air can interfere with the mVOC peaks because they frequently appear in low concentrations inside the air (Wessén & Schoeps, 1996). Microorganisms primarily produce carbon dioxide and microbial biomass through the breakdown of complex organic molecules into simpler components. A large variety of by-products can be created during this process; some are specific to microorganisms, some can also be natural elements of construction materials, and others are the same as those that can come from outdoor sources, such as engine exhausts. These volatile substances will fall under any volatile class because of their extremely varied chemical characteristics (Wessén & Schoeps, 1996).

The volatile organic compounds (VOCs) generated by microorganisms have attracted the attention of researchers in the domains of microbiology, biotechnology, medicine, and agriculture in recent years. Most VOCs are lipophilic substances with tiny molecular weights (on average around 300 Da), high vapour pressure, and low boiling temperatures. Bacterial VOCs can be found in many different chemical forms, including terpenoids, alcohols, ketones, and compounds that contain sulphur and alkenes, among others. The entire population of microbes shares some compounds, whereas specific strains are the only ones that can produce others. It is discovered that the multifunctional and complex process of microbial VOC creation involves the composition of elements such as culture growth stage, pH, medium feeding, and aeration (Sidorova et al., 2021). Some exposure to mVOCs, which are industrial substances as well, can reveal health impacts. Since industrial exposure levels are often much higher than those observed in the mVOC scenario, the lowest doses that are now accessible are high when compared to levels of microbiological origin. Concerns about mVOC exposure have primarily focused on indoor surroundings in homes (Korpi et al., 2009). The produced mVOCs could undergo additional chemical transformations into different chemicals in the environment. For instance, alcohols are quickly transformed into aldehydes and ultimately into carboxylic acids by the reaction of ketones with hydroxyl radicals in the air.

11.2 MICROBIAL VOLATILE ORGANIC COMPOUNDS

Bacteria use quorum sensing, a technique that generates diffusible signals, for cell-to-cell communication. Chemical communication via diffusible autoinducers occurs under situations of close cell proximity and high concentrations of signalling chemicals. However, studies have shown that in addition to these cell-to-cell signals, the microbes also produce a variety of organic and inorganic volatile chemicals, which may be found in the gas phase (Weisskopf et al., 2021).

11.2.1 Occurrence of Microbial Volatile Organic Compounds

Because of their great affinity for absorption by building materials, some mVOCs have been hypothesized to represent emissions from an earlier, or even prior, microbial contamination. The primary source of mVOCs is fresh, metabolically active microbial contamination. The level of mVOCs is influenced by the proportion of production levels, the uptake by and release from furnishings and building materials, as well as ventilation (Wessén & Schoeps, 1996). Construction of buildings in wet environments promotes the proliferation of microorganisms. Research on microbial volatile organic compounds suggests that the production of volatile organic compounds by growing microorganisms may have an effect on the quality of indoor air (mVOCs).

From the studies conducted by Wessén and Schoeps (1996), it was learned that there are 23 varied compounds of mVOCs which arise from unique microbial sources. The compounds can be divided into classes A and B, where the occurrence frequency of class A is comparatively higher than that of class B. Compounds of class B were further learned to be correlated to microbial impact with the indoor environment. Experiments showed that the most frequently occurring compound of mVOC was detected to be 2-hepatanone, followed by borneol and geosmin. Furthermore, most of the mVOC compounds were found to have similar frequencies of occurrence (Ryan & Beaucham, 2013).

11.2.1.1 Factors Affecting the Occurrence of mVOCs

As a consequence of microbial metabolism, mVOC production is influenced by the same variables that control microbial growth. These include substrates and nutrients (for instance, a deficiency in a nutrient causes terpene emissions, and the presence of a deficiency in a certain amino acid in the substrate causes sulphide and nitrogen compounds to be produced), moisture conditions (water content and humidity levels), temperature, and microorganism species and strains (which serve as the basis for methods for species identification based on mVOC profile) (Korpi et al., 2009).

In order to collect chemicals from the atmosphere above a culture medium that has been infected with the desired bacterium or microorganisms, microbial volatiles are typically collected. The growing medium itself has a major impact on the production and operation of volatiles, in addition to the changes caused by the use of different methods for analysis and collection. Other environmental factors, such as temperature, humidity, pH, and oxygen levels, in addition to the content of the medium, are likely to affect the composition of microbial volatiles. These factors may have an impact on the volatiles' physical features directly (temperature and humidity, for instance, will have an impact on molecular volatility) or inadvertently (for example, by changing metabolism and growth rate) (Weisskopf et al., 2021). It is expected that the presence or absence of oxygen will significantly affect how bacteria synthesize volatile compounds, specifically depending on whether they prefer fermentation (which is favourable in micro-oxic and anoxic settings) or respiration (favourable in oxic conditions).

11.2.2 Physiochemical Properties of Microbial Volatile Organic Compounds

An individual mVOC cannot be connected to a particular microbial species. The same mVOC can be produced by a wide range of microorganisms, comprising bacterial and fungal species. This is expected given that mVOC synthesis in all microbial species depends on identical growth and metabolic conditions. A compound's volatility can be estimated from its molecular weight, boiling temperature, vapour pressure, and the number of carbon atoms. Bacterial emissions of volatile chemicals include a diverse spectrum of structural characteristics and include substances from many different chemical classes, such as alcohols, carbonyl compounds, aromatic compounds, hydrocarbons, and compounds containing sulphur and nitrogen (Schulz & Dickschat, 2007). Compounds like geosmin, 2-methylisoborneol, and thujopsene are found to have moderate volatile compounds according to their vapour pressure date at 25°C (Leinonen & Kultamaa, n.d.), whereas compounds like 2-isopropyl-3-methoxy-pyrazine are considered volatile compounds and α-Pinene and β-Pinene are very volatile compounds.

According to a detailed analysis of the presence of volatiles in bacteria, they can be categorized into common, group, and unique compounds. Common compounds can be found in many different bacterial strains, oftentimes randomly. They comprise all of the fundamental fermentation by-products and molecules generated from the primary metabolism, including certain sulphur compounds. However, these chemicals alone can create volatile mixtures that are strain- or species-specific, and the mixture can then have particular biological consequences (Weisskopf et al., 2021). This group of substances includes geosmin and 2-methylisoborneol, which have only ever been found in actinobacteria, myxobacteria, and cyanobacteria. This category also includes 2,5-Bis(isopropyl)pyrazine, which is produced by only a few species of myxobacteria, including *Chondromyces crocatus* (Dickschat et al., 2005) and *Paenibacillus polymyxa* (Blom et al., 2011). Last but not least, some chemicals, such as streptopyridine in *Streptomyces* species and cyclohexenyl esters and salinilactone in *Salinispora* species, are generated by only one or a few closely related strains. Their biosynthesis is linked to that of particular bigger metabolites or signalling molecules (Groenhagen et al., 2014; Weisskopf et al., 2021).

Due to the small number of functional groups, many volatiles are polar molecules with low water solubility; yet, this solubility is enough to permit diffusion into the aqueous solution. Algal pheromones, which are typically C11 hydrocarbons and generally polar molecules with low water solubility, are a good example. Due to the absence of the hydration sphere, these chemicals are predicted to diffuse especially quickly in the water phase, allowing for a variety of chemical signals. This occurs as a result of their few functional groups, tiny effect sizes, and quick diffusion rates. Even small molecules, such as amino acids, carry a hydration sphere in water, which eventually results in slower diffusion. Amino acids are more polar and therefore more nonvolatile than other small molecules (Weisskopf et al., 2021). Bacterial volatiles display dramatically different structural variations as a result of their extensive metabolic diversity. Primary metabolism, fermented by-products, sulphur metabolism, the fatty acid biosynthesis pathway, and the terpene pathway are major metabolic sources for microbial volatiles (Schulz & Dickschat, 2007).

From their basic metabolism, many bacteria emit volatiles. The origins of these volatiles include acids such as 2-methyl butanoic, 3-methyl butanoic, and methyl propanoic, as well as their reduction products and associated alcohols. Derivatives of valine, leucine, and isoleucine are also present. Other volatile substances, such as simple alcohols, esters, acetic acid, hydrogen cyanide, or ammonia, are also possible to produce (Weisskopf et al., 2021). Another group of rather common molecules are aromatic compounds, which mainly come from the shikimate pathway. Particularly 2-phenyl ethanol and 2-phenyl indole, along with several other aromatic compounds are particularly prevalent and generated by a variety of microorganisms. Compounds of sulphur are frequently found in microbial volatiles. They primarily come from methanethiol and hydrogen sulphide, two substances that are challenging to find using conventional gas chromatography–mass spectrometry () techniques. The latter quickly dimerizes to dimethyl disulphide by oxidation, which is considerably simpler to detect (Brock et al., 2014).

The fatty acid biosynthesis route is the source of a significant group of bacterial volatiles, which results in a variety of molecules with different chain lengths, numbers and positions of methyl branches, and levels of oxidation in the functional groups. Compounds with the terminal methyl-branching result from variations in the starter units that take the place of the standard acetyl-CoA starter (for instance, with isovaleryl-CoA). When a ketocarboxylate intermediate is decarboxylated, methyl ketones, which are frequently present in bacteria, are formed, although alcohols and esters are also frequently created (Dickschat et al., 2004). Terpenes are a group of diversely structured bacterial volatiles. There are numerous ways to produce terpenes, and numerous terpene synthases have indeed been found. The enzymes required for the creation of the additional volatiles mentioned thus far, however, have gotten less attention.

11.2.3 Toxicokinetics of Microbial Volatile Organic Compounds

Similar to any chemical molecule, the absorption rate, distribution, biotransformation, and excretion affect the body burden of mVOCs. The most frequent way to be exposed to mVOCs is through breathing because they are by definition volatile. It is safe to suppose that the dermal uptake of vapour is not much more than a few per cent of the pulmonary uptake and that the respiratory uptake of mVOCs is frequently critical when it comes to organic solvents, some of which may also be referred to as mVOCs. Small, largely neutral chemicals called mVOCs readily diffuse through biological membranes. They are therefore readily transported throughout the blood and other tissues as well as between alveolar air and blood. Distribution of mVOCs within the body is determined by their tissue blood partition coefficients (Korpi et al., 2009).

At concentrations much greater than those that are relevant in this situation, mVOCs have several toxic systemic consequences. VOC exposure is known to cause sensory discomfort, and mVOCs also have this effect. The interior environment contains four main kinds of organic chemicals that have the potential to irritate the senses. The groups are (a) chemically stable, nonreactive organic compounds (such as octane, toluene, and butanol); (b) chemically reactive organic compounds, such as alkenes, which can react with ozone by themselves or with nitrogen dioxide in the presence of light to produce new oxygenated products; and (c) organic compounds that form chemical bonds to receptor sites in the mucous membranes.

11.3 ANTIMICROBIAL ACTIVITY OF MICROBIAL VOLATILE ORGANIC COMPOUNDS

Bacteria and fungi regularly create and release VOCs into the environment. These substances may have an impact on plants and microorganisms and are distinguished by their high vapour pressure and low molecular weight. According to reports, bacteria can release a variety of volatile chemicals that have substantial biological effects on a wide range of plant diseases. Small molecules with high vapour pressure and low boiling temperatures are known as volatile compounds. Due to their varied chemical configurations, the molecules can move between roots and microorganisms across long distances. They also belong to several groupings, such as tiny aliphatic and aromatic molecules (Safara et al., 2022).

Rhizobacteria produce VOCs which do have antibacterial and growth regulator properties and engage in relationships with host plants and plant-pathogenic microbes. Inhibition of the quorum-sensing cell-to-cell communication system, which controls the production of antibiotics, pigments, exoenzymes, and toxins, may be present in some of the VOCs created by the *Pseudomonas* and *Serratia* strains. (Popova et al., 2014). The volatile organic chemicals generated by *Pseudomonas fluorescens* WR-1 and *Bacillus amyloliquefaciens* SQR-9 can inhibit *Ralstonia solanacearum*'s growth and pathogenicity. *R. solanacearum* cells underwent morphological alterations as a result of VOCs generated by *Bacillus* strains (Tahir et al., 2017). *Agrobacterium*'s crown-gall disease may be avoided by using VOCs generated by soil-borne *Serratia plymuthica* variant IC1270 and *P. fluorescens* strain B-4117. It was discovered that tomato plant stem tissues released the volatile alkyl sulphide chemical dimethyl disulphide after being exposed to *S. plymuthica* strain IC1270 (DMDS) (Popova et al., 2014). DMDS is the primary headspace volatile produced by this strain of the bacterium. The biocontrol action of strain IC1270 against crown-gall disease was mediated by this VOC, as shown by the inhibition of *Agrobacterium* growth in plate experiments. These results imply that some bacterial volatiles may contribute to the promotion of hostile behaviours in bacterial strains associated with plants (Popova et al., 2014).

The production of several volatile, dynamic chemicals by microbes is well recognized to protect plants against illness. As a result of increased research interest in this topic, there have been several papers on the manufacture of mVOCs by diverse bacteria. GC-MS is widely used to chemically profile the volatile chemicals produced by bacteria. They can be classified as alcohols, aldehydes, ketones, acids, esters, ethers, hydrocarbons, and sulphur-containing chemicals, among other things (Agisha et al., 2019). Bacterial VOCs can be seen as important components of the complex interactions that occur between and among bacteria as well as between bacteria and other species, including eukaryotes, in natural habitats. In this study, we investigated the effects of VOCs generated by several strains of *Serratia* and *Pseudomonas*, mostly soil-borne and rhizospheric isolates from various geographical locations. Numerous species (bacteria, fungi, *Drosophila*, and nematodes), some of which are harmful to agricultural plants, have been discovered to be eradicated or to have their growth reduced by the overall pool and specific VOCs produced by these bacteria (Popova et al., 2014).

11.4 MICROBE INTERACTIONS IN VOCs

11.4.1 Bacteria-Bacteria Interactions

Other microorganisms are directly harmful towards bacteria through bacterial VOCs. Examples include the production of dimethyl disulphide by two rhizospheric bacteria, *P. fluorescens* and *S. plymuthica*, which had bacteriostatic impacts on the plant viral pathogens *Agrobacterium tumefaciens* and *Agrobacterium vitis*, and the sesquiterpene albaflavenone by *Streptomyces albidoflavus*, which had action against *Bacillus* (Dandurishvili et al., 2011). The synthesis of VOCs as well as their antibacterial activity is controlled by using several species of *Pseudomonas* and *Bacillus* as plant pathogens. Studies have revealed that *R. solanacearum*, the causing bacteria for wilt disease, is strongly inhibited by the VOCs produced by *Bacillus* species, including benzaldehyde, 1,2-benzothiazole-3 (2 H)-one, and 1,3-butadiene (Schulz-Bohm et al., 2017; Tahir et al., 2017).

The type III secretion system, the generation of extracellular polysaccharides (EPS), and the global virulence regulator PhcA are only a few examples of the genes whose transcriptional expression levels were changed by the mVOCs. Additionally, the mVOCs made plants resistant to wilt illness through systemic resistance. Additionally, VOCs produced by specific bacteria might promote the development of nearby bacteria in the rhizosphere. When *Collimonas pratensis* and *S. plymuthica*'s VOCs are present, *P. fluorescens* Pf0–1 can thrive, according to research by Garbeva et al. from 2014. These VOCs boosted *P. fluorescens* Pf0–1's production of secondary metabolites that had an antibiotic effect on *Bacillus* and encouraged the expression of genes that are responsible for motility. This shows that *C. pratensis* and *S. plymuthica* may be drawing in and encouraging *P. fluorescens*' growth in an effort to boost its own chances of defeating various bacterial rivals or fungal infections from soil (Garbeva et al., 2014).

In addition to their negative effects on bacteria, VOCs can impact the behaviour of other bacteria and their resistance to antibiotics. Ammonia, trimethylamine, hydrogen sulphide, nitric oxide, and 2-amino-acetophenone are examples of bacterial volatiles that may affect the development or spread of biofilms or modify bacterial motility (Audrain et al., 2015). Additionally, bacteria may use their mobility to relocate to regions with better resources and fewer rivals. *Streptomyces venezuelae* possesses a novel mode of development that permits nonmotile bacteria to reach locations with greater resources, according to recent research (Jones et al., 2017; Schulz-Bohm et al., 2017).

11.4.2 Bacteria-Fungi Interactions

In long-distance fungus-bacterial interactions, fungal VOCs can be significant and can influence the phenotypic reactions of the interacting partners. For instance, *P. fluorescens* that generates 2,4-diacetyl phloroglucinol had up-regulation of the biocontrol gene (phlA) when *Trichoderma atroviride* VOCs were present (Lutz et al., 2004). Under varying nutrient concentrations and during various growth phases, diverse fungal and oomycetal soil strains emit volatiles, and soil bacterial strains phenotypically react to these volatiles (Schmidt et al., 2015). Out of the evaluated phenotypical reactions, such as growth alteration, antimicrobial activities, biofilm formation, or motility, fungal and oomycetal VOCs had a substantial impact, either positive or negative, on the motility of bacteria (both swimming and swarming). The fungus may thus be using this observation as a potential strategy to control the movement of mutualistic bacteria in order to attract them to it and chase away the competition (Piechulla et al., 2017).

Numerous soil bacteria are capable of producing VOCs that have antifungal properties, and as a result, they can influence the phenomenon known as soil fungi stasis, which hinders the growth or germination of fungal propagules (Garbeva et al., 2014; Schulz-Bohm et al., 2017). *Rhizoctonia solani* is resistant to VOCs produced by *Streptomyces* species, according to recent findings by Cordovez et al., 2015, which may aid in the management of plant diseases. Another recent investigation found that while these *Lysobacter* strains produced nonantagonistic VOCs when grown on a medium rich in sugar, they produced antagonistic VOCs when grown on a medium rich in protein (Lazazzara et al., 2017).

11.4.3 Fungi-Fungi Interactions

Conspecific mycelial growth and spore germination are controlled by specific fungal VOCs during population formation, which is done in a concentration-dependent manner (Schulz-Bohm et al., 2017). Additionally, fungi interact negatively and are inhibited by fungal VOCs. For instance, endophytic fungi *Muscodor albus* and *Oxyporus latemarginatus* can effectively prevent the growth of certain pathogenic fungi that affect plants, like *Botrytis cinerea* and *R. solani* (Sears et al., 2001). The VOCs generated by *Trichoderma* species have a significant impact on plant pathogenic fungi such as *Fusarium oxysporum*, *R. solani*, *Sclerotium rolfsii*, *Sclerotinia sclerotiorum*, and *Alternaria brassicicola* (Amin et al., 2010; Schulz-Bohm et al., 2017).

When exposed to VOCs produced by the nonpathogenic fungus *F. oxysporum* CanR-46, such as 5-hexenoic acid, limonene, octanoic acid, and 3,4–2H-dihydropyran, 14 different fungi, including the pathogenic *Verticillium dahlia*, were unable to establish their mycelium (Zhang et al., 2015). Recent proteomic research suggested that essential metabolic pathways may be blocked by fungus-derived VOCs, inhibiting the growth of fungi (Fialho et al., 2016). Some fungi have the ability to degrade the antifungal substances that their microbial rivals create. For instance, *Trichoderma harzianum*'s poisonous 6-pentyl-alpha pyrone can be detoxified by *Fusarium graminearum* (Cooney et al., 2001; Schulz-Bohm et al., 2017).

11.5 QUORUM SENSING

Our understanding of quorum sensing (QS) stems from the notion of bacteria engaging in intercellular communication to orchestrate collective behaviour. The underlying mechanism of biofilm development and quorum sensing are synchronized on an intricate level. Bacterial cells interact with the residential members of the community by establishing a complex circuit of density-dependent methods broadly classified as quorum sensing. As the bacterial cells are enclosed in the matrix, the development of biofilm acts as a 'security blanket' in stressful settings (Velmourougane et al., 2017). The major domains of specialization that fall under the regulation of the complex network include sporulation, regulation of gene expression, provision of defensive mechanisms against unpredictable harsh environmental conditions, synthesis and expression of virulence factors, biofilm formation, and bioluminescence. along with antibiotic production (Mooney et al., 2018). Bacterial cells involve themselves in amalgamating the bottled-up information incorporated in numerous autoinducers to regulate gene expression and establishing a universal dialect which paves the way for interspecies communication (Bassler & Losick, 2006; Waters & Bassler, 2005). Depending on the cellular consistency, QS flips a switch that causes bacterial cells to convert from their existing lifestyle to a biofilm lifestyle. The autoinducers act as signalling molecules secreted by the cells of the colony to establish a complex chemical network among cell members in the vicinity (Waters & Bassler, 2005).

Bacteria have diverse biofilm-forming behaviours, depending on their quorum-sensing status. As a consequence of aggregation and recognition of autoinducer molecules, *Pseudomonas aeruginosa* produces biofilms at high cell density (HCD), whereas *Staphylococcus aureus* and *Vibrio cholerae* develop biofilm at low cell density (LCD) (Bronesky et al., 2016; de Kievit & Iglewski, 2000). The intricate QS mechanism of the bacterial cells is built around the production, release, and identification of signalling compounds termed autoinducers (AI) (Whiteley et al., 2017). Autoinducers are constitutively synthesized at the elementary level and the gradual accumulation of the signalling molecules during microbial growth results in the formation of a positive feedback loop (Lal, 2009). The receptor protein controls the transcriptional activity of the signalling molecule it perceives which in turn regulates the synthesis of the receptor protein itself; hence they are correctly referred to as autoinducers (Lee & Zhang, 2015). QS allows bacteria to assess the density of the surrounding neighbourhood by monitoring the amount of a particular signalling chemical secreted by members of the community. Only when the cellular density attains a higher level is the signal accumulation in the surrounding environment adequate to turn on the response. QS signalling compounds are characterized by low molecular weight and are categorized as acyl-homoserine lactones (AHLs), furanosyl borate diesters (AI2), cis-unsaturated fatty acids (DSF family signals), and peptides (Solano et al., 2014).

The proper execution of a QS system relies on the synchronization of three-step mechanisms. First is the bacterial community's production of signalling molecules (AIs). When the cell density is low, lesser concentrations of AIs are generated, significantly lower than the AI concentration necessary for recognition. The satisfactory response is generated by a greater concentration of AI which is produced when the cell density achieves a higher level. Second is the proper recognition of the chemical signalling molecules. AIs are identified by receptors only when they exceed a certain threshold. Third is gene expression and establishment of the autoinduction loop. After the successful detection of signalling molecules, gene expression takes place, along with additional AI synthesis via the forwarding feed mechanism (Rutherford & Bassler, 2012). When the quorum-sensing lifestyle is activated, autoinducer-receptor complexes turn on the genetic coding of the autoinducer synthase, which ultimately increases the concentration of the extracellular autoinducer molecules. This feedback control autoinduction loop is considered to synchronize bacterial population behaviour (Mukherjee & Bassler, 2019).

The microbial cell density correlates to the external concentration of AI. Bacteria monitor AI concentrations in order to regulate gene expression and detect any modifications in the cellular population (Rajamanikandan et al., 2017). Gram-negative bacteria use acylated homoserine lactone (AHL) to do this, while Gram-positive bacteria utilize peptides (AIP) (Waters & Bassler, 2005). There are two main QS systems: (a) LuxI/LuxR type acyl homoserine lactone (AHL) QS system; and (b) two-component autoinducing peptide (AIP) QS system (Swem et al., 2008). Aside from the QS systems that are functional in separate groups of bacteria, there is another kind of QS system which is operational in Gram-positive bacteria as well as in Gram-negative bacteria. AI-2 serves as a medium for inter-species and intraspecies communication. There are three types of AIs found in bacteria: Acyl-homoserine lactone (AHL) (Gram-negative bacteria) and autoinducer peptide (AIP) (Gram-positive bacteria) for intraspecies communication and autoinducer-2 (AI-2) (Gram-positive and Gram-negative bacteria) for interspecies communication.

TABLE 11.1
Various QS Systems in Different Groups of Bacteria

Bacterial species	QS system	AI molecule	Reference
Bacillus subtilis	ComX	ComP/ComA	(Oslizlo et al., 2014)
Streptococcus pneumoniae	CSP (competence stimulating peptide)	AgrC/AgrA	(Banerjee & Ray, 2017)
Agrobacterium tumefaciens	TraI/TraR	N-(3-oxo octanoyl)-HS	(Pallaval Veera Bramhachari et al., 2018)
B. subtilis	CSF (competence stimulating factor)	ComP/ComA	(Bareia et al., 2018)

11.5.1 Modes of Quorum Sensing

11.5.1.1 Gram-Negative Bacteria

The LuxI enzyme generates N-acylated homoserine lactones that serve the role of signalling molecules (AI) in Gram-negative bacteria. These signalling molecules penetrate the cell membrane of the bacteria, and the bacterial population density is determined by the number of growing cells. When the LuxR receptor protein reaches the threshold concentration, it induces the transcription of effector genes to commence. N-acylated homoserine lactones function as diffusible communication signalling compounds, influencing distinct and diverse physiological activities dependent on cellular population. In this context, we can talk about the QS system present in the Gram-negative bacteria *P. aeruginosa*. The quorum sensing mechanism is controlled by LasI/LasR and RhlI/RhlR, which are considered the LuxI/LuxR homologues. The countenance of factors responsible for virulence like alkaline phosphatase protease, elastase, and exotoxin A falls under the administration of the quorum-sensing molecular circuit. To state another instance, the production of bioluminescence in *Vibrio fischeri* can be considered where the mechanism of QS circuitry is overseen and modulated by the lux AB genes and the lux CDE genes. lux CDE genes encode enzymes responsible for the production of substrates for luciferase, which leads to the expression of bioluminescence (Nazzaro et al., 2013).

11.5.1.2 Gram-Positive Bacteria

Gram-positive bacteria deploy oligopeptide autoinducer molecules and two-component systems which comprise transcription factors that regulate any changes taking place in the dominion of gene expression and transmembrane histidine sensor kinase receptors (Papenfort & Bassler, 2016). QS occurs in Gram-positive bacteria due to peptide secretion via the ATP-binding cassette transporter. The peptide signals are disinterred by sensor kinases. AIP is secreted from Gram-positive bacteria, and once it attains a sufficient quantity, it interacts with kinase receptors. Phosphorylation occurs as a result of kinase receptor activation. Such dual-component managerial systems, which consist of a detector and a counter-regulator protein, transmit information via phosphorylation (Miller, 2001). To state an example of a QS system regulated by the synthesis of oligopeptide we can take the case of *S. aureus*. Here the regulation of gene expression takes place through the Agr QS system. Ribosomes generate these AI peptides, which then undergo post-translational modification to become active. This Agr system regulates and oversees the synthesis of biofilm along with the exhibition of virulence factors like exotoxins. The cell membrane–associated ATP-binding cassette transporter stimulates autoinducer peptide secretion (Verbeke et al., 2017).

11.6 QUORUM QUENCHING

QS pathways lead to the development of virulence factors and endurance traits in bacteria. As a result, the interest of global researchers has been piqued in deconstructing the core principle of the quorum sensing mechanism, which may contribute to a better understanding of the mechanism of autoinducer-mediated cell-to-cell interaction and aid in the process of designing quorum signalling modulators. Fluorescence, mass chromatography, spectroscopy, bioluminescence, chemiluminescence, and colourimetry are some of the methods used to evaluate QS/QQ. They are all capable of measuring the qualitative and quantitative properties of QS/QQ molecules. Several strategies go along with the process of quorum quenching: (i) interfering with Lux-I synthase proteins to inhibit signal molecule synthesis (Lade et al., 2014); (ii) enzyme-mediated demolishment of signalling molecules by AHL acylase and AHL lactonase (Fuentes et al., 2016; Lade et al., 2014); (iii) meddling with analogue compounds of signalling receptors (Koch et al., 2005); and (iv) jamming the AHL-LuxR complex (Chen et al., 2011; Rampioni et al., 2014). (See following sections.) These strategies can be put to use for achieving the required inhibition algorithm of the AHL-regulated QS network in Gram-positive bacteria as well as AIP-regulated Gram-negative bacteria.

11.6.1 Inhibiting AI Synthesis/Development of AI Adversary Molecules

Gram-negative bacteria implement QQ based on AHL inhibition, wherein enzymes such as acylase, oxidoreductase, and lactonase can ideally block AHL, limiting AHL buildup outside the cellular environment and, as a result, repressing QS-related gene expression (Smith et al., 2003; Suga, 2003). The three types of enzymes that may degrade AHL are lactonases, which can hydrolyze the lactone moiety, AHL-acylase, which can hydrolyze the amide link, and AHL oxidoreductase, which can oxidize the third carbon in AHL molecules (Dong et al., 2002). Another approach to disrupting bacterial transmission is the utilization of antagonist molecules. These compounds can attach to the target by interacting with inductors for the similar binding site, or they can participate in noncompetitive binding with the receptor to impede transmission of signals into the cell which is accomplished by the help of inductor molecules (Bodede et al., 2018). Biofilm development can also be avoided by using AHL analogues that disrupt QS communication. Chow et al. proved the efficacy of this method against *Acinetobacter baumanii* (Chow et al., 2014).

TABLE 11.2
Quorum-Quenching Enzymes

QS signal molecules	Working mechanism	Reference
AHL lactonase	Hydrolyzes the lactone moiety and opens the HSL ring	(Joachimiak et al., 2002)
AHL acylase	Cuts the acyl side chain and hydrolyzes the amide bond	(Lin et al., 2003)
Oxidoreductase	Facilitates the acyl side chain's redox reaction	(Chowdhary et al., 2007)

11.6.2 Inhibition of the Autoinducer-Receptor Binding

Designing an inhibitor molecule which can mimic an inhibitor of competitive nature and combat with AHL to occupy the similar binding site on the surface of the receptor is another method for interfering with the biofilm signalling cascade (Petersen, 2016; Shiner et al., 2006). The small molecule inhibitor named savrin inhibits the transcriptional regulator AgrA. It binds to DNA, preventing the development of RNAIII, which, together with AgrA, is accountable for generating a multitude of virulence factors (Sully et al., 2014). For instance, halogenated furanones synthesized by the marine macroalga *Delisea pulchra* were the first anti-QS compounds discovered, with action focused on competitive binding to LuxR-type proteins. This action mechanism promotes the enzymatic breakdown of LuxR-type proteins (Asfour, 2018).

11.6.3 Degradation of Autoinducers

The most widely known procedure of quorum quenching is the enzyme-mediated degradation of AHL molecules. They are catalyzed by four different types of enzymes: lactonases and amylases, respectively, hydrolyze HSL ring and amide linkage, whereas reductases and oxidases modify but do not disintegrate AHL (Rehman & Leiknes, 2018). The AHL-lactonase enzyme generated by *B. cereus* VT96 has a direct impact on the production of virulence factors such as biofilm development, synthesis of exopolysaccharide, and production of pyocyanin in *P. aeruginosa* PAO1 (Rajesh & Rai, 2016). *A. tumefaciens P. aeruginosa* and *K. pneumoniae* all produce enzymes which possess the ability to digest the AHL molecules. MomL lactonase, isolated from *Muricauda olearia* Th120, is capable of degrading both long and short chain AHL, reducing pathogenicity in many pathogenic bacteria (Wang et al., 2019).

11.6.4 Inhibition of Surface Adhesion Initiation

Some surfaces have features that make microbial adhesion easier. Bacterial adhesion to hydrophobic and nonpolar surfaces has a higher binding strength than adherence to hydrophilic surfaces because the repelling force between the surface and the bacterium is lessened (Jamal et al., 2018). Surface structure influences cellular pioneering as well. Since the amount of surface area in contact between bacteria and surface material is necessary in the attachment stage, abrasive surfaces are more likely to be colonized than smooth surfaces due to lower contact surface areas. On rough surfaces, forces are also reduced, preventing adhering cells from separating (Song et al., 2015). The changes in the surface topography inhibit attachment of bacteria to the surface, reducing the possibility of biofilm synthesis (Estrela et al., 2009). One approach to preventing bacterial adhesion to specific surfaces is to coat them with antimicrobial compounds (metal-based nanoparticles) and surfactants (such as quaternary ammonium salts) (Paluch et al., 2020; Piecuch et al., 2016).

11.7 QUORUM QUENCHING BY VOCs

Animal and plant volatiles have been extensively researched in the past, but microbial volatiles (i.e., bacteria and fungi) have gotten little focus for many years. Volatile chemicals have the potential to disrupt the QS regulation mechanism. Multiple virulence factor regulators (MvfR) and other global virulence regulators have been demonstrated to be regulated by the volatile 2-aminoacetophenone, leading to the aggregation of quorum-sensing mutants and a change in the character of the infection (from acute to chronic) in *P. aeruginosa* (Kesarwani et al., 2011). Kesarwani and colleagues propose the possibility of QS-regulated airborne communication among bacteria because 2-AA is both a QS-regulated chemical and a microbial volatile signalling compound (Schulz & Dickschat, 2007). 2-AA, a small volatile agent synthesized by *P. aeruginosa* and *Burkholderia thailandensis*, was shown to be a QS-controlled compound. It also acts as an essential regulator of both acute and chronic virulence factors. The results reveal that the 2-AA molecule, which is characterized by low molecular weight, silences the MvfR regulons with the establishment of a negative feedback loop. This feedback loop is generated with the utilization of transcriptionally and post-transcriptionally repressed synthesis of MvfR ligands which ultimately leads to the restriction of HAQ-mediated QS signalling.

When the signalling cascade is induced, pathogenic factors that cause chronic and acute infection are stimulated. The researchers discovered that 2-AA suppresses MvfR via regulation of the feedback loop and improves in vitro and in vivo cell viability in flies, most likely by disrupting the cell lysis mechanism and triggering alterations in lasR (Kesarwani et al., 2011). Some microbial volatiles (such as dimethyl disulphide and others) have the ability to impede quorum sensing, which might be a potential approach to volatile compounds because these molecules would impede only virulence and not growth, thus resulting in diminished emergence of resistance in the intended microorganisms. VOCs generated by rhizospheric bacteria, on the other hand, have been demonstrated to impede quorum sensing. The property to disrupt quorum sensing networks, referred to as quorum quenching (QQ), is an important bacterial competing technique that can grant one bacterial species an advantage over another.

Bacteria commonly produce a variety of volatile sulphur-containing molecules known as dimethyl trisulphide (DMTS) and dimethyl disulphide (DMDS) (Schulz & Dickschat, 2007). The most prevalent mVOC in the headspace of *S. plymuthica* IC1270 was determined to be DMDS, and it was discovered to potentially limit the production of *Agrobacterium* strains (Dandurishvili et al., 2011). Plant species like *Allium* and *Brassica,* as well as *Burkholderia ambifaria* (Groenhagen et al., 2013), have been demonstrated to generate DMDS (Kyung & Lee, 2001). Recently, it was shown that bacterial VOCs, particularly DMDS, exhibit significant quenching activity of the QS network (Chernin et al., 2011). The opportunistic pathogens *Pseudomonas chlororaphis* 449, *P. aeruginosa* PAO1, and *P. fluorescens* 2–79 were administered with DMDS, which greatly decreased AHL production. Strong correlations were seen when these bacteria were exposed to a group of *S. plymuthica* IC1270–produced VOCs, with DMDS serving as the primary headspace volatile (Dandurishvili et al., 2011). It was noted that AHLs containing both long and short chains along with substituted and unsubstituted derivatives experienced a decline. The findings indicated that the VOCs were effective at preventing the transcription of the relevant AHL synthase-encoding genes. It was determined that there must be additional factors associated with the suppression of AHL formation because there was no quantitative association established between the degree of decrease in transcription of AHL-encoding genes and the suppression of AHL synthesis (Chernin et al., 2011). Enhancements in the formation of phenazines and exotoxins as well as the formation of biofilms are caused by 2,3-butanediol's effects on the *P. aeruginosa* OS system and activation of the LasI/LasR regulatory system (Venkataraman et al., 2014). A volatile compound, 2,3-butanediol, synthesized by *Enterobacter aerogenes* promoted the production of additional antibiotics by *P. aeruginosa* when it was cultivated alongside that organism.

Dimethyl disulphide (DMDS) manufactured by ketones (2-nonanone, 2-heptanone) and *S. plymuthica* IC1270 formed by *Pseudomonas* strains are two instances of distinct VOCs that interfered with the function of bacterial QS systems (Chernin et al., 2011; V.A. Plyuta et al., 2014). Chernin and colleagues found that volatiles produced by the rhizobial microbes *S. plymuthica* and *P. fluorescens* may decrease N-acyl homoserine lactone–mediated quorum sensing by *Pectobacterium, Chromobacterium, Agrobacterium* and *Pseudomonas*. When AHL-producing bacteria were exposed to VOCs synthesized by *S. plymuthica* and *P. fluorescens*, this quorum-quenching action was discovered. This approach curtailed bacterial AHL synthesis and inhibited AHL synthase gene transcription (Chernin et al., 2011). A study conducted by Chernin et al. (2011) discovered that VOCs generated by rhizospheric strains of bacteria *P. fluorescens* B-4117 and *S. plymuthica* IC1270 may be exploited to disrupt the QS network coordinated by N-acyl homoserine lactone (AHL) signal molecules. A quorum-quenching effect was seen when AHL-producing bacteria were exposed to VOCs generated by strains IC1270 and B-4117 or to the key volatile compounds formed by strain IC1270, dimethyl disulphide (DMDS). In their investigation, they demonstrated the antibacterial effect of rhizospheric volatiles, which can be employed to control several plant diseases as well as act as potent inhibitors of AHL-regulated QS networks.

According to LC-MS/MS analyses, treating strains *P. aeruginosa* PAO1, *P. chlororaphis* 449, or *P. fluorescens* 2–79 with VOCs emitted by DMDS or strain IC1270 considerably lowers the number of AHL molecules synthesized by those specific bacteria. The primary headspace volatile compound released by strain IC1270 is alkyl sulphide dimethyl disulphide (DMDS) (Dandurishvili et al., 2011). As a result, the QQ activity of this molecule was investigated. DMDS was evaluated at concentrations of up to 200 μmol. The QS response was inhibited even at low dosages (10–20 μmol), but the most prominent impact was shown at levels of 50–100 μmol. In similar investigations, 1-undecene, the key volatile compound produced by *P. fluorescens* B-4117, failed to exhibit QQ action even at high doses of up to 750 μmol. *P. chlororaphis* 449 produced volatile organic compounds that hindered the inducing activity of the quorum sensing mechanism, indicating a significant antagonistic association between AHL molecules and VOCs among the similar strain of species. The ability of strain 449 to manufacture VOCs had no impact on the AHL signalling of the same cells, which demonstrates autoimmunity of the QS response to volatile compounds synthesized by a similar host. When *P. chlororaphis* 449 is exposed to VOCs produced by B-4117, IC1270, DMDS or 449 itself, it results in the inhibition of the transcription of the AHL synthase genes (phzI and csaI) (Chernin et al., 2011).

VOCs synthesized by *B. subtilis* inhibit swarming in *Burkholderia glumae, E. coli, P. polymyxa* and *P. aeruginosa* (Kim et al., 2013). 1-butanol and indole decrease *E. coli* and *P. aeruginosa* cell motility across solid surfaces, but 3-hydroxy-2-butanone and 2-butanone promote *P. aeruginosa* cell motility (Létoffé et al., 2014). *P. chlororaphis* 449, *P. fluorescens* B-4117, *S. plymuthica* IC1270, and *Serratia proteamaculans* 94 total volatile pools, as well as unique individual VOCs formed by these bacteria (the ketone 2-pentanone, 2-heptanone, 2-undecanone, and the DMDS), inhibited formation of biofilm by plant-pathogenic agrobacterium of multiple origin (*A. tumefaciens* strains Chry5, C58, and Sh-1). These findings imply that particular VOCs have the ability to either kill or inhibit the growth of plant pathogenic bacteria, the virulence of which is linked to biofilm production on diseased plant species (V. Plyuta et al., 2016). Considering the ubiquity of bacterial VOCs and AHLs in the ecosystem, no studies have been conducted to demonstrate actual QQ activities of bacterial VOCs against plant diseases, and the issue of whether such interactions occur under normal conditions remains unanswered. The underlying mechanisms of VOCs' impact on the regulatory procedure of quorum sensing are not yet explored properly, and that leaves a void for achieving a true understanding of the technique underneath. QS systems have been used to monitor a set of cellular activities, taking into consideration the management of bacteria's antagonistic characteristics, ability to colonize particular unique ecological niches, and pathogenicity. Modulation of quorum sensing regulation, along with the hindering effect of VOCs on bacterial proliferation and ability to survive, can provide a wealth of study options (Chernin et al., 2011; V.A. Plyuta et al., 2014).

11.8 CONCLUSION

The chemical properties and complexities of mVOCs that are produced by different microorganisms have been studied since the past few decades. It has been concluded that these small, odorous molecules are capable of modifying or promoting their growth along with their neighbouring organisms. Previous research on mVOCs has mostly focused on the unidirectional effects of microbial VOCs on the creation of a unique organism and reactions of that specific organism that recognize themselves without mVOCs mediated communication. Additionally, microbial interactions that occur underground involve many species and are much more complicated than simple one-on-one interactions, which have a big impact on mVOC emission. The interactions that take place among bacteria, fungi, and plants are the belowground mVOC-mediated interactions that have received the greatest attention in the scientific literature as of late. Because protists, archaea, and other rhizosphere creatures, including nematodes and earthworms, are currently under-studied in this area, there is a paucity of knowledge regarding the synthesis of VOCs by these groups (Schulz-Bohm et al., 2017). Studies conducted by (Popova et al., 2014) have shown VOCs synthesized by strains of *Serratia* and *Pseudomonas* that have been primarily isolated from plant rhizospheres are potent growth inhibitors of a variety of microorganisms, including cyanobacteria and plant pathogenic bacteria and fungi. 1-undecene, DMDS, and ketones were identified by LC-MS/MS analysis as the primary headspace VOCs released by the examined bacterial strains.

Furthermore, conclusions have been made regarding the localization and interpretation of QS signals interactions in plants and microbes that profoundly affect the health of the plant. It can control and manipulate these signals and provide efficient ways to reduce diseases and protect plant health. QS regulates useful properties in beneficial bacteria (such as the generation of antifungal chemicals) and aggressive ones in infections, so care must be taken while using QQ enzymes, which break down a wide variety of AHLs, mostly because the ratios of benefits to drawbacks may fluctuate based on a community's organization and the environment (Helman & Chernin, 2015). The use of QS antagonists is a method for QS manipulation. QS antagonists, which have a limited spectrum of activity compared to AHL-degrading enzymes, may be used to target particular infections. The drawback of this process is that the persistence of these compounds in the rhizosphere may be brief and highly reliant on the environment. However, modifying QS pathways offers a potentially effective strategy to change the relationship between plants and microorganisms in the plants' favour. This is especially important given the close correlation between virulence and QS in pathogenic bacteria and the environmental input of synthetic plant protection agents, which may contaminate foods and have adverse health effects.

REFERENCES

Agisha, V. N., Kumar, A., Eapen, S. J., Sheoran, N., & Suseelabhai, R. (2019). Broad-spectrum antimicrobial activity of volatile organic compounds from endophytic *Pseudomonas putida* BP25 against diverse plant pathogens. *Biocontrol Science and Technology*, *29*(11), 1069–1089. https://doi.org/10.1080/09583157.2019.1657067

Amin, F., Razdan, V. K., Mohiddin, F. A., Bhat, K. A., & Sheikh, P. A. (2010). Effect of volatile metabolites of trichoderma species against seven fungal plant pathogens in-vitro. *The Journal of Phytology*, 2, 34–37.

Asfour, H. (2018). Anti-quorum sensing natural compounds. *Journal of Microscopy and Ultrastructure*, *6*(1), 1. https://doi.org/10.4103/JMAU.JMAU_10_18

Audrain, B., Farag, M. A., Ryu, C.-M., & Ghigo, J.-M. (2015). Role of bacterial volatile compounds in bacterial biology. *FEMS Microbiology Reviews*, *39*(2), 222–233. https://doi.org/10.1093/femsre/fuu013

Banerjee, G., & Ray, A. K. (2017). Quorum-sensing network-associated gene regulation in Gram-positive bacteria. *Acta Microbiologica et Immunologica Hungarica*, *64*(4), 439–453. https://doi.org/10.1556/030.64.2017.040

Bareia, T., Pollak, S., & Eldar, A. (2018). Self-sensing in Bacillus subtilis quorum-sensing systems. *Nature Microbiology*, *3*(1), 83–89. https://doi.org/10.1038/s41564-017-0044-z

Bassler, B. L., & Losick, R. (2006). Bacterially speaking. *Cell*, *125*(2), 237–246. https://doi.org/10.1016/j.cell.2006.04.001

Blom, D., Fabbri, C., Connor, E. C., Schiestl, F. P., Klauser, D. R., Boller, T., Eberl, L., & Weisskopf, L. (2011). Production of plant growth modulating volatiles is widespread among rhizosphere bacteria and strongly depends on culture conditions: Volatile-mediated impact of bacteria on Arabidopsis thaliana. *Environmental Microbiology*, *13*(11), 3047–3058. https://doi.org/10.1111/j.1462-2920.2011.02582.x

Bodede, O., Shaik, S., Chenia, H., Singh, P., & Moodley, R. (2018). Quorum sensing inhibitory potential and in silico molecular docking of flavonoids and novel terpenoids from Senegalia nigrescens. *Journal of Ethnopharmacology*, *216*, 134–146. https://doi.org/10.1016/j.jep.2018.01.031

Bramhachari, P. V., Yugandhar, N. M., Prathyusha, A. M. V. N., Mohana Sheela, G., Naravula, J., & Venkateswarlu, N. (2018). Quorum sensing regulated swarming motility and migratory behavior in bacteria. In P. V. Bramhachari (Ed.), *Implication of quorum sensing system in biofilm formation and virulence* (pp. 49–66). Springer. https://doi.org/10.1007/978-981-13-2429-1_5

Brock, N. L., Menke, M., Klapschinski, T. A., & Dickschat, J. S. (2014). Marine bacteria from the Roseobacter clade produce sulfur volatiles via amino acid and dimethylsulfoniopropionate catabolism. *Organic & Biomolecular Chemistry*, *12*(25), 4318. https://doi.org/10.1039/c4ob00719k

Bronesky, D., Wu, Z., Marzi, S., Walter, P., Geissmann, T., Moreau, K., Vandenesch, F., Caldelari, I., & Romby, P. (2016). *Staphylococcus aureus* RNAIII and its regulon link quorum sensing, stress responses, metabolic adaptation, and regulation of virulence gene expression. *Annual Review of Microbiology*, *70*(1), 299–316. https://doi.org/10.1146/annurev-micro-102215-095708

Chen, G., Swem, L. R., Swem, D. L., Stauff, D. L., O'Loughlin, C. T., Jeffrey, P. D., Bassler, B. L., & Hughson, F. M. (2011). A strategy for antagonizing quorum sensing. *Molecular Cell*, *42*(2), 199–209. https://doi.org/10.1016/j.molcel.2011.04.003

Chernin, L., Toklikishvili, N., Ovadis, M., Kim, S., Ben-Ari, J., Khmel, I., & Vainstein, A. (2011). Quorum-sensing quenching by rhizobacterial volatiles: Quorum quenching by volatile compounds. *Environmental Microbiology Reports*, *3*(6), 698–704. https://doi.org/10.1111/j.1758-2229.2011.00284.x

Chow, J. Y., Yang, Y., Tay, S. B., Chua, K. L., & Yew, W. S. (2014). Disruption of biofilm formation by the human pathogen acinetobacter baumannii using engineered quorum-quenching lactonases. *Antimicrobial Agents and Chemotherapy*, *58*(3), 1802–1805. https://doi.org/10.1128/AAC.02410-13

Chowdhary, P. K., Keshavan, N., Nguyen, H. Q., Peterson, J. A., González, J. E., & Haines, D. C. (2007). *Bacillus megaterium* CYP102A1 oxidation of acyl homoserine lactones and acyl homoserines. *Biochemistry*, *46*(50), 14429–14437. https://doi.org/10.1021/bi701945j

Cooney, J. M., Lauren, D. R., & di Menna, M. E. (2001). Impact of competitive fungi on trichothecene production by *Fusarium graminearum*. *Journal of Agricultural and Food Chemistry*, *49*(1), 522–526. https://doi.org/10.1021/jf0006372

Cordovez, V., Carrion, V. J., Etalo, D. W., Mumm, R., Zhu, H., van Wezel, G. P., & Raaijmakers, J. M. (2015). Diversity and functions of volatile organic compounds produced by Streptomyces from a disease-suppressive soil. *Frontiers in Microbiology*, *6*. https://doi.org/10.3389/fmicb.2015.01081

Dandurishvili, N., Toklikishvili, N., Ovadis, M., Eliashvili, P., Giorgobiani, N., Keshelava, R., Tediashvili, M., Vainstein, A., Khmel, I., Szegedi, E., & Chernin, L. (2011). Broad-range antagonistic rhizobacteria Pseudomonas fluorescens and Serratia plymuthica suppress Agrobacterium crown gall tumours on tomato plants: Biocontrol of Agrobacterium by bacterial antagonists. *Journal of Applied Microbiology*, *110*(1), 341–352. https://doi.org/10.1111/j.1365-2672.2010.04891.x

de Kievit, T. R., & Iglewski, B. H. (2000). Bacterial quorum sensing in pathogenic relationships. *Infection and Immunity*, *68*(9), 4839–4849. https://doi.org/10.1128/IAI.68.9.4839-4849.2000

Dickschat, J. S., Reichenbach, H., Wagner-Döbler, I., & Schulz, S. (2005). Novel Pyrazines from the Myxobacterium *Chondromyces crocatus* and marine bacteria: Novel pyrazines from myxobacteria and marine bacteria. *European Journal of Organic Chemistry*, *2005*(19), 4141–4153. https://doi.org/10.1002/ejoc.200500280

Dickschat, J. S., Wenzel, S. C., Bode, H. B., Müller, R., & Schulz, S. (2004). Biosynthesis of volatiles by the myxobacterium myxococcus xanthus. *ChemBioChem*, *5*(6), 778–787. https://doi.org/10.1002/cbic.200300813

Dong, Y.-H., Gusti, A. R., Zhang, Q., Xu, J.-L., & Zhang, L.-H. (2002). Identification of Quorum-quenching *N*-acyl homoserine lactonases from *Bacillus* species. *Applied and Environmental Microbiology*, *68*(4), 1754–1759. https://doi.org/10.1128/AEM.68.4.1754-1759.2002

Estrela, A., Heck, M., & Abraham, W.-R. (2009). Novel approaches to control biofilm infections. *Current Medicinal Chemistry*, *16*(12), 1512–1530. https://doi.org/10.2174/092986709787909640

Fialho, M. B., de Andrade, A., Bonatto, J. M. C., Salvato, F., Labate, C. A., & Pascholati, S. F. (2016). Proteomic response of the phytopathogen Phyllosticta citricarpa to antimicrobial volatile organic compounds from Saccharomyces cerevisiae. *Microbiological Research*, *183*, 1–7. https://doi.org/10.1016/j.micres.2015.11.002

Fuentes, J., Garbayo, I., Cuaresma, M., Montero, Z., González-del-Valle, M., & Vílchez, C. (2016). Impact of microalgae-bacteria interactions on the production of algal biomass and associated compounds. *Marine Drugs*, *14*(5), 100. https://doi.org/10.3390/md14050100

Garbeva, P., Hordijk, C., Gerards, S., & de Boer, W. (2014). Volatile-mediated interactions between phylogenetically different soil bacteria. *Frontiers in Microbiology*, *5*. https://doi.org/10.3389/fmicb.2014.00289

Groenhagen, U., Baumgartner, R., Bailly, A., Gardiner, A., Eberl, L., Schulz, S., & Weisskopf, L. (2013). Production of bioactive volatiles by different burkholderia ambifaria strains. *Journal of Chemical Ecology*, *39*(7), 892–906. https://doi.org/10.1007/s10886-013-0315-y

Groenhagen, U., Maczka, M., Dickschat, J. S., & Schulz, S. (2014). Streptopyridines, volatile pyridine alkaloids produced by *Streptomyces* sp. FORM5. *Beilstein Journal of Organic Chemistry*, *10*, 1421–1432. https://doi.org/10.3762/bjoc.10.146

Helman, Y., & Chernin, L. (2015). Silencing the mob: Disrupting quorum sensing as a means to fight plant disease: Quorum quenching as a means to fight plant disease. *Molecular Plant Pathology*, *16*(3), 316–329. https://doi.org/10.1111/mpp.12180

Jamal, M., Ahmad, W., Andleeb, S., Jalil, F., Imran, M., Nawaz, M. A., Hussain, T., Ali, M., Rafiq, M., & Kamil, M. A. (2018). Bacterial biofilm and associated infections. *Journal of the Chinese Medical Association*, *81*(1), 7–11. https://doi.org/10.1016/j.jcma.2017.07.012

Joachimiak, A., Zhang, R., Pappas, T., Brace, J., Miller, P., Bashkin, J., & Winans, S. (2002). Structure of a bacterial quorum-sensing transcription factor complexed with autoinducer-type pheromone and DNA. *Acta Crystallographica Section A Foundations of Crystallography*, *58*(s1), c274. https://doi.org/10.1107/S0108767302095934

Jones, S. E., Ho, L., Rees, C. A., Hill, J. E., Nodwell, J. R., & Elliot, M. A. (2017). Streptomyces exploration is triggered by fungal interactions and volatile signals. *ELife*, *6*, e21738. https://doi.org/10.7554/eLife.21738

Kesarwani, M., Hazan, R., He, J., Que, Y., Apidianakis, Y., Lesic, B., Xiao, G., Dekimpe, V., Milot, S., Deziel, E., Lépine, F., & Rahme, L. G. (2011). A quorum sensing regulated small volatile molecule reduces acute virulence and promotes chronic infection phenotypes. *PLoS Pathogens*, *7*(8), e1002192. https://doi.org/10.1371/journal.ppat.1002192

Kim, K., Lee, S., & Ryu, C.-M. (2013). Interspecific bacterial sensing through airborne signals modulates locomotion and drug resistance. *Nature Communications*, *4*(1), 1809. https://doi.org/10.1038/ncomms2789

Koch, B., Liljefors, T., Persson, T., Nielsen, J., Kjelleberg, S., & Givskov, M. (2005). The LuxR receptor: The sites of interaction with quorum-sensing signals and inhibitors. *Microbiology*, *151*(11), 3589–3602. https://doi.org/10.1099/mic.0.27954-0

Korpi, A., Järnberg, J., & Pasanen, A.-L. (2009). Microbial volatile organic compounds. *Critical Reviews in Toxicology*, *39*(2), 139–193. https://doi.org/10.1080/10408440802291497

Kyung, K. H., & Lee, Y. C. (2001). Antimicrobial activities of sulfur compounds derived from *S*-Alk (En) YL-L-Cysteine sulfoxides In *Allium* and *Brassica*. *Food Reviews International*, *17*(2), 183–198. https://doi.org/10.1081/FRI-100000268

Lade, H., Paul, D., & Kweon, J. H. (2014). Quorum quenching mediated approaches for control of membrane biofouling. *International Journal of Biological Sciences*, *10*(5), 550–565. https://doi.org/10.7150/ijbs.9028

Lal, A. (2009). How bacteria talk to each other: regulation of gene expression by quorum sensing. *Current Opinion in Microbiology*, *2*(6), 582–7.

Lazazzara, V., Perazzolli, M., Pertot, I., Biasioli, F., Puopolo, G., & Cappellin, L. (2017). Growth media affect the volatilome and antimicrobial activity against Phytophthora infestans in four Lysobacter type strains. *Microbiological Research*, *201*, 52–62. https://doi.org/10.1016/j.micres.2017.04.015

Lee, J., & Zhang, L. (2015). The hierarchy quorum sensing network in Pseudomonas aeruginosa. *Protein & Cell*, *6*(1), 26–41. https://doi.org/10.1007/s13238-014-0100-x

Leinonen, E. N. R., & Kultamaa, B. K. A. (n.d.). *Environmental properties of chemicals, 2*, 1170.

Létoffé, S., Audrain, B., Bernier, S. P., Delepierre, M., & Ghigo, J.-M. (2014). Aerial exposure to the bacterial volatile compound trimethylamine modifies antibiotic resistance of physically separated bacteria by raising culture medium pH. *MBio*, *5*(1), e00944–13. https://doi.org/10.1128/mBio.00944-13

Lin, Y.-H., Xu, J.-L., Hu, J., Wang, L.-H., Ong, S. L., Leadbetter, J. R., & Zhang, L.-H. (2003). Acyl-homoserine lactone acylase from Ralstonia strain XJ12B represents a novel and potent class of quorum-quenching enzymes: Quorum-quenching AHL-acylase. *Molecular Microbiology*, *47*(3), 849–860. https://doi.org/10.1046/j.1365–2958.2003.03351.x

Lutz, M. P., Wenger, S., Maurhofer, M., Défago, G., & Duffy, B. (2004). Signaling between bacterial and fungal biocontrol agents in a strain mixture. *FEMS Microbiology Ecology*, *48*(3), 447–455. https://doi.org/10.1016/j.femsec.2004.03.002

Miller, M. B. (2001). Quorum sensing in bacteria. *Annual Reviews of Microbiology*, *55*, 165–99.

Misztal, P. K., Lymperopoulou, D. S., Adams, R. I., Scott, R. A., Lindow, S. E., Bruns, T., Taylor, J. W., Uehling, J., Bonito, G., Vilgalys, R., & Goldstein, A. H. (2018). Emission factors of microbial volatile organic compounds from environmental bacteria and fungi. *Environmental Science & Technology*, *52*(15), 8272–8282. https://doi.org/10.1021/acs.est.8b00806

Mooney, J. A., Pridgen, E. M., Manasherob, R., Suh, G., Blackwell, H. E., Barron, A. E., Bollyky, P. L., Goodman, S. B., & Amanatullah, D. F. (2018). Periprosthetic bacterial biofilm and quorum sensing: Periprosthetic bacterial biofilm and quorum sensing. *Journal of Orthopaedic Research*, *36*(9), 2331–2339. https://doi.org/10.1002/jor.24019

Mukherjee, S., & Bassler, B. L. (2019). Bacterial quorum sensing in complex and dynamically changing environments. *Nature Reviews Microbiology*, *17*(6), 371–382. https://doi.org/10.1038/s41579-019-0186-5

Nazzaro, F., Fratianni, F., & Coppola, R. (2013). Quorum sensing and phytochemicals. *International Journal of Molecular Sciences*, *14*(6), 12607–12619. https://doi.org/10.3390/ijms140612607

Oslizlo, A., Stefanic, P., Dogsa, I., & Mandic-Mulec, I. (2014). Private link between signal and response in *Bacillus subtilis* quorum sensing. *Proceedings of the National Academy of Sciences*, *111*(4), 1586–1591. https://doi.org/10.1073/pnas.1316283111

Paluch, E., Rewak-Soroczyńska, J., Jędrusik, I., Mazurkiewicz, E., & Jermakow, K. (2020). Prevention of biofilm formation by quorum quenching. *Applied Microbiology and Biotechnology*, *104*(5), 1871–1881. https://doi.org/10.1007/s00253-020-10349-w

Papenfort, K., & Bassler, B. L. (2016). Quorum sensing signal–response systems in Gram-negative bacteria. *Nature Reviews Microbiology*, *14*(9), 576–588. https://doi.org/10.1038/nrmicro.2016.89

Petersen, J. (2016). Heap leaching as a key technology for recovery of values from low-grade ores – A brief overview. *Hydrometallurgy*, *165*, 206–212. https://doi.org/10.1016/j.hydromet.2015.09.001

Piechulla, B., Lemfack, M. C., & Kai, M. (2017). Effects of discrete bioactive microbial volatiles on plants and fungi: Discrete bioactive mVOCs. *Plant, Cell & Environment*, *40*(10), 2042–2067. https://doi.org/10.1111/pce.13011

Piecuch, A., Lamch, Ł., Paluch, E., Obłąk, E., & Wilk, K. A. (2016). Biofilm prevention by dicephalic cationic surfactants and their interactions with DNA. *Journal of Applied Microbiology*, *121*(3), 682–692. https://doi.org/10.1111/jam.13204

Plyuta, V., Lipasova, V., Popova, A., Koksharova, O., Kuznetsov, A., Szegedi, E., Chernin, L., & Khmel, I. (2016). Influence of volatile organic compounds emitted by *Pseudomonas* and *Serratia* strains on *Agrobacterium tumefaciens* biofilms. *APMIS*, *124*(7), 586–594. https://doi.org/10.1111/apm.12547

Plyuta, V. A., Popova, A. A., Koksharova, O. A., Kuznetsov, A. E., & Khmel, I. A. (2014). The ability of natural ketones to interact with bacterial quorum sensing systems. *Molecular Genetics, Microbiology and Virology*, *29*(4), 167–171. https://doi.org/10.3103/S0891416814040077

Popova, A. A., Koksharova, O. A., Lipasova, V. A., Zaitseva, J. V., Katkova-Zhukotskaya, O. A., Eremina, S. Iu., Mironov, A. S., Chernin, L. S., & Khmel, I. A. (2014). Inhibitory and toxic effects of volatiles emitted by strains of *Pseudomonas* and *Serratia* on growth and survival of selected microorganisms, *Caenorhabditis elegans*, and *Drosophila melanogaster. BioMed Research International*, *2014*, 1–11. https://doi.org/10.1155/2014/125704

Rajamanikandan, S., Jeyakanthan, J., & Srinivasan, P. (2017). Molecular docking, molecular dynamics simulations, computational screening to design quorum sensing inhibitors targeting LuxP of vibrio harveyi and its biological evaluation. *Applied Biochemistry and Biotechnology*, *181*(1), 192–218. https://doi.org/10.1007/s12010-016-2207-4

Rajesh, P. S., & Rai, V. R. (2016). Inhibition of QS-regulated virulence factors in Pseudomonas aeruginosa PAO1 and Pectobacterium carotovorum by AHL-lactonase of endophytic bacterium Bacillus cereus VT96. *Biocatalysis and Agricultural Biotechnology*, *7*, 154–163. https://doi.org/10.1016/j.bcab.2016.06.003

Rampioni, G., Leoni, L., & Williams, P. (2014). The art of antibacterial warfare: Deception through interference with quorum sensing–mediated communication. *Bioorganic Chemistry*, *55*, 60–68. https://doi.org/10.1016/j.bioorg.2014.04.005

Rehman, Z. U., & Leiknes, T. (2018). Quorum-quenching bacteria isolated from red sea sediments reduce biofilm formation by pseudomonas aeruginosa. *Frontiers in Microbiology*, *9*, 1354. https://doi.org/10.3389/fmicb.2018.01354

Rutherford, S. T., & Bassler, B. L. (2012). Bacterial quorum sensing: Its role in virulence and possibilities for its control. *Cold Spring Harbor Perspectives in Medicine*, *2*(11), a012427. https://doi.org/10.1101/cshperspect.a012427

Ryan, T. J., & Beaucham, C. (2013). Dominant microbial volatile organic compounds in 23 US homes. *Chemosphere*, *90*(3), 977–985. https://doi.org/10.1016/j.chemosphere.2012.06.066

Safara, S., Harighi, B., Bahramnejad, B., & Ahmadi, S. (2022). Antibacterial activity of endophytic bacteria against sugar beet root rot agent by volatile organic compound production and induction of systemic resistance. *Frontiers in Microbiology*, *13*, 921762. https://doi.org/10.3389/fmicb.2022.921762

Schmidt, R., Cordovez, V., de Boer, W., Raaijmakers, J., & Garbeva, P. (2015). Volatile affairs in microbial interactions. *The ISME Journal*, *9*(11), 2329–2335. https://doi.org/10.1038/ismej.2015.42

Schulz, S., & Dickschat, J. S. (2007). Bacterial volatiles: The smell of small organisms. *Natural Product Reports*, *24*(4), 814. https://doi.org/10.1039/b507392h

Schulz-Bohm, K., Martín-Sánchez, L., & Garbeva, P. (2017). Microbial volatiles: Small molecules with an important role in intra- and inter-kingdom interactions. *Frontiers in Microbiology*, *8*, 2484. https://doi.org/10.3389/fmicb.2017.02484

Sears, J., Dirkse, E., Markworth, C., & Strobel, G. A. (2001). Volatile antimicrobials from Muscodor albus, a novel endophytic fungus. *Microbiology*, *147*(11), 2943–2950. https://doi.org/10.1099/00221287-147-11-2943

Shiner, E. K., Terentyev, D., Bryan, A., Sennoune, S., Martinez-Zaguilan, R., Li, G., Gyorke, S., Williams, S. C., & Rumbaugh, K. P. (2006). Pseudomonas aeruginosa autoinducer modulates host cell responses through calcium signalling. *Cellular Microbiology*, *8*(10), 1601–1610. https://doi.org/10.1111/j.1462-5822.2006.00734.x

Sidorova, D. E., Plyuta, V. A., Padiy, D. A., Kupriyanova, E. V., Roshina, N. V., Koksharova, O. A., & Khmel, I. A. (2021). The effect of volatile organic compounds on different organisms: Agrobacteria, plants and insects. *Microorganisms*, *10*(1), 69. https://doi.org/10.3390/microorganisms10010069

Smith, K. M., Bu, Y., & Suga, H. (2003). Library screening for synthetic agonists and antagonists of a Pseudomonas aeruginosa autoinducer. *Chemistry & Biology*, *10*(6), 563–571. https://doi.org/10.1016/S1074-5521(03)00107-8

Solano, C., Echeverz, M., & Lasa, I. (2014). Biofilm dispersion and quorum sensing. *Current Opinion in Microbiology*, *18*, 96–104. https://doi.org/10.1016/j.mib.2014.02.008

Song, F., Koo, H., & Ren, D. (2015). Effects of material properties on bacterial adhesion and biofilm formation. *Journal of Dental Research*, *94*(8), 1027–1034. https://doi.org/10.1177/0022034515587690

Suga, H. (2003). Molecular mechanisms of bacterial quorum sensing as a new drug target. *Current Opinion in Chemical Biology*, *7*(5), 586–591. https://doi.org/10.1016/j.cbpa.2003.08.001

Sully, E. K., Malachowa, N., Elmore, B. O., Alexander, S. M., Femling, J. K., Gray, B. M., DeLeo, F. R., Otto, M., Cheung, A. L., Edwards, B. S., Sklar, L. A., Horswill, A. R., Hall, P. R., & Gresham, H. D. (2014). Selective chemical inhibition of AGR quorum sensing in staphylococcus aureus promotes host defense with minimal impact on resistance. *PLoS Pathogens*, *10*(6), e1004174. https://doi.org/10.1371/journal.ppat.1004174

Swem, L. R., Swem, D. L., Wingreen, N. S., & Bassler, B. L. (2008). Deducing receptor signaling parameters from in vivo analysis: LuxN/AI-1 quorum sensing in vibrio harveyi. *Cell*, *134*(3), 461–473. https://doi.org/10.1016/j.cell.2008.06.023

Tahir, H. A. S., Gu, Q., Wu, H., Niu, Y., Huo, R., & Gao, X. (2017). Bacillus volatiles adversely affect the physiology and ultra-structure of Ralstonia solanacearum and induce systemic resistance in tobacco against bacterial wilt. *Scientific Reports*, *7*(1), 40481. https://doi.org/10.1038/srep40481

Velmourougane, K., Prasanna, R., & Saxena, A. K. (2017). Agriculturally important microbial biofilms: Present status and future prospects. *Journal of Basic Microbiology*, *57*(7), 548–573. https://doi.org/10.1002/jobm.201700046

Venkataraman, A., Rosenbaum, M. A., Werner, J. J., Winans, S. C., & Angenent, L. T. (2014). Metabolite transfer with the fermentation product 2,3-butanediol enhances virulence by Pseudomonas aeruginosa. *The ISME Journal*, *8*(6), 1210–1220. https://doi.org/10.1038/ismej.2013.232

Verbeke, F., De Craemer, S., Debunne, N., Janssens, Y., Wynendaele, E., Van de Wiele, C., & De Spiegeleer, B. (2017). Peptides as quorum sensing molecules: Measurement techniques and obtained levels in vitro and in vivo. *Frontiers in Neuroscience*, *11*. https://doi.org/10.3389/fnins.2017.00183

Wang, J., Lin, J., Zhang, Y., Zhang, J., Feng, T., Li, H., Wang, X., Sun, Q., Zhang, X., & Wang, Y. (2019). Activity improvement and vital amino acid identification on the marine-derived quorum quenching enzyme MomL by protein engineering. *Marine Drugs*, *17*(5), 300. https://doi.org/10.3390/md17050300

Waters, C. M., & Bassler, B. L. (2005). Quorum sensing: cell-to-cell communication in bacteria. *Annual Review of Cell and Developmental Biology*, 21, 319–46.

Weisskopf, L., Schulz, S., & Garbeva, P. (2021). Microbial volatile organic compounds in intra-kingdom and inter-kingdom interactions. *Nature Reviews Microbiology*, *19*(6), 391–404. https://doi.org/10.1038/s41579-020-00508-1

Wessén, B., & Schoeps, K.-O. (1996). Microbial volatile organic compounds—what substances can be found in sick buildings? *The Analyst*, *121*(9), 1203–1205. https://doi.org/10.1039/AN9962101203

Whiteley, M., Diggle, S. P., & Greenberg, E. P. (2017). Progress in and promise of bacterial quorum sensing research. *Nature*, *551*(7680), 313–320. https://doi.org/10.1038/nature24624

Zhang, Q., Yang, L., Zhang, J., Wu, M., Chen, W., Jiang, D., & Li, G. (2015). Production of anti-fungal volatiles by non-pathogenic Fusarium oxysporum and its efficacy in suppression of Verticillium wilt of cotton. *Plant and Soil*, *392*(1–2), 101–114. https://doi.org/10.1007/s11104-015-2448-y